CAMBRIDGE CLASSICAL TEXTS
AND COMMENTARIES

60

GARGILIUS MARTIALIS: THE AGRICULTURAL FRAGMENTS

GARGILIUS MARTIALIS

THE AGRICULTURAL FRAGMENTS

JAMES L. ZAINALDIN

Harvard University, Massachusetts

CAMBRIDGE
UNIVERSITY PRESS

CAMBRIDGE
UNIVERSITY PRESS

University Printing House, Cambridge CB2 8BS, United Kingdom

One Liberty Plaza, 20th Floor, New York, NY 10006, USA

477 Williamstown Road, Port Melbourne, VIC 3207, Australia

314–321, 3rd Floor, Plot 3, Splendor Forum, Jasola District Centre, New Delhi – 110025, India

79 Anson Road, #06–04/06, Singapore 079906

Cambridge University Press is part of the University of Cambridge.

It furthers the University's mission by disseminating knowledge in the pursuit of education, learning, and research at the highest international levels of excellence.

www.cambridge.org
Information on this title: www.cambridge.org/9781108499897
DOI: 10.1017/9781108759489

First published 2020

Printed in the United Kingdom by TJ International Ltd, Padstow Cornwall

A catalogue record for this publication is available from the British Library.

Library of Congress Cataloging-in-Publication Data
NAMES: Gargilius Martialis, Quintus, author. | Zainaldin, James L., editor.
TITLE: Gargilius Martialis, the agricultural fragments / James L. Zainaldin.
OTHER TITLES: Hortis | Cambridge classical texts and commentaries.
DESCRIPTION: New York : Cambridge University Press, 2019. | Series: Cambridge classical texts and commentaries | Includes bibliographical references and index. | English and Latin.
IDENTIFIERS: LCCN 2019044727 (print) | LCCN 2019044728 (ebook) | ISBN 9781108499897 (hardback) | ISBN 9781108759489 (epub)
SUBJECTS: LCSH: Agriculture – Early works to 1800.
CLASSIFICATION: LCC PA6389.G5 A28 2019 (print) | LCC PA6389.G5 (ebook) | DDC 630–dc23
LC record available at https://lccn.loc.gov/2019044727
LC ebook record available at https://lccn.loc.gov/2019044728

ISBN 978-1-108-49989-7 Hardback

For Ingrid, Jamil, and Carl

CONTENTS

The plate section can be found between pp 206 and 207

PLATES

FIGURES

TABLES

PREFACE

Gargilius Martialis' agricultural fragments contain hardly a sentence without a clue which, if traced to its conclusion, yields something that fascinates, delights or enlightens. In puzzling over these many problems, it became clear to me that in spite, or rather because, of the obscurity of the author, a full-scale commentary would be useful for calling attention to the many interesting questions raised by the work. In partial justification for the length of my treatment of Gargilius' short text I offer two remarks. First, in proportion to the vast number of ancient Greek and Roman works pertaining to what were usually called τέχναι or *artes*, there are few complete modern studies (still fewer in English). The fullness of the present undertaking is meant as a tribute to and confirmation of the efforts others have made to reclaim for us this extraordinary sphere of ancient thought. Second, Gargilius' intellectual engagement with the rich tradition before him has the consequence that his writings can only be appreciated in light of those of previous authors. A dearth of extensive commentaries on the Greek and Roman agricultural authors has therefore made it necessary to go into greater detail on traditional problems that bear on Gargilius' text than would be otherwise required.

It is a pleasant duty to express my gratitude here for the help so many have generously rendered to me while I was preparing this commentary. For various kinds of aid and encouragement I am grateful to George Conklin, Adam Gitner, Jared Hudson, Nicholas Kahn, Oliver Marjot, Jonathan Master, Gregory Mellen, Marco Romani Mistretta, Vivian Nutton, Jeremy Rau, Mark Schiefsky and the participants in a seminar on Roman Gardens taught by Kathleen Coleman in the Harvard Department of the Classics in 2015. Werner Jobst and Bahadır Yıldırım helped me locate a hard-to-find detail of the Great Palace Mosaic in Istanbul, and Örgü Dalgıç generously provided me with a photograph of it. Li Lihua graciously hosted me in the beautiful campus of Sun Yat-sen University in Guangzhou, China, for three weeks in the fall of 2018, affording me the opportunity to complete many crucial revisions. Bobbi Angell provided the exquisite illustrations of the quince, peach, almond and chestnut included in this volume.

The Harvard Department of the Classics supported this project, not least by providing on numerous occasions undisturbed research time and generous material support. I thank especially Alyson Lynch and Teresa Wu for their help in these respects. Grants and fellowships from the Harvard Graduate School of Arts and Sciences further enabled me to pursue this project. I am grateful also to the staff and researchers at the Thesaurus Linguae Latinae, especially Manfred Flieger, Michael Hillen and my Redaktorinnen, Marijke Ottink and Josine Schrickx, for hosting me in Munich in the summer of 2016. Besides acquiring a deeper appreciation of the invaluable work being completed there, I benefited enormously from access to the treasures of the library and *Zettelarchiv*.

James Adams offered criticism on an early draft of a portion of the text and commentary which proved fundamental for guiding the development of the project. I am grateful for his continued encouragement thereafter and for sharing in advance parts of his forthcoming work. Garth Tissol read portions of this work in different stages of its composition and always provided stimulating feedback, besides offering much guidance and support of a more general nature. For these benefits, as for his friendship and unstinting kindness over the years, I am deeply indebted.

I thank the Series Editors for including this volume among the Cambridge Classical Texts and Commentaries, and I am particularly grateful to Stephen Oakley and Michael Reeve for reading the entire typescript. Stephen Oakley's meticulous criticisms greatly strengthened many of the arguments offered here; how fertile were the suggestions of Michael Reeve may perhaps best be grasped by glancing at the apparatus criticus.

I thank also Richard Tarrant, from whose patient instruction I had previously benefited, for his help on various details of the text and commentary. It will not surprise anyone if I admit that what I have learned from the comments of these scholars extends far beyond the present undertaking. I am indebted to Michael Sharp and the publication staff at Cambridge University Press, including Hal Churchman and Bethany Johnson, for so smoothly guiding the entire process and seeing the work through to publication. I wish further to express my appreciation to my copyeditor Malcolm Todd for his expert eye and keen attention to detail.

There is finally no adequate way to express my profound gratitude to Kathleen Coleman, who as adviser, mentor and friend supported and guided this work. It would be insufficient to acknowledge her unfailing advice and incisive criticism on countless versions in draft, were I not also to add that it was with her encouragement that the idea for a commentary on Gargilius' agricultural writings was first conceived. On numerous occasions her aid was also critical in securing opportunities for funding and research time without which this project could not have been completed as swiftly as it was.

In this work, as in everything else, I have been inspired by my wife, Katherine van Schaik, on whose μαιευτικὴ τέχνη I had continual occasion to rely. Only she can know how much this work owes to her.

I must acknowledge that in spite of the kind help I have received there will no doubt remain errors and deficiencies. I alone am responsible for these, but I hope that I may divert Columella's statement to address my own trepidation: *cum aut magnitudinem totius rei quasi quandam uastitatem corporis aut partium eius uelut singulorum membrorum subtilitatem dispicio, uereor ne supremus ante me dies occupet quam uniuersam disciplinam ruris possim cognoscere.*

INTRODUCTION

I GARGILIVS MARTIALIS: LIFE AND WORK

(a) Life

Of our author Gargilius Martialis (G.), little is known with certainty. Several recent and accessible overviews report the facts and the few deductions that we can make, and so the account here can be summary.[1] G.'s references to the Quintilii brothers (d. *c.*183) and Galen (d. early third century), and the extensive use made of his own work by Palladius (*fl.* late fourth or fifth century), outline very roughly the time when he lived and worked.[2] Servius and Cassiodorus mention him as an agricultural author, both with high esteem, and the latter testifies also to his medical authorship (below, §I(b)).[3] The *Historia Augusta* twice mentions a Gargilius (Martialis) *auctor*; if we accept a plausible identification with our G. (the *nomen* Gargilius is very rare), then we can locate at least part of his life in the early third century, for there is attributed to him (*Alex. Seu.* 37.9) a

[1] The fundamental study is Mazzini (1977); the latest comprehensive survey of questions of his identity, biography and oeuvre is Maire (2002*a*) ix–xxvi (documentary materials reproduced at cvii–cxiii). See also among recent work Condorelli (1978) xi–xiv, Tapper (1980) 1–7, Riddle (1984) 408–13, Önnerfors (1993) 264–5, Stok (1993), Fischer (1997*a*), Schulze (2002) 91–3; older but still useful are Cichorius (1887), S–H III 222–4, *RE* (1) 7.760.37–762.53 s.v. 'Gargilius' (Stadler), *PIR*[2] G no. 82.

[2] Quintilii mentioned 6x in *Pom.* (below, §II), Galen mentioned 17x in *Med.* (*CG* s.v.); Palladius refers to G. by name 13x (below, §VI).

[3] Serv. *ad* Verg. *G.* 4.148 *ALIIS: Gargilium Martialem significat* (Verg. *uerum haec* (sc. horticulture) *ipse equidem spatiis exclusus iniquis / praetereo atque aliis post me memoranda relinquo*); Cassiod. *Inst.* 1.28.5 *quod si huius studii* (sc. horticulture) *requirantur auctores, de hortis scripsit pulcherrime Gargilius Martialis, qui et nutrimenta holerum et uirtutes eorum diligenter exposuit, ut ex illius commentarii lectione praestante Domino unusquisque et saturari ualeat et sanari; quem uobis inter alios codices reliqui.*

work on the dietary habits of Severus Alexander (d. 235).[4] The ascription of such a biography to G. is maintained implicitly elsewhere in the *HA* (*Prob.* 2.7), when he is included among certain earlier historians who wrote 'less elegantly than truthfully'.[5] There is good reason to doubt the veracity of these attributions,[6] but we need not reject the historical evidence for the time when G. lived. Regardless of whether a biography of Severus is really to be ascribed to him, G. must still be accounted a polymath for the erudition of the agricultural and medical writings that have survived; nor was he an entirely bookish one, for although his writings show the fruit of meticulous research, he also appeals at times to his own experience to settle a dispute or vouch for a piece of advice.[7]

We know more about G. if we accept a further reasonable identification with the Q. Gargilius Martialis named on two inscriptions from Auzia, Mauretania,[8] which bear witness to a decorated career as a soldier and provincial official of North Africa and will give a more definite idea of the dates of G.'s life, for one of them testifies to his death on 25 March 260 in

[4] *Alex. Seu.* 37.9 *habuit cottidie et mul<si> sine pipere sextarios quattuor, cum pipere duo, et, ne longum sit omnia inserere quae Gargilius eius temporis scriptor singillatim persecutus est*, etc.

[5] *Prob.* 2.7 *Marium Maximum, Suetonium Tranquillum, Fabium Marcellinum, Gargilium Martialem, Iulium Capitolinum, Aelium Lampridium ceterosque qui haec et talia non tam diserte quam uere memoriae tradiderunt.*

[6] See Champlin (1981) 190, Syme (1971) 47, 278, (1983) 100–1, 103, Chastagnol (1994) cx–cxi, 560; judicious summary conclusions at Maire (2002*a*) xv–xvi.

[7] Note the first-person verbs at *Pom.* 1.1.7, 2.11.2, and G.'s appeal to *experimenta* at 2.1.5; among G.'s numerous appeals to personal experience in *Med.* (Maire (2002*a*) lxiii–lxiv), cf. 2.11 *ut experimento meo nossem*; 25.3 *intellegi datum est . . . sicuti postea experimenta docuerunt*; 32.8 *nobis expertum est*; 53.16 *uehemens hoc esse etiam domesticis in uxore seruata experimentis probaui.*

[8] *CIL* VIII 20751, *CIL* VIII 9047 = *ILS* 2767.7, with which see the interpretative notes of Mazzini (1977) 101–2, 115–18 (synopsis at 108). The former (*CIL* VIII 20751) is a funerary inscription commissioned by Q. Gargilius Martialis in honour of his parents, Q. Gargilius Martialis and Iulia Prima; the latter (*CIL* VIII 9047 = *ILS* 2767.7) a stele erected at public expense to Q. Gargilius Martialis (the younger) after his death, recounting his distinguished military and political career.

a Berber uprising.[9] This G.'s tribe was Quirina, and his father of the same name (Q. Gargilius Martialis) was also a veteran.[10] There is no reason to suppose that a successful career in the military and in public life, culminating in the titles of *decurio* for two colonies and *patronus prouinciae*, would have kept G. from his technical writings.[11] In view of the probability that our author G. is to be identified with this Q. Gargilius Martialis, it is interesting to note that at one place in his horticultural writings (*Pom.* 4.1.2–3) G. seems to assume an Italian perspective for his instructions, criticizing the Carthaginian author Mago for his ignorance about the chestnut and expecting that Celsus, 'a man very experienced in the Italian practice' (4.1.3n. *Italicae disciplinae peritissimum*), would correct him.[12] But if this shows a preoccupation with Rome and its peninsular environs, it has less definite biographical significance: Italian land was the epicentre, real or imagined, of the written Roman agronomic tradition, and G. would thus have had a good reason to adopt this perspective. Similarly, Columella, although he does not forget his native Spain when he writes,[13]

9 It will be noted that this date would give him a *floruit* in agreement with that suggested by the *HA*.

10 It is generally, and not unreasonably, supposed that it is the younger and more illustrious Gargilius with whom our G. is to be identified; I follow this assumption here, but it cannot be established with complete certainty that the elder Gargilius was not the author.

11 Proofs to the contrary are offered, for example, by Velleius Paterculus; by Pliny the Elder, who besides his *Historia naturalis* produced historiographic writings (lost) and a manual on spear-throwing (*De iaculatione equestri liber unus*, lost); and especially by Sex. Iulius Frontinus, whose voluminous technical output included works on land-surveying (now fragmentary), the military art (lost), stratagems (*Strategemata*, surviving) and water management (*De aquaeductu urbis Romae*, surviving).

12 On the north- and east-Mediterranean provenance of the chestnut (hence G.'s complaint), see 4.1.2–3n.

13 Baetica is recalled at, e.g., 2.10.35, 3.12.6, 5.1.5, 5.8.5, 7.1.2, 7.2.4. He also mentions often the advice of his uncle M. Columella (e.g. 2.15.4, 5.5.15, 7.2.4, 12.21.4–5, 12.40.2), a farmer of the same province (5.5.15 *eruditus ac diligentissimus agricola Baeticae prouinciae*); but these recollections must be in some part family loyalty.

adds the same caution about reconciling agricultural advice from North Africa with the Italian climate.[14]

(b) Work

There are only two sets of writing transmitted to the present that can with certainty be attributed to G. The first comprises 60 chapters containing advice about the therapeutic use of fruits, vegetables and herbs. Travelling in numerous manuscripts, always anonymously or under the name of Pliny (and attached to the pseudo-Plinian *Medicina*), it was restored to G. in 1875 by V. Rose, who also furnished it with its present utilitarian title *Medicinae ex holeribus et pomis* (*Med.*).[15] The second set of writings, transmitted in a single, early (sixth-century) manuscript **N**,[16] comprises four substantial but fragmentary chapters on the cultivation of fruit trees: quince (1 section; abbreviated henceforth *cydon.*), peach (12 sections; *persic.*), almond (8; *amygd.*) and chestnut (7; *cast.*). For these agricultural fragments, which are the subject of this volume, I adopt the traditional and utilitarian title *De arboribus pomiferis* (*Pom.*), but they have also been called *De hortis*.[17] **N** further contains, immediately following

[14] Correcting the errors of the North African agronomists had a good pedigree among the Roman farmers, going back at least as far as Tremelius Scrofa (first century BCE): Col. 1.1.6 *ceterum non dissimulanda erunt agrorum cultori praecepta rusticationis quae cum plurima tradiderint Poeni ex Africa scriptores, multa tamen ab his falso prodita coarguunt nostri coloni, sicut Tremelius, qui querens id ipsum tamen excusat, quod Italiae et Africae solum caelumque diuersae naturae nequeat eosdem prouentus habere.*

[15] Rose (1875), where it is still printed following three books of the *Medicina Plinii*. For *Med.*, the standard text is now the Budé of Maire, which furthermore offers a detailed introduction, notes and bibliography. There have been identified a dozen surviving manuscripts from the ninth century on (and extracts of *Med.* made it into the eighth-century codex of abbey medicine called the 'Lorscher Arzneibuch', Bamberg med. 1): for the transmission of *Med.*, see Maire (2002*a*) lxxiii–lxxxviii, Zainaldin (forthcoming 1).

[16] For full discussion of **N**, see below, §VII.

[17] The title *De arboribus pomiferis* is Mai's. Mazzini calls the fragments *De hortis*, arguing ((1988) 17) that it is the only title with ancient precedent; this is true

the horticultural advice for the quince (*Pom.* 1.1), a truncated section that opens with the words *Medicina ex cydoneis* and provides another eight sentences detailing therapeutic uses of the quince. Crucially, these sentences are also transmitted by the manuscripts of *Med.*, where they in fact constitute the beginning of G.'s chapter on quince medicine (*Med.* 43.1–8). This small area of overlap between **N** and the medical manuscripts (printed in this volume as Appendix 1) not only allows G. to be identified as the author of *Med.*, but also provides us with an important clue to the original form of his work.

The most probable conclusion on the evidence of **N** is that *Med.* and *Pom.*, despite their largely independent transmissions, represent two parts of what was once a single larger work. This treatise may itself have been called *De hortis* and, as it seems, would have described plant by plant both the cultivation and medical use of garden produce.[18] We may form some idea of the shape of this original *De hortis* if we imagine a work in which each of the chapters of *Med.* on various fruits, vegetables and herbs (60 in total) was preceded by a chapter dealing with the horticulture of the relevant plant organized according to the agenda that can be reconstructed from the fragments of *Pom.* (for which see below, §III(a)).[19] Further confirmation for such a hypothesis is found in the remarks of Cassiodorus, which

(see Cassiodorus quoted above, n. 3), but I prefer to reserve that name for the larger, original work (see below, this section). (Mai had previously used the title *De hortis* in order to describe the combination of *De arboribus pomiferis* and *De pomis seu Medicina ex pomis*, the latter a work of dubious authenticity: see below, n. 28.) Scotti calls the fragments *De re hortensi.* Condorelli has the simple but descriptive *fragmenta ad holera arboresque pertinentia.*

18 On this original work, the existence of which is not doubted by modern scholars, see Condorelli (1978) xxv–xxvi, Riddle (1984) 412–13, 427–9, Mazzini (1988) 16–17, 135, Condorelli (1995–8) 241–52, Maire (2002*a*) xxvi, lx–lxiv.

19 Hence it may be inferred that we are missing from *Pom.* (and the original *De hortis*) 56 agricultural chapters dealing with the remaining plants found in *Med.* (for a list of which, see Maire (2002*a*) c–civ), unless for some reason we suppose that G. did not furnish every plant with notes on its cultivation. G. alludes in *Pom.* to one such missing chapter, that on the pomegranate: see Table 1 and below, n. 22.

strongly suggest that the agricultural and medical writings were joined,[20] and in the agreement of *Pom.* and *Med* in the chapter ordering of the plants, as summarized in Table 1.[21]

As has been persuasively argued, the original *De hortis* would have taken the form of a manual for the *paterfamilias* teaching how both to cultivate and to administer the plants that would nourish and treat members of his household.[23] Cato's *De agri cultura* is thus perhaps the closest extant precedent that we have,

Table 1: The Order of the Chapters

Plant	Chapter number in *Pom.*	Chapter number in *Med.*
Pomegranate	Uncertain, but before quince[22]	41
Quince	1	43
Peach	2	44
Almond	3	53
Chestnut	4	56

[20] See above, n. 3. The key details are: *et nutrimenta holerum et uirtutes eorum diligenter exposuit, ut ex illius commentarii lectione . . . unusquisque et saturari ualeat et sanari.*

[21] It must be duly noted that in ordering the chapters of *Pom.* there is a choice in where to place *persic.* (below, §IX), whether after *cydon.* or after *cast.*; but this mild circularity does not vitiate the more general agreement of chapter order. For the chapter order of *Med.*, see Maire (2002*a*) xcix–civ. The uneven gaps between chapters in *Pom.* and *Med.* can be explained by rearrangement of material in *Med.* at the time of or after its excerpting, if not simply by disturbance in **N** itself.

[22] The chapter on the pomegranate can be placed before that on the quince because G. refers to the pomegranate in *cydon.* on the grounds that he has already discussed it: cf. 1.1.13n. *Consentiunt aliqui posse* (sc. *cydonea*) *et scrobibus et in doleis eodem modo quo punica supra uel alia m<ala> seruari.*

[23] The reconstruction of the intention and audience of the original work rests not only on the *prima facie* argument from the paired chapters of agriculture and medicine, but also from persuasive observations about the non-specialist character of G.'s language and advice, especially of *Med.*: in his reworking of his sources, G. avoids excessively abstruse medical language, strives for literary effect (sometimes to the point of obfuscating his instructions) and prioritizes simple and practical therapies. See Maire (1997), esp. 316–18, (2000), esp. 162–4, (2002*a*) lx–lxiv, (2003*b*), esp. 366–7. Stok (1993) places the original *De hortis* in a 'vein of encyclopaedism' ('un filone enciclopedico' (229)) represented by Cato and Pliny the Elder (i.e. and not Celsus or Varro).

and it is worth noting the suggestive statement found in G.'s report on Cato's medical advice for the cabbage (*Med.* 30.3): *ceterum militares uiri gloriosas cicatrices gratuito holere curabant, eodem horto usi ad salutem dum illos pascit et sanat.*[24] It is hard not to think that G. reflects here on his own work as well as Cato's.

There are a handful of other works that have been attributed to G.[25] As mentioned above, a biography of Severus Alexander has been hypothesized on the strength of the remarks in the *Historia Augusta*, but these are hardly to be trusted.[26] The short veterinary work *Curae boum* was once attributed to G., but is now rightly separated from his name.[27] Several fragments dealing with the medical use of fruits and vegetables have been attached to him, but none securely.[28] Finally, there has been

[24] Cato is named 3x in G.'s writings, all in *Med.* 30.1–14, where he contributes many medical uses for the cabbage.

[25] For synoptic treatments, see Mazzini (1977) 108–15, Stok (1993) 220–4, Maire (2002*a*) xiv–xxiv.

[26] For strong grounds to doubt the *HA* here, see above, n. 6. Önnerfors (1993) 272 thinks that G.'s taste for prose rhythm (see below, §v(a)) might suggest a 'schriftstellerische Tätigkeit als Historiker', because 'die Geschichtsschreibung zählte ja zu den schönen, der Poesie nahestehenden Gattungen der antiken Literatur'.

[27] The work, edited by Lommatzsch, comprises 23 sections of veterinary medicine under the heading *Curae boum ex corpore Gargili Martialis.* They are transmitted in a single manuscript **L** (Voss. lat. f. 71, *anno* 1537) following the four books of Vegetius' *Digesta artis mulomedicinae.* Authenticity can be precluded on stylistic among other grounds (regarding style, an absence of prose rhythm, monotonous language and lack of G.'s characteristically dense source-citation): see Mazzini (1977) 111–13, (1988) 133–4, Fischer (1997*b*), Zainaldin (forthcoming 2). Condorelli (1985) 1025 n. 1 allows that there may be a (remote) Gargilian substrate to the work.

[28] The most substantial of these is the spurious work confusingly called both *De oleribus Martialis* (when edited by Rose (1864–70) II 131–50) and *De pomis seu Medicina ex pomis* (when edited by Mai (1831) 418–26). Mazzini (1977) 114 identifies the work *De oleribus / De pomis* as being in truth the nucleus of books 1 and 3 of the pseudo-Hippocratic *Dynamidia*; the first part of it is a copy of the Latin translation of the Hippocratic περὶ διαίτης, the second a reworking of some chapters of *Med.* For the sources of the *Dynamidia*, see further Ferraces Rodríguez (1999) 23–55 and *passim.*

hope that genuinely Gargilian writings can be recovered from various late Latin pharmacological writings.[29]

II GARGILIVS IN THE AGRICULTURAL TRADITION

To judge by *Pom.*, if G.'s agricultural writings had survived in full they would have been by far the most detailed extant treatment of the cultivation of fruit trees, and probably many other edible or medically useful plants, from Greco-Roman antiquity. Although in some important respects they would also have been radically different from previous extant writings, they can only be understood in terms of the Greek and Roman agronomic tradition to which they belong. For the remainder of the Introduction, I will restrict my remarks about G.'s agricultural writings to *Pom.*, that is, to the surviving chapters dealing with the arboriculture of quince, peach, almond and chestnut.[30]

Pom. stands out among the extant agricultural writings from Rome both in form and content. Its chapter-by-chapter treatment of fruits is not found in earlier treatises, and it is the first work that attempts to discuss systematically and completely the cultivation of individual trees from the stage of soil preparation, seed selection and planting through to harvesting and preservation.[31] Moreover, the wealth of detail found in *Pom.* is unprecedented, as is the thoroughness with which G. searches Greek and Roman agricultural authorities for alternative

[29] Ferraces Rodríguez, for example, has argued that the pseudo-Dioscoridean *De herbis femininis* is based on a Latin translation of Dioscorides' *Materia medica* prepared by G.: see Ferraces Rodríguez (1999), esp. 173–224; cf. also Ferraces Rodríguez (1994), (2000). But in reviewing the arguments of Ferraces Rodríguez, Maltby (2008) offers reasonable grounds for doubting the hypothesis.

[30] This narrow focus is of course dictated by the loss of the other chapters, but the opening fragment also indicates a specific demarcation of the discussion of fruits: cf. 1.1.1 ****uerum de fructibus pauca nobis dicenda sunt* (with 1.1n.).

[31] On the uniqueness of *Pom.* in this respect, see further below, §III(a).

techniques of cultivation. But despite its novelty, *Pom.* remains deeply indebted to earlier agricultural writings and must be studied as belonging to that tradition: thus, Columella and Pliny the Elder had both discussed fruit trees in detail in separate books within their works (*De re rustica* 5, *Historia naturalis* 17), and, whatever its authorship, the *De arboribus* if it circulated independently by G.'s time could have supplied a precedent for a treatment of arboriculture not embedded in a comprehensive or encyclopaedic framework.[32] An even more obvious indication of G.'s debt to the earlier agronomists is the remarkable density of named authorities appearing in his work, and the great extent to which, even where he leaves a source unnamed, his account can be traced back to strands of advice found in earlier writings. Before discussing further the structure of *Pom.*, it will therefore be useful to survey in brief G.'s Greek and Roman precedents in arboriculture in order to understand what models for composition he may have had before him and what traditional constraints and opportunities they might have offered (although our knowledge of many sources, and *a fortiori* of G.'s use, is limited). In the following, I will discuss in chronological order (1) those authors G. mentions by name in *Pom.* (e.g. Columella, Celsus and the Quintilii), regardless of whether their works still survive; and (2) those authors whom G. does not mention by name in *Pom.* (e.g. Theophrastus, Cato and Varro), but whose extant writings appear to bear witness to the source of G.'s language or advice, if they are not the source itself.[33]

[32] Thus, what is relevant for our purposes is not the content of *Arb.* (which in many cases duplicates that of Columella's *De re rustica* 5) but the fact that such a book was valued enough to be transmitted apart from the work that originally contained it. On *Arb.*, see further below, this section.

[33] The list makes no claim to exhaust the possible sources of influence on G.: for a sampling of agricultural authors excluded here, consider the lengthy lists at Varro, *Rust.* 1.1.7–10, Col. 1.1.7–14; also the list of the sources of the *Geoponica* compiled by Dalby (2011) 36–49.

One of the earliest authorities on whom G. relies is the Punic agronomist **Mago** (fl. unknown, probably sometime fourth to early second century BCE), whose rustic work in 28 books exerted a profound influence on subsequent Greek and Roman writers on agricultural themes.[34] It is probably this work, or rather, the numerous translations and abridgements of it that were made into both Greek and Latin (cf. Varro, *Rust.* 1.1.10), that provided the model for the systematic approach to agriculture adopted by later Roman authors. This is unsurprising in view of the high esteem in which Mago was held: Varro (*ibid.*) gives him the place of honour among agronomists – *hos* (sc. *scriptores*) *nobilitate Mago Carthaginiensis praeteriit* –, a judgement with which Columella (1.1.13) agrees, calling him *rusticationis parens*. Pliny and Columella tell us that his work was so important that a translation of his 28 books was made into Latin *ex senatus consulto* (Col. *ibid.*, Plin. *HN* 18.22), which has been dated to as early as 140 BCE. Although G. obviously departs from Mago's comprehensive approach, he uses his arboricultural advice extensively (named 5x: 3.1.2, 3.1.4, 3.3.4, 4.1.2, 4.1.3), whether in an apparently direct form, that is, through a translation or epitome, or indirectly, that is, through quotations or paraphrase in later agronomists.[35] But

34 Pliny (*HN* 18.22) gives us the only scrap of information we have about Mago's life when he lists him next to Xenophon as a *dux* who wrote on agricultural topics, but we may suspect the identification to be too pat. Indeed, scholars have doubted whether Mago is anything more than a 'venerable and distinguished name associated with a large corpus of information from Punic Africa, the accumulation of centuries of development' (White (1973) 475; cf. Mahaffy (1889) 32) – a Hippocrates of agronomy, as it were. Whether or not this is the case, Mago is consistently represented by the Romans as a historical figure and the progenitor of much agricultural wisdom. For elucidation of the shadowy figure of Mago and of the influence of Punic agriculture on Rome, see Martin (1971) 43–52, White (1973), esp. 470–5, Heurgon (1976), Ameling (1993) 259–60, Greene and Kehoe (1995), Santini (2000), Cataudella (2002) 41–6, Domínguez Petit (2004), Krings (2008) 24–7. Speranza (1974) 75–119 collects the fragments, to which from the *agrimensores* add Lachmann 348.19–350.16.

35 Mago is tied to Celsus at 3.1.2 (Mago and Celsus agree), 3.1.4 (they disagree), 4.1.2–3 (they agree); it is not unreasonably inferred that Celsus was

in spite of Mago's great stature, G. does not slavishly transmit his instructions: he quotes them where he approves, but usually engages the opinion of another authority, and will revise his recommendations where he believes that they might mislead (cf. 4.1.3 *quidquid Magonis ignorantiam fugerat*). To anticipate the discussion below, it may be said that while the influence of the great Carthaginian agronomist is still significant in *Pom.*, later Roman authorities had firmly established their own dominance in the tradition by G.'s time.

G. mentions **Aristotle** (384–322 BCE) once, referring to a work called *georgica* (3.7.4 *Aristoteles in georgicis*). It is not at all certain what work G. means. We may mention three treatises on plants or agriculture that have been connected with the name of the great philosopher. The first is the γεωργικά, a work whose title is found in an appendix of pseudepigrapha attached to the list of Aristotle's writings following the *Vita Menagiana*.[36] The second is the extant treatise in two books called περὶ φυτῶν that travels with the *corpus Aristotelicum*. This work, which survives to the present not in the original Greek but in Arabic, Hebrew, Latin, Syriac and a Greek back-translation, was in fact penned by the first-century BCE Peripatetic Nicolaus of Damascus.[37] Third, there is the hypothesis of a genuine work by Aristotle called περὶ φυτῶν, inferred from certain references in Aristotle's own writings and from the comments of some ancient authors; it is often assumed as the basis for Nicolaus' treatise.[38] Aristotle certainly shows an interest in plants in his writings, but there

the medium for Mago's advice in some or all of these cases (cf. Santini (2000) 377). At 3.3.4, Mago appears with Diophanes (they agree); here G. may have had Diophanes' Greek translation/epitome in his hands.

36 See Moraux (1951) 266; and for further discussion of evidence for the work, Rose (1863) 268–75. Rose later (1886 (209)) seems to consider the γεωργικά identical to the περὶ φυτῶν attributed to Aristotle.

37 See concisely Moraux (1973) 487–9; more fully the edition of Drossaart Lulofs and Poortman.

38 For ancient references to an Aristotelian περὶ φυτῶν, see concisely Moraux (1973) 490–1, Drossaart Lulofs and Poortman (1989) 1 n. 1; material is collected also at Rose (1886) 209–14.

is no definitive evidence that he wrote such a treatise.[39] Finally, we must countenance the possibility that works on plants or agriculture besides these were circulating under Aristotle's name, any one of which G. might have had before him. The only source that can be excluded with confidence is the περὶ φυτῶν of Nicolaus, which, as we have it, does not contain the advice that G. reports. It may be relevant to note that the earlier Roman agronomists who include Aristotle in their lists of authors writing on rustic matters (Varro, *Rust.* 1.1.8, Col. 1.1.7) seem to have an eye not to any botanical books he wrote, but rather his zoological works.[40]

Though G. never mentions Aristotle's successor **Theophrastus** (371/0–287/6 BCE) by name,[41] there are numerous places where he remains the oldest and sometimes only extant witness to a tradition of agricultural advice surfacing in *Pom.* His *De causis plantarum* and *Historia plantarum* are not concerned with providing agronomic advice *per se*, but they can provide a window onto the Greek didactic agricultural tradition insofar as Theophrastus relied on contemporary practice for his data. For example, when he wishes to illustrate a claim

[39] For judicious remarks, see Moraux (1973) 490–1, Hardy and Totelin (2016) 7–8; also Hugonnard-Roche (2003). Drossaart Lulofs and Poortman (1989) xvi, 1 have no doubts about the existence of an Aristotelian περὶ φυτῶν behind that of Nicolaus. Senn (1929) argues vigorously, but not decisively, against its existence, maintaining that Aristotle's references to such a work are instead to Theophrastan material; *contra* Senn, see Regenbogen (1937), noting disagreements between Aristotle and Theophrastus on plants.

[40] Consider thus that the doctrines that appear attached to Aristotle's name in Varro (*Rust.* 2.1.3, 2.5.13) and Columella (7.3.12, 9.3.1) deal exclusively with non-plant life, e.g. bees, horses and bulls, and stem from his biological writings, not from a treatise on husbandry *per se*. A particularly clear example is Varro, *Rust.* 2.5.13 (on the sex of cows), which is based on the remarks at Arist. *Hist. an.* 583b2–9 and *Gen. an.* 763b28–764a2 (see further Flach *ad loc.*).

[41] For a concise overview of Theophrastus' life, see Diggle (2004) 1–3. For the botanical writings, see Wöhrle (1985), Hardy and Totelin (2016) 8–10 and *passim*. All of Theophrastus' botanical writings now benefit from the new editions and translations of Amigues (*Hist. pl.* (1988–2006), *Caus. pl.* (2012–17)), which also offer introductions and notes.

about a tree's capacity to undergo a change in nature (φύσις), he refers to the farmers' technique of bleeding the almond of its noxious sap in order to improve the produce; the advice, recycled by Pliny, eventually finds its way into *Pom.*[42] In other cases, Theophrastus' botanical theory itself may reflect concerns that underpin, however remotely, G.'s advice. Such could be the case with G.'s claim that the almond when cropped by animals becomes bitter (3.5.2 *pecorum . . . dente contacta in amarum saporem ex dulcibus transeunt*). G.'s closest source seems to be Pliny,[43] but it is ultimately Theophrastus who theorizes the dangers presented to the plant that is browsed or cut back: the reason (τὸ αἴτιον) is the same as for many other changes in plants, ὅτι τῆς ἀρχῆς ἀλλοιουμένης, συναλλοιοῦται καὶ τὸ τέλος (*Caus. pl.* 5.17.5).[44]

Two perhaps surprising omissions from *Pom.* are the names of the famous Republican agricultural writers **Cato** (234–149 BCE) and **Varro** (116–27 BCE),[45] although their absence may simply be an accident of transmission. Still, it is true that these authors' influence on G. can be appreciated more readily in his agricultural language than in his arboriculture.[46] The scarcity of specific influence in the agricultural advice is no doubt a result of the fact that, while both Cato's *De agri cultura* and Varro's *De re rustica* played a profound role in the development of the written agricultural tradition at Rome, neither of their works treats the cultivation of fruit trees in an especially detailed or sophisticated

[42] For full discussion, see 3.7.2n. [43] For details, see 3.5.2n.

[44] For the same reason adduced in another context, see 3.1.1n.

[45] For a remarkable introduction to Cato and Varro the agronomists, see White (1973). For Cato in particular, see also Brehaut (1933) xiii–xlv, Thielscher (1963), Goujard (1975) vii–xlv, Astin (1978), esp. 182–210, Diederich (2007) 11–22, 156–72; and for Varro, Skydsgaard (1968), Diederich (2007) 22–53, 172–209.

[46] That is, a high proportion of the technical terminology that G. employs has its observable origin in the agricultural language of Cato and Varro (as noted *passim* in the commentary), although in most cases it seems to reach G. only indirectly, with Columella or Pliny (and probably other later authors) intervening. A stark counter-example to this situation however is G.'s *hiasco* (3.1.1), which is attested elsewhere only in Cato: see 3.1.1n. *hiascere coeperunt*.

way.[47] Both Cato and Varro remain important for understanding G.'s work, but more as forming a part of the linguistic and intellectual backdrop to *Pom.* than as palpable sources for it.

The obscure **Diophanes of Nicaea** (first century BCE) is mentioned by G. twice for his advice on almonds (3.3.4, 3.4.3).[48] At 3.3.4, he is paired with Mago on the grounds that they report the same advice for the almond;[49] this appearance fits with what little we know of him, that he rendered Cassius Dionysius' epitome of Mago (already supplemented by material from the Greek agronomists) into a handy six books, apparently for his patron Deiotarus of Galatia (Varro, *Rust.* 1.1.8, 1.1.11, Col. 1.1.10).

G. quotes from the *Eclogues* of **Virgil** (70–19 BCE) once, on the chestnut (4.1.1 *mea quas Amaryllis amabat* = Verg. *Ecl.* 2.52). The reference comes unfortunately in a lacunose sentence, and it is therefore not absolutely certain to what end G. is using it; it may simply be a rhetorical flourish in order to describe an especially desirable crop of chestnuts.[50] Elsewhere, Virgilian resonances in *Pom.* tend to be with the *Georgics*,[51] as one might expect given

[47] The important exceptions are of course the olive and vine (the latter usually handled by the agronomists along with trees). The light handling of fruit trees can be explained in part by the nature of the Roman economy in the second and first centuries BCE and in part by the simple fact that many of the trees were foreign in origin and would not be established in Roman farms and markets until much later. In the former respect, it is telling that Cato urges cultivation of quince (*Agr.* 7.3) and almond (*Agr.* 8.2) only for those holdings near a city, and that Varro has his eyes trained on more lucrative enterprises, such as the *pastio uillatica* of *Rust.* 3. In the latter respect, one should note that the peach – a highly developed subject of discussion in *Pom.* – was unknown to Varro and Cato, and still uncommon at the time when Columella wrote: see White (1970) 258 n. 82, Sadori et al. (2009) 50, Zohary et al. (2012) 144–5.

[48] For what few deductions we can make about his life, see *RE* (1) 5.1049.17–33 s.v. 'Diophanes' (Wellmann).

[49] See further 3.3.4n. *post eum Diophanes*.

[50] As suggested at 4.1.1n.

[51] Whether these are direct or not, it is hard to say; for examples, see 2.5.7n., 2.9.2n. *sic . . . imperatur*.

both that work's subject matter and the high esteem in which it was held among later agronomists.[52]

With **Iulius Atticus** (early first century CE) we enter the early Imperial period, a fertile time for agricultural writing at Rome.[53] Atticus is mentioned by G. three times for his advice on the chestnut (4.1.6, 4.6.4, 4.7.2). That the references to him appear only on this subject is no accident: he is known through Columella (1.1.14) for the single book on viticulture that he wrote, in which he would doubtless have had occasion to discuss the cultivation of timber crops for use as vine supports and frames. The chestnut was highly valued for this purpose, and although G. wrote for those interested in cultivating the tree for its fruit, he does not scruple to report the opinions of those dealing with the chestnut as a timber crop (*satio caedua*).[54]

Though Atticus wrote before both Columella and Pliny, for whom he is an important source, it is not certain whether he also predates **Cornelius Celsus** (early first century CE), whose encyclopaedic *Artes* contained five books devoted to the discussion of agriculture (composed prior to 39 CE).[55] Celsus' books were employed by Columella and Pliny, the former thinking them praiseworthy for their elegance and restraint (9.2.1). Celsus

52 For Virgil as an agricultural authority, see Christmann (1982), Spurr (1986*a*), Doody (2007).

53 For Iulius Atticus, see Reitzenstein (1884) 27–30, 54, S–H II 791, *PIR*² I no. 183, Richter (1981–3) III 587–8, 620, *NP* s.v. 'Iulius (IV 3)'.

54 See further 4.1–7n.

55 For an introduction to Celsus (with an inevitable focus on his medical writings), see Schulze (2001); also Krenkel (1973), Langslow (2000) 41–8, Gautherie (2017) 42–9. Reitzenstein (1884) 30–41, 55–6 investigates the contents of Celsus' lost agricultural books, for which see also Marx (1915) 5–13 (testimonia and fragments), Krenkel (1973) 22, Schulze (2001) 15–16, 85–6. On the scant evidence available, Reitzenstein (1884) 27, 33 suggests that Atticus was an older contemporary of Celsus and that the latter may have used his work. At all events Celsus wrote the agricultural books of the *Artes* before 39 CE, but it is difficult to be more precise than this; they may have been composed before 21 CE (implied by the tentative suggestion of Langslow (2000) 44 for the date of *De medicina*), around 25/26 CE (Cichorius (1922) 411–17) or after 27 CE (Reitzenstein (1884) 31).

evidently devoted substantial space to arboriculture in the *Artes* (whether he had a separate book on trees is uncertain), as is apparent both from the breadth and frequency of G.'s references to him (cited 7x, for the peach, almond and chestnut: 2.3.4, 2.4.1, 3.1.2, 3.1.4, 3.8.1, 4.1.3, 4.6.2). Indeed, G. mentions his name more often than that of any other author; although he once gently reproves him (4.1.3), he also calls him *Italicae disciplinae peritissimus* (*ibid.*), and the tone of his references in general shows that he values Celsus' advice highly. Were Celsus' agricultural books to survive, we might find that they were a source for much more of the advice in *Pom.* than at first appears, judging not least by the fact that G. can be seen to draw extensively on Columella and Pliny even where he does not mention them by name.

Of course, we are in a position to gauge the full extent of G.'s debt to earlier agronomists only when their writings are extant. If we restrict our scope to these surviving works, the *Res rustica* of **Columella** (*fl. c.*50 CE) is the single most influential work on *Pom.* as regards both language and content.[56] (Only Pliny the Elder's *Historia naturalis* could compete for the title; see below.) This situation is hardly surprising, for Columella's treatise, a masterly synthesis of traditional agricultural advice with the observations and experience of an inquisitive farmer and savvy businessman, has come down as the preeminent achievement of the Roman technical tradition of agronomy. It is relevant for *Pom.* in particular that Columella discusses at length and with great sophistication nearly every aspect of the cultivation of fruit trees, from preparation of the soil and propagation of the plants to the selection and preservation of mature fruits, basing his advice on general and usually sound agronomic principles. Book 5, most of which is devoted to arboriculture, is the key text for *Pom.*, but G. draws extensively from

[56] For general discussions of Columella's person and his work, see White (1970) 26–8, Martin (1971) 289–373, Richter (1981–3) III 588–650, Martin (1985), Flach (1990) 198–204, Marcone (1997) 26–30, Diederich (2007) 53–69, 209–58, Fögen (2009) 152–200, Reitz (2013).

many other parts of the *Res rustica* (book 11, for example, on preservation) to fill out his discussions of particular fruits. This wide-ranging use naturally suggests that G. read Columella's arboricultural advice in the context of the larger *Res rustica*, that is, in its place within book 5 of that work, but it cannot be wholly ruled out that a copy of the so-called *liber de arboribus* was also in his hands at some time.[57] The *De arboribus*, which is transmitted in manuscripts of the *Res rustica* but is evidently an excerpt from a different work, contains advice on viticulture and arboriculture, and overlaps extensively in its material with Col. 5, sharing many whole passages nearly verbatim. It was long thought to be a work of Columella's youth, but its authenticity has been doubted in recent decades; if it is spurious, it will probably have been assembled sometime by the second or third centuries, still early enough for G. to have known it.[58] It is in any event apparent that G. relies heavily on Columella throughout *Pom.*, quoting him directly once (3.1.4) and referring to him by name in four other places (3.1.3, 3.3.5, 4.6.5, 4.7.2). These figures may however give a less than accurate impression of the extent of Columella's presence in the text, for G.'s more common practice is to reproduce Columella's advice tacitly (i.e. without attribution), reshaping it in some important respects but often retaining tell-tale elements of the original diction and syntax.[59] Besides the debt owed to his advice, *Pom.* also shows

[57] Because Col. 5 and *Arb.* are so close in language and advice in their respective sections on arboriculture, it would be hard to find conclusive evidence for G.'s use of *Arb.*; there are a handful of echoes of statements found in *Arb.* but not Col. (see e.g. 2.5.7n. *largus umor*; 3.1.4n. *radices exitum moliuntur*), but these slight connections are inconclusive in view of the uncertainty surrounding G.'s other sources.

[58] The case for its inauthenticity is made by Richter (1972) (summarized at (1981–3) III 600–1), who is joined by the editor of the most recent and authoritative edition (Rodgers (2010) xvi). See also Hentz (1983). Goujard (1979), (1986*a*) 10–11, (1986*b*) defends its authenticity. For a conspectus of commentators' opinions, before and after Richter, see Fögen (2009) 158 n. 17; and for a very concise overview of the issue, see Hine (2011) 635 n. 39.

[59] This situation is very readily shown by consulting 1.1. Although Columella is never named here, his advice seems to underpin G.'s writings as a remote

significant linguistic influence from Columella: even where the instructions are not based on him, G. uses words and phrases that are distinctly Columellan in their pedigree.

Pliny the Elder (23/24–79 CE) is second among surviving authors only to Columella in his influence on *Pom.*[60] How well acquainted G. was with the *Historia naturalis* (*HN*) is perhaps best illustrated by his use of that work in *Med.*: of its 60 chapters, all but seven are carefully based on Pliny's voluminous writings.[61] The *HN* is not as overwhelmingly present in *Pom.*, but it is still an important source. That G. would have found much of relevance for his project in the *HN* is hardly surprising, for Pliny compiles, organizes and comments on a great deal of agricultural information in his work: before G.'s time, book 17 of the *HN* can be matched only by Columella's *Res rustica* 5 for the detailed information it provides on the cultivation of fruit trees.[62] There are many places where Pliny's instructions overlap with Columella's, presumably where they have a common source (e.g. Mago, although it often remains unidentified); G. comments on such agreement in one case (4.6.5n.), but it is more typical for him to produce a silent composite of their writings:[63] in these instances, if their respective contributions can be disentangled at all (often they cannot), it is only by close study of the language and details of the advice. The *HN* also contains much advice employed by G. that is found in no other

or proximate source in several places: see 1.1.4–5n., 1.1.8n., 1.1.10n. (the advice also in Pliny, but the language points to Columella), 1.1.11n., 1.1.13n.

60 For Pliny as an agricultural writer, see White (1970) 28–9, Martin (1971) 375–85, Frederiksen (1980), Beagon (1992) 161–77 and *passim*, Marcone (1997) 31–4, Doody (2007), Hardy and Totelin (2016) 20–1.

61 See Maire (2002*a*) lvii–lx, (2002*b*). As Maire observes, however, G.'s use of the *HN* in *Med.* is hardly passive.

62 Most of G.'s references to Pliny can accordingly be traced to this book, but he also read elsewhere in it for information relevant to his purposes, such as 15 (on preservation of fruits). For an instance of statements collected even further afield, from *HN* 23 and 24, see 1.1.16n.

63 A very remarkable case of this is 4.5 (n.), a whole section that seems to be a tightly fitting join of parallel passages in Columella and Pliny.

extant work and on some occasions thus provides a crucial source for reconstructing the text of *Pom.* or interpreting its advice.[64] G. refers to Pliny by name only twice (4.1.5, 4.6.5), which gives a far from adequate idea of his importance; but he often suppresses Pliny's name in *Med.*, too, and he may have had good authorial reasons for doing so. The direct references to Pliny in *Pom.* do not clearly reveal G.'s attitude to his instruction, but the extensive if tacit use made of it both here and in *Med.* suggests high esteem.

The final datable authors whom G. employs are the **Quintilii** brothers (both d. 183 CE), Sex. Quintilius Condianus and Sex. Quintilius Valerius Maximus.[65] The brothers were famous for doing everything together: they shared the consulship in 151 and were both victims of Commodus. Together they wrote an agricultural work in Greek.[66] G. cites them frequently and in every chapter of *Pom.* (6x altogether: 1.1.14, 2.1.2, 2.4.3, 3.3.1, 4.1.9, 4.6.6). If their work had survived we might have seen how much the language and instruction of *Pom.* owes to them, but we are hardly in a position to conclude with White (1970) 29 that 'much of Gargilius Martialis probably comes directly from them'. Palladius never cites the Quintilii, but they are referred to often (19x) in the *Geoponica*.[67]

Of the remaining figures whom G. cites, **Curtius Iustus** and **Iulius Fronticus**, virtually nothing is known. Unless the agronomist Curtius Iustus is to be identified with the senator of the Antonine period C. Curtius Iustus (*cos.* 150 or 151 CE), or with his son,[68] there is no trace of his existence except for the

[64] For example, Pliny's text supports the emendation of *Pom.* at 3.7.2 from *pruna* (**N**) to *pituita*: see 3.7.2n.

[65] For the Quintilii, see Boll (1911), *PIR*[2] Q no. 21 (Sex. Quintilius Condianus), 27 (Sex. Quintilius Valerius Maximus), *RE* (1) 24.984.27–985.13 s.v. 'Quintilius', 22 (Hanslik), 986.59–987.3 s.v. 'Quintilius', 27 (*id.*), White (1970) 29.

[66] For the evidence see Gemoll (1883) 192–3, but there are otherwise few indications of its original form.

[67] For the appearances, see Dalby (2011) 46.

[68] For C. Curtius Iustus, see Birley (2005) 254–5.

references in *Pom.* (named 4x: 2.1.2, 2.4.2, 2.7.4, 4.1.7). A Iulius Fronticus, as the name is transmitted in the manuscript (4.1.8), is likewise unknown, and indeed, the name Fronticus appears nowhere else in the Latin language.[69] It has been suggested that the reference should be to Sex. Iulius Frontinus, who wrote a number of technical treatises, including one on land-surveying – does G.'s statement that 'Fronticus' gave advice *soli de genere* (4.1.8n.) point in this direction? – , but is not known to have penned anything on agriculture.[70] Another suggestion for Fronticus is Fronto. An author by that name is mentioned as a source for agricultural instruction in the *Geoponica*, but because nothing further is known of him, there would be no certainty in the alteration.[71]

We are fortunate that we know as much as we do about G.'s sources, and that several of the principal ones, especially Columella and Pliny, survive intact. But the uncertainties that remain, and the loss of so many other sources, place severe limitations on our attempts to assess G.'s precise debt to his agricultural predecessors. Which did G. use directly? Which indirectly? In what form did he find the instructions he reports? When he adds details, are they his own, or those of an intermediary? Even in cases where there seems to be a transparent parallel with advice found in Columella or Pliny, it is difficult to be sure that these are the writers upon whom G. relied: the similarity could as easily derive from a common source such as Mago, Diophanes, Celsus or indeed another author unknown to us. Besides, how many of these earlier authors might have been filtered through those later, such as the Quintilii, before reaching

[69] It is absent from Kajanto (1965).

[70] For the identification with Frontinus, see Scotti (1833) 130 n. 1, *RE* (1) 10.603.42–5 s.v. 'Iulius', 243 (Kappelmacher), Condorelli (1978) xix n. 7. For Frontinus' writings on land-surveying, see concisely Campbell (2000) xxvii–xxxi.

[71] For the identification with Fronto, see Scotti (1833) 130 n. 1, *RE* (1) 7.112.50–61 s.v. 'Fronto', 13 (Wellmann); for Fronto as a source in the *Geoponica*, see Dalby (2011) 43.

G.? In some cases the strength of the linguistic parallels with a specific source may persuade us of a direct relationship; more often, we must be content with a reconstruction necessarily tentative and hypothetical. One among many possible examples will suffice to illustrate the challenge. In *amygd.*, G. reports a technique of slowly filling in the planting hole for the almond in proportion to the amount of growth of the young tree: 3.2.3n. *Hanc paulatim adobruunt singula arboris incrementa numerantes, tantumque radicibus cumuli largiuntur quantum altitudinis truncus adiecerit.* The technique is reported by Columella (5.5.4 = *Arb.* 5.4), who in turn puts it in the mouth of Mago (though *Arb.* does not); Columella goes on to critique and revise Mago's technique. Since G. does not tell us where he found the advice, it is up to us to guess: Mago? Columella? *De arboribus*? (Not to mention other authors in the Magonian tradition, such as Diophanes.) There is a further complication: we in fact find the substance of the recommendation already in Theophrastus (*Caus. pl.* 3.4.2). So it seems that we must put on the table G.'s elusive *Graeci auctores*,[72] plus any of their epitomators and the early Roman agronomists who may have read the Greeks on this point. In order to stop this speculative regress, it is useful to invoke the cautionary remarks made by White (1973) 470 apropos of Varro's relationship to Mago:[73]

> Did writer A use source Q directly, or only through an epitomator or translator? Did A misunderstand or mistranslate Q, and so on? This is a legitimate form of inquiry only if identifiable remains of the earlier writer are available for comparative study; we cannot conjure up a text from later citations, and then attempt to define the relationship of any particular writer to our putative source.

There is a value in inquiring into G.'s sources where we can, not only for the history of agriculture but also in order to better

[72] G. mentions the *Graeci* (sc. *auctores*) 7x in *Pom.* (2.1.3, 2.4.5, 2.11.2, 2.12.5, 4.2.1, 4.3.1, 4.7.3), never specifying any members of this group by name.

[73] Cf. also White (1973) 469–70, 474–5 and *passim*, Skydsgaard (1968) 64–88, esp. 66.

appreciate his authorial strategies, but our ignorance about many of his sources will always impose limits on the questions to which we can find satisfactory answers.

This juncture brings us to a final and important aspect of G.'s relationship with previous authors. When studying G.'s sources, we must always be aware that his use of them is neither naïve nor passive; on the contrary, it is often inflected by specific rhetorical aims that characterize the literary and didactic project of *Pom.* (see below, §III(b)).[74] Each of G.'s decisions in handling earlier authorities – to name a source or not; to suppress, augment, revise or restructure their advice; to adopt or discard the original language – should be considered with a view to the range of potential authorial motivations. For example, disagreeing with or correcting an authority would be a way for G. to display erudition and ensure that his own name remains connected with the agronomic tradition.[75] Eliding mention of a source might work towards a similar end, by muting intellectual rivals or else giving the impression of G.'s independent control of material. Besides, he may have been inclined to quote a specific author among other reasons in order to make a didactic point about the quality or transmission of advice, in order to lend the weight of authoritative consensus to a particular instruction,[76] or even for *uariatio*, so as to demonstrate his wide reading and avoid the appearance of relying too much on a single author.[77] If these considerations offer another stumbling block to the reconstruction of G.'s sources, they are also valuable for the light that they shed on him as an independent voice in the agricultural tradition.

[74] Maire (2002*a*) liv–lx, (2002*b*) has stressed the same point about G.'s use of medical authorities in *Med.*

[75] This strategy is common among technical authors of all stripes. For agriculture, consider Pliny's polemic against Virgil: Bruère (1956), Howe (1985) 563, 570–2, Doody (2007).

[76] This aspect is discussed below, §III(b), under the rubric of G.'s didactic strategy of 'extensive layering'.

[77] For remarks on literary *uariatio* in *Pom.*, see below, §V(a).

III STRUCTURE AND METHOD

(a) Structure

As mentioned already (§II), *Pom.* is unique both in its form and in the scope of its instruction. Earlier authors such as Columella and Pliny had devoted significant space to the cultivation of fruit trees, but *Pom.* differs from them in its evidently specialized orientation: whereas the reader of Columella and Pliny had to jump within or between books in order to gather all of the information germane to a single fruit tree,[78] the chapters of *Pom.* seem designed for those who want in front of them all at once the full information regarding the cultivation of a given species. Thus, the most distinctive feature about the structure of *Pom.* is the relatively self-sufficient and comprehensive treatment offered by each chapter (i.e. for each tree), which presents in a systematic fashion and under the rubric of a single species the horticultural 'history' of a plant from seed to fruit.[79] This specialized chapter-structure must be motivated, at least in part, by the much greater volume of information that G. controls in comparison with earlier writers on arboriculture: whereas Columella, for example, had only devoted two lines, about 30 words, to the peach (5.10.20 planting time, 11.2.11 grafting time), a new fruit at Rome in his day,[80] G.'s chapter *persic.* (although truncated at both ends and lacunose in the middle) runs to 12 sections, a little under

[78] Consider the example of the pomegranate: Columella's reader would find general information on preparing a nursery for fruit trees at 5.10.1–9; specific advice for the pomegranate including planting time, remedies for improving the produce and a technique for preserving at 5.10.15–16; general information on grafting fruit trees at 5.11; and then very detailed advice for preserving pomegranates at 12.46. Pliny's treatment in book 17 is also diffuse, and valuable information on preserving is reserved for other books (e.g. 15). Of course, moving between books, chapters and even sections would not have been easy in a papyrus roll.

[79] This consideration does not necessarily mean that G. included no discussion of important general aspects of agriculture, such as manuring or grafting, elsewhere in his work, although that may be the case: see below, this section.

[80] See above, n. 47.

1,100 words. The scale of G.'s achievement stands out across all of Greco-Roman antiquity, and can be illustrated by consulting Palladius, who, writing at least a century after G. and using him extensively as a source,[81] devotes fewer than half the words to the peach in his corresponding chapter (12.7.1–8). G. seems to have wished to excel his predecessors not only in convenience, but also in detail: *Pom.* would supersede previous writings and offer a complete guide to the arboricultural state of the art.

What did a complete chapter of *Pom.* look like? Although no single chapter survives in its entirety, the surviving fragments overlap enough to allow us to reconstruct its typical shape. We are least confident about the form that the opening of a chapter would have taken, for no chapter survives from its beginning. G. may have had some remarks of a more general character to make, but in any case, he would quickly have turned to issues of climate, soil, planting time and preparation of the plants. The internal organization of each chapter varies in some respects (see below, this section), but they are on the whole consistent enough to describe a general structure, as given in Table 2.

Table 2: The Structure of a Chapter

Subject	Components of subject	Representative sections
Preparation	Planting time	2.1–3; 4.1–3
	Seed storage and preparation	2.2–3; 4.2–4
	Climate and situation	2.3–4; 4.6
	Soil quality	2.4; 4.5
	Preparation of soil	3.1, 3.3; 4.7
Planting	Propagation, by seeds or slips	2.5; 3.1, 3.3
Cultivation (of the young plant in the nursery)	Watering, weeding, hoeing, etc.	2.6; 3.1–2
	Transplantation and establishment in orchard	2.6–7; 3.2, 3.4

[81] On Palladius' use of G., see below, §VI.

Subject	Components of subject	Representative sections
Cultivation (of the mature plant in the orchard)	Pruning of branches and roots; maintenance and preventive measures	2.8; 3.5
	Improvement of produce	2.9; 3.6–7
[Grafting][82]	Grafting proper	2.10
	Budding	2.11
Harvesting and Preservation	Gathering	3.8
	Preservation	1.1; 2.12; 3.8
	Transportation to market[83]	1.1

According to our hypothesis about the original form of the work (above, §1(b)), and as indicated by the portion of *Med.* transmitted immediately following 1.1 in our manuscript **N**,[84] after harvesting and preservation the chapter would have led seamlessly into a discussion of the therapeutic uses of the fruit, that is, into the corresponding chapter of *Med.*

The chapter structure of *Pom.* as presented in Table 2 corresponds remarkably well to the old sixfold division of the agricultural year that Varro has Stolo propose in the *De re rustica* (1.37.4). The *gradus* of Stolo's division are based both on the seasons of the year and on the natural growth of the plant: *est altera . . . temporum diuisio coniuncta quodam modo cum sole et luna* [*quin* del. Goetz] *sexpertita, quod omnis fere fructus quinto denique gradu peruenit ad perfectum ac uidet in uilla dolium ac modium, unde sexto prodit ad usum. primo praeparandum, secundo serendum, tertio nutricandum, quarto legendum, quinto condendum, sexto promendum.*[85]

82 Grafting is bracketed here because it is not found in *amygd.* (for a discussion of the omission, see 3.1–4n.), and it is thus uncertain whether it was a regular part of other chapters; possibly it was not. (*cydon.* and *cast.* are too lacunose to be informative on the question.)

83 For the oblique reference to transportation of fruit, see 1.1n.

84 For the text of the portion shared between **N** and the MSS of *Med.*, see Appendix 1.

85 On Stolo's division, see further Skydsgaard (1968) 21–5.

Stolo's natural division of the agricultural year is at the same time a natural division and ordering of the subject matter for treatment in speech or writing, in the sense that it takes the operations involved in cultivating a plant as they are imagined following from one another on a physical or biological clock, as it were. The natural structure indeed imposes a great deal of order on the chapter-level organization of *Pom.* It does not, however, always account as neatly for the sequence of thought within and sometimes across individual sections. This is hardly surprising, for different plants have their own peculiar and often variable developmental schedules, which do not always overlap with the ordinary agenda of the seasonal calendar; what is more, didactic considerations, such as the need to explain unusual or potentially confusing instructions, also pull G. from time to time away from the strict subject of discussion. Thus, more often than not the organization of the material within a particular section, that is, on a sentence-to-sentence level, follows a fuller, 'mixed' style, which takes its cue partly from the expected seasonal calendar, partly from the peculiarities of cultivating a specific plant, and partly from didactic exigency. The varying demands of these agendas not infrequently lead G. to reject a strict rubric for organization: he digresses in order to treat questions germane to a topic that he has introduced, and sometimes shows an almost scholarly (some might say pedantic) interest in correcting the advice of others or appending relevant information.

This full, mixed style can be illustrated by a brief overview of the structure of the first section of *persic.* (2.1). At the outset of the section, G. is reporting the planting times recommended by a number of agricultural authorities. Some (Curtius Iustus and the Quintilii; perhaps others lost in the lacuna) moot a winter planting time (2.1.2), but the Greeks unanimously recommend autumn (2.1.3). G. thinks to supply the reason for this fairly wide difference of opinion, which is the observation that peaches that have fallen from the tree sprout of their own accord around autumn (2.1.3 *ratione naturali . . . siquidem*, etc.).

This observation leads into a short digression on planting technique: G. explains that, for the same natural reason (2.1.3 *ideoque*), many believe that the entire peach fruit, not simply the stone, ought to be planted, so that the flesh can nourish and sweeten the seed. There follows yet another digression apropos of the mention of the fruit, in which G. says that even in his day some erroneously believe that peach stones which have touched someone's teeth become sterile or degraded. The digression concludes only in 2.2 (n.), which, with the introduction of a new planting time (2.2.1), reinstates the strict subject of discussion. The example offered by 2.1 is a relatively stark one, but it gives a good idea of the flexible and discursive development of G.'s ideas within the typical structure of the chapter outlined above.[86]

If we can feel reasonably confident in our appreciation of the scope and structure of a chapter in *Pom.*, there nevertheless remain some fundamental questions about the comprehensiveness of the treatise as a whole. It has already been suggested (§1(b)) that in the original *De hortis* there would have been many more chapters (probably 56[87]) covering additional fruits, vegetables and herbs, as well as corresponding discussions of their therapeutic use. This does not tell us everything we would like to know about *Pom.*, however. The most important question for understanding the work's advice and didactic goals is whether anywhere in the treatise there was a systematic discussion of the most general aspects of arboriculture, from the establishment of farm facilities and irrigation systems to methods for digging, manuring, grafting, establishing the trees in the orchard and so on. The question is appropriate on both intrinsic and extrinsic grounds, that is, not only because many

[86] In the introductory notes to individual sections and groups of sections in the commentary, I attempt to trace the arc of each chapter according both to the structure of a 'typical' chapter presented above and to G.'s tendency for explanatory digression. Cf. also 3.5–7n.

[87] See above, n. 19.

of G.'s recommendations presuppose a prior familiarity with such topics if they are to be intelligible,[88] but also because every other surviving agricultural work from Greco-Roman antiquity, including Palladius' *Opus agriculturae*, which owes much to *Pom.* for its arboricultural advice, contains some general treatment of those issues.

Although the lacunose state of the treatise does not allow certainty in answering these questions, there are two possibilities that we should entertain. The first is that *Pom.* (in the context of the original *De hortis*) contained no such general discussion. The strongest argument for this position is the *a priori* one from the structure of the work, namely, that there seems to be no point in the apparent chapter organization where this discussion would fit. On the basis of the specialized orientation of the work hypothesized above (this section), it could be further argued that G. might have conceived of *Pom.* as a kind of supplement for those who were already knowledgeable about agriculture but desired a more detailed investigation of the peculiar character of different fruit trees. In this case, he might simply have assumed a certain body of agricultural information that would be known to his readers from personal experience and other authors' works and dispensed with discussion of points that were not contentious. It is thus instructive to compare the tenor of Pliny's remark at the beginning of his discussion of trees: *HN* 17.9 *non uolgata tractabimus nec quae constare animo aduertimus, sed incerta atque dubia, in quibus maxime fallitur uita.*[89]

The strength of the *a priori* argument should not be overstated, however. There is of course no reason why G. could not

[88] For one example among many, see the remarks on grafting at 2.10n. For further reflection on the difficulties of appreciating G.'s advice, see below, §IV.

[89] But it should be noted that Pliny's remark is not an advertisement of utilitarian brevity: consider for example the elaborate overview of grafting at *HN* 17.99–123, which includes a detailed natural history of the technique.

have inserted general remarks as prefatory or concluding matter in the original treatise. (The absence of such a section in *Med.* is no conclusive argument against its existence.[90]) In this vein it should be noticed that the opening line of our manuscript, 1.1.1 ****uerum de fructibus pauca nobis dicenda sunt*, appears to be a kind of transitional statement that would have been at home in a discursive section formally different from the four chapters of *Pom.* that we now have. Furthermore, and the unique form of G.'s treatise notwithstanding, there are generic precedents for an organization of agricultural material that places general remarks in sections preceding or following the discussion of specific plants.[91] But even were we to dismiss the hypothesis of a more general section external to the chapters on specific plants, G. could easily have digressed within (sequentially earlier?) chapters to insert a more extended treatment of an issue of wider relevance; Palladius occasionally does so,[92] and by all indications G. is even more prone to digress from a strict presentation of material than he is. Although in the final analysis we cannot have a sure answer to our questions about *Pom.*'s treatment of general agricultural advice, there is a pragmatic solution that can at least make us better readers of what does survive: that is, to assess in each case the generic precedents and substantive *comparanda* from surviving Greek and Roman agricultural writing in order to form an idea of the information 'missing' from the work (for the sources helpful in such reconstruction, see further below, §IV). From this perspective, we may set aside the question whether 'missing' means lost in the vagaries of transmission or omitted from the original treatise itself.

[90] Indeed, Maire (2002*a*) xxviii–xxxi has argued vigorously that we should suppose an original preface for *Med.* lost in the excerpting or subsequent transmission of the work.

[91] These precedents are discussed at length at 3.1–4n. as they relate to grafting.

[92] See, for example, Palladius' overview of patch-budding (*emplastratio*) in the chapter on fruits treated under the month of July (7.5.2–4).

(b) Didactic Method

Theophrastus laments the difficulty of the scientific study of plants owing to their variable parts and growth;[93] consensus in the practical study of their cultivation may have seemed in some respects more attainable, but here too there was a wealth of potentially conflicting observations and techniques that the cultivator had to handle for himself and, if he purported to teach the art, also for his audience. In order to show how G. navigates the challenges involved in evaluating and communicating agricultural knowledge, it will be useful to describe in brief compass the didactic method of *Pom.* Characterizing G.'s didactic method not only allows us to situate him in the technical tradition of agriculture, but also presents a key to better understanding the presentational logic of advice within *Pom.* itself.

One of the principal challenges to the reader of *Pom.* is the number of different instructional 'styles' or voices that coexist in the text. In particular, *Pom.* alternates among three distinct modes of presenting agricultural facts and advice: (1) seemingly uncritical report of advice (whether from named or unnamed authorities), (2) forthright approbation or rejection of advice in G.'s own voice, and (3) critical but oblique approbation or rejection of advice. All of these instructional styles can be well illustrated by *cydon.* (1.1n.): the section opens (1.1.6) with an unspecified *aliqui* who recommend one method for preserving the quince (this is the first style); G. counters (1.1.7) in the first person (*nos*) with a different recommendation (second style); next he presents a slew of alternative techniques (1.1.8–13), none of which he explicitly endorses or rejects, attributed to unspecified groups of authorities (*quidam, sunt qui, multi, aliqui*; all first style); next he reports a method espoused by the Quintilii (1.1.14), again without passing judgement (first style); and, finally, he introduces a further general statement about quince

[93] Cf. e.g. *Hist. pl.* 1.1.10 ὅλως δὲ πολύχουν τὸ φυτὸν καὶ ποικίλον, καὶ χαλεπὸν εἰπεῖν καθόλου, κτλ.

preservation (1.1.15) to which he attaches oblique approval (cf. *Diligentioribus rectissime uisum est*; thus, third style).[94]

Besides these three styles of presentation, there are moreover stretches of the text where the voice seems especially ambiguous, so that G.'s own thoughts or opinions may become indistinguishable from those of the authorities whom he reports, whether or not they are named. This occurrence can be concisely illustrated by a sentence from *amygd.*: 3.4.3 *Bene Diophanes ait amygdalas in transitu esse ponendas, quoniam non semel maturant et per partes se fructus eius accommodant, nec possint ita fures dispoliare*, etc. After attributing to Diophanes the recommendation to plant almonds along a crosspath, G. adds an explanation in the *quoniam*-cause that is subsequently amplified by the words *nec possint*, etc. (The basic idea here is to avoid thievery of individual ripe almonds on the tree before the owner is ready to harvest the entire lot of them.[95]) Now, is the explanation of the slow maturation of the almonds (*quoniam . . . accommodant*) Diophanes', or has G. supplied it himself in order to explain (and endorse) Diophanes' advice?[96] The same question can be asked about the elaboration *nec possint*, etc., which offers an explanation of a slightly higher order. Although it may be unimportant in the present instance to assign credit precisely (except perhaps from the point of view of *Quellenforschung*), what should be observed is how difficult it is to disentangle the voices of Diophanes and G. even within a single sentence. The phenomenon is typical of many other places in *Pom.*, and the interpretative difficulties

[94] For the approval conveyed by *diligentioribus* and *rectissime*, see nn. *ad loc.*

[95] For further discussion, see 3.4.3n.

[96] The rule that if a causal clause contains an alleged or attributed reason we would expect the subjunctive (see 2.1.3n. *siquidem . . . uideamus*) is not decisive here: *quoniam* reason-clauses with the subjunctive are rare (Pinkster 650), and even if Diophanes had offered the reason for the advice, it is possible that G. would have used the indicative if he thought the reason true or considered it an obvious fact (as when modal assimilation to the subjunctive is avoided 'if doubt might arise as to the factual status of the content of the clause' (Pinkster 668)).

are only exaggerated as G. carries over advice from sentence to sentence and section to section.

With all of these different voices and potential ambiguities in the text, what message is the reader supposed to gather from a chapter? Is there a single 'story' about the cultivation of each fruit tree that the reader ought to extract from G.'s writings? It may be reasonable to doubt whether these questions are even appropriate to ask of *Pom.* If the work was intended as a specialized and comprehensive treatment of the included trees (and it seems clear that it was), then G. might have been content simply to gather and present as much information as he thought worthy of mention. His role would thus be one of *compiling* and *describing* the methods of cultivation that he knew of for each tree, pre-selecting and curating but not necessarily judging the value of every piece of advice. It would be up to the (more or less expert) reader to weigh a given set of instructions for himself and decide whether and how to adapt it for his own purposes. This interpretation of *Pom.* has some advantages, and, following it, we may wish to call *Pom.* 'encyclopaedic' in a very qualified sense.[97] But the approach is not entirely adequate. For one thing, it cannot explain the second and third instructional styles discussed above, except as arbitrary interventions into the text on G.'s part. Even more significantly, however, it does not account for the ways in which the very pre-selection and curation of information can be tacitly informed by didactic aims.

As I argue, the different voices in *Pom.* can in fact be understood in terms of a specific authorial technique which I will call 'didactic layering'. By 'didactic layering', I mean a purposive repetition, or 'layering', of facts and instructions that gradually and implicitly establishes consensus around a particular piece of advice. I distinguish two such kinds of layering in *Pom.*, which I

[97] It must be heavily qualified because the word is controversial in its application to Greek and Roman writings: see e.g. Doody (2009), (2010) 11–23, König and Woolf (2013*b*), esp. 1–5, König and Woolf (2013*c*); and more generally, the articles collected in König and Woolf (2013*a*).

will call 'extensive' and 'intensive'. The easiest way to appreciate how these extensive and intensive techniques of didactic layering structure the information that G. presents is to consider a few carefully chosen and representative examples. I describe first 'extensive' didactic layering. Extensive layering refers to the way that the repetition of a strand of advice in *successive sections* within a chapter, or 'extensively', lends implicit priority to that advice. That is, by selectively repeating advice from earlier sections, G. invites the reader to share his assumption of *this* recommendation rather than *that* one, for it is only on the basis of *this* recommendation that the following instructions can be understood.

Consider from this perspective the case of the planting time and method for the peach (2.1–3). In 2.1, we hear of two possible planting windows: January (in warm climates, 2.1.2 *calidioribus locis*) and autumn (2.1.3).[98] The autumn planting (*autumnalis satio*) supposes that the cultivator will plant not only the pips, but the entire peach fruits (*poma ipsa matura integra*), taking a cue from the fact that fallen peaches naturally sprout in this season. After a short digression on this subject, G. returns in 2.2 to the question of planting time and reports that spring (2.2.1 *ab exortu Fauoni*) has also been recommended for the peach; in order to make the planting viable, however, one must preserve the seeds in the interim to avoid rot (2.2.2). In the next section, 2.3, G. has yet another word for the planting time, mentioning autumn again and commending it in the case of a warm climate (2.3.2 *In calido merito sub ipsum tempus autumni ossa merguntur*). In the remainder of 2.3 he downplays fear of the winter cold and reports Celsus' technique for preparing the seeds for planting by soaking them in water (2.3.3–5). In the movement across sections 2.1–3 summarized briefly here, we notice that 2.3 restores both the warm climate (2.1.2 *calidioribus locis* / 2.3.2 *calido*) and the autumn planting (2.1.3 *autumnalis satio* / 2.3.2 *sub ipsum tempus autumni*) mentioned in 2.1 and assumes these as the

[98] It is possible that additional planting times were offered in the lacuna at the beginning of the section: see 2.1n.

basis for further instruction. This assumption is not a wholly neutral one, but rather evinces the method of 'extensive' didactic layering. In the first place, it deemphasizes Curtius Iustus' recommendation for a January planting in warm climates, and perhaps also the Quintilii's recommendation to plant apricots in the same month. In the second place, the report in 2.3 that it is the stones that are planted (2.3.2 *ossa merguntur*) deemphasizes the Greeks' earlier recommendation to plant the ripe fruits whole (2.1.3 *non ossa sed poma ipsa matura integra*). It is not that G. formally renounces the planting times and methods aired in 2.1, but rather that the selection and repetition of certain details encourages the readers to adopt a particular strand of advice if they are to follow the discussion of *Pom.* further.

A word of qualification is in order here. I do not mean to suggest that the method of didactic layering should be understood as completely overriding the descriptive aspect of *Pom.* discussed above by muting all other strands of advice so as to establish one right method of cultivation. This would be an equally inadequate approach to the text, for there are many places where ambiguity and alternatives in instruction are not controversial and do not force a choice in techniques of cultivation on the reader. Furthermore, even where a strand of advice is deemphasized (such as the January planting time or the method of planting whole peaches given in 2.1.3), it is not entirely removed as an option for the reader; readers may still form their own opinion of a recommendation, even if it is downplayed by G. It may therefore be best to speak of didactic layering in terms of the establishment of different didactic 'strata' of varying epistemic authority in the text, rather than in terms of firm prescription: the selection, repetition and placement of information guide our understanding of the relative *applicability* of different methods of cultivation. Those methods which are repeated and assumed for further instruction are more widely *applicable* in the specific sense that they are given a more prominent role in the configuration of different techniques for cultivating trees. On the other hand, those methods which are not repeated are not necessarily expunged as possibilities, but rather become relatively less prominent among the multiplicity of possible methods for cultivation.

With this important qualification in mind, we may now turn to the second kind of didactic layering, which I called 'intensive'. Whereas 'extensive' didactic layering describes repetition in successive sections within a chapter, 'intensive' layering describes repetition within a *single section*. That is, just as the prominence of a strand of advice can be established extensively by selecting and reusing it in the course of a chapter, so too can it be established intensively, by selecting and concentrating within a brief space the opinions of several different authorities. These overlapping and authoritative opinions establish a weight of consensus around some particular fact or piece of advice. For a good illustration of intensive layering, we may consider the opening section of *cast.* (4.1n.), where G. reports the opinions of no fewer than eight authorities on the question of the best time to plant the chestnut. Although their opinions vary considerably, one dominant recommendation is discerned: five of the eight authorities prefer a planting time in the winter.[99] No other timeframe receives as much support, and the effect of the repetition of the winter planting time is to establish its prominence within the range of recommendations. There is a secondary didactic stratum (in the sense of the term suggested above) that is established by the agreement of Celsus and Mago, who both prefer a planting time for the chestnut that is the same as the almond; G. however intervenes to censure their choice, rebuking Celsus for not correcting Mago.[100] When

[99] Columella (4.1.4), Pliny (4.1.5), Iulius Atticus (4.1.6), Curtius Iustus (4.1.7) and the Quintilii (4.1.9).

[100] 4.1.3n. *Cornelium tamen Celsum, Italicae disciplinae peritissimum, decuit aliquatenus edocere quidquid Magonis ignorantiam fugerat; sed et ipse castaneas amygdalorum exemplo reiectat.* It is not certain which planting time Mago has in mind when he says 4.1.2 *castaneam . . . sicut Graecam nucem* (= *amygdalum*) *serito*, because opinions about the planting time of the almond varied (for a conspectus of the surviving recommendations, see 3.1n.).

readers leave this section, they may thus feel that a winter planting time is the soundest choice for the almond, even though G. never explicitly joins his voice to those of Columella, Pliny, Iulius Atticus, Curtius Iustus and the Quintilii. This persuasion would be the result of the 'intensive' didactic layering within the section.[101] This kind of layering can be further manipulated and reinforced by G.'s decision to name authorities or to group and report them with more or less colourless expressions such as *sunt qui*, *multi*, *aliqui*, *quidam*, etc.:[102] by flattening and anonymizing the perhaps numerous authorities behind a set of instructions, these expressions allow G. to control when he will or will not marshal the weight of earlier writers.

In sum, it is possible to make sense of the multitude of voices and styles of presentation in *Pom.* both descriptively and prescriptively. That is, it is no doubt correct up to a point to understand G.'s goal as that of compiling and presenting advice to an informed audience that could make its own decisions about the usefulness of this information. But on the other hand, we have suggested that this descriptive aspect of *Pom.* must be qualified by an appreciation for the way in which indirect didactic strategies, especially that of 'didactic layering' proposed here, allow G. to implicitly evaluate and give prominence to certain of the recommendations that he reports. This prominence is not necessarily to be understood in terms of strict injunction, but rather as the establishment of epistemic strata in the text that reflect the relative applicability of the advice. Of course, we must further consider the many places where G. intervenes directly or indirectly to approve or reject advice (these were the second and third instructional styles discussed above). It is only

[101] It is not necessarily the final word, however: after packaging the opinions of 4.1, G. goes on to discuss additional planting times in 4.2–3, and in 4.3 elaborates the spring planting that was also mentioned in connection with Columella at 4.1.4.

[102] These are extremely common in *Pom.*: see the *index nominum* of Condorelli (1978). Maire (1997) 309–10 catalogues the *uariatio* in G.'s presentation of sources in *Med.*

when we take into consideration the full range of both descriptive and prescriptive aspects of *Pom.* that we can understand the didactic method that informs the work.

IV UNDERSTANDING THE AGRICULTURE OF *DE ARBORIBVS POMIFERIS*

The agricultural instruction of *Pom.* is not always easy to interpret. G.'s recommendations frequently come without explanation, whether because he gave greater detail in a part of his writings now lost or because he assumed a certain measure of background knowledge from his audience (on the question, see above, §III(a)). Furthermore, our ability to appreciate these recommendations on their own terms is impeded by several factors. Not only do localized issues, such as textual corruption, hinder us from understanding crucial details of *Pom.*, but we must also reckon with more profound historical changes that shift the way we as readers approach the work: the present dominance of mechanized and scientifically based agriculture; the prevailing but tacit acceptance of its biochemical underpinnings (i.e. soil and plant science); and the great remove of most citizens of developed countries from agricultural activities (something historically unprecedented until recent times). G. wrote for an audience who, whatever their level of expertise, would have known at least something about the fundamental work on a Roman farm, from manuring and irrigation to transplanting and preservation. Furthermore, G. and his audience would have shared (up to a point) a similar set of beliefs and experiences that would have provided a framework for understanding the value and relevance of the issues addressed in *Pom.*[103] For us, so many centuries later, the questions raised by

[103] I do not mean to downplay the sophistication of much ancient agricultural advice, which has been readily recognized by the most informed authorities (for a classic appreciation, see White (1970)). For an investigation into the Romans' shared set of 'scientific' beliefs, see Lehoux (2012).

G.'s advice can range from the unfamiliar to the puzzling: What are the best spells and rituals to ensure an abundant harvest? Where does a peach tree feed from, and how does that affect planting techniques for it? Do roots and weeds smell, and, if so, why would that matter? Why must one align the almond seed with the cardinal directions? Does the quince really taint other fruit stored with it?[104] In the commentary, I have not only attempted to provide background for general topics in ancient agriculture as they become relevant to the instructions of *Pom.*, but also to reconstruct a basis for understanding the more puzzling aspects of G.'s advice. Here I offer a brief overview of the ancient contexts which aid in understanding *Pom.*

The most significant source for appreciating the advice of *Pom.* is that of the **written tradition** of agricultural advice at Rome. Through consultation of the writings of the agronomists before G.'s time (see above, §II), as well as those writing after him but in the same tradition (see below, §VI), we can obtain a reasonably good idea of ancient attitudes towards trenching and preparing soil for the nursery; propagating plants through seeds, cuttings and grafts; manuring; irrigating; transplanting from the nursery; arranging the plants in the orchard; pruning, culling and nourishing the trees; and gathering and preserving the fruit for a variety of purposes.[105] Against the backdrop of the ancient consensus (or disagreement) about these issues, G.'s own recommendations acquire their full significance as a response or contribution to a venerable intellectual tradition;

[104] For discussion of these issues, see, respectively, 3.6.2–3n. (magic); 2.7.1n. with 2.7.1n. *contra . . . alimentum* (peach nutrition); 2.6.1n. *tenero germini noceant flatus herbarum* and 2.8.1n. *inhalantibus fibris* (roots and weeds); 3.1.4n. *aciem . . . Aquiloni* (almond seed); 1.1.15n. (quince storage). (For explanation of cross-reference to notes in this work see below, §IX.)

[105] See, respectively, 3.3.1n. *omnem . . . fodiunt* (trenching); 2.5.7n. (propagating from cuttings) and 2.10–11n. (grafting); Appendix II (manuring); 3.3.2n. *quo . . . immoretur* (irrigating); 2.5.6n. *transferant* (transplanting); 3.4.2n. *uiginti . . . pronuntiant* (arranging in orchard); 2.9.2n. *quotiens . . . imperatur* (culling); 3.5.1n. *ramos . . . praefractos* (pruning); 1.1n. (preserving fruit).

furthermore, his advice can be restored to a concrete sphere of agricultural activity in the Roman world. There are more specific challenges in *Pom.* that we can tackle with the help of the agronomic tradition, too: in addition to the puzzles mentioned above, we can hazard answers to questions such as why a watery climate is said to make peach sprouts fond of manure; why G. makes no mention of grafting in the case of the almond; what a 'Tarentine' nut is, and how to produce it; whether or not the chestnut is transplanted with special difficulty; and so on.[106] Certain of these problems may seem trivial to us now, but it would be unwise to dismiss them out of hand when we consider that in many cases the issues at stake were discussed and debated for well over a thousand years in the Greek and Roman agricultural traditions, from the time of Aristotle and Theophrastus to the Byzantine *Geoponica* (tenth century CE) and beyond.

As we consider the durability of the written tradition of ancient agronomy, we must be careful not to ignore the role that **practice** and **experience** played in its development. Neither *Pom.* nor any other Roman agricultural work reflects the reality of rural life in a straightforward way, of course; but G. and many of the other writers in the tradition were practising agriculturalists, and it is inconceivable that their personal experience and expertise did not influence their instructions. Indeed, this influence is seen most clearly in *Pom.* in the first-person statements with which G. introduces his own opinions into the text.[107] Sometimes our methods of investigation must be more indirect, however: we may, for example, draw on the archaeological and textual evidence for Roman agricultural implements in order to understand better the methods that G. mentions for pruning and hoeing, propping up plants and storing fruits;

[106] See, respectively, 2.4.2n. *sic . . . germina* (peach sprouts); 3.1–4n. (omission of grafting for the almond); 3.3.4–5n. (Tarentine nut); 4.4.2–3n. (transplantation of chestnut).

[107] See above, n. 7.

or we may adopt an anthropological approach that illuminates ancient techniques by comparison with contemporary subsistence farming practices in the Mediterranean.[108] While it is *ipso facto* impossible to recover the full experiential context for G.'s advice, adopting a practical perspective can help to balance our interpretation of textual sources.

Pom. also shows the influence in some places of agricultural **magic** and **ritual**: hang the shoe of a pack animal from a peach tree and the fruit will not drop; insert a honeyed wedge of wood into the tree and the almonds will become sweet.[109] It is hardly surprising to find G. purveying magical advice, which was always present to some degree in Greco-Roman agronomy.[110] What rather deserves emphasis is that such advice often freely combines what we think of as 'unscientific', superstitious elements with 'scientific' principles of agronomy. It is not always possible to tease apart these influences; nor, however, is it necessarily desirable, for the distinction is a false one from the standpoint of the ancient agronomists:[111] for the purposes of understanding G.'s advice, we must occasionally be willing to allow spells and ritual to coexist with his plant science.

This juncture brings us to the impact of **natural philosophy** on *Pom.* As one might expect, the direct influence of philosophical writing on the work is slight. Nevertheless, there are some cases in which G.'s advice can be best understood in light of the biological theories elaborated most fully

[108] For the material technology of Roman agriculture, see esp. White (1967), (1975), (1984) 58–72; for the archaeology of Roman produce gardens and orchards in general, see Jashemski (2018*a*), esp. 140–3, (2018*b*). For the comparative approach to the study of Mediterranean farming, see Halstead (2014).

[109] See, respectively, 2.9.3n. (pack-animal shoes); 3.6.2–3n. (wedge of wood).

[110] But in fact, *Pom.* shows a relatively low proportion of magical instruction, certainly in comparison with later Greek and Roman agricultural writings (Palladius, *Geoponica*). In general, we never see in agronomy the vigorous attempts to expurgate the magical and supernatural that are found in many (but not all) circles of ancient medicine (for which see Lloyd (1979) 10–58).

[111] Cf. Houston (1989) 79–80.

by the natural philosophers. Thus, it may be that we can find a clue in Theophrastus to understanding why digging around the almond will cause it to lose its flowers, or why pig manure applied to its roots is an especially effective remedy for a mature tree; or that we can appreciate the advice to plant the peach in shallow soil if we know something about the general theory of plant growth and nutrition elaborated by Aristotle and Theophrastus.[112] Of course, these theories were not the exclusive province of φυσικοί, and nothing compels us to believe that G. must have found them in philosophical works; but a reasonable scepticism here should not diminish the potential evidentiary value of such writings, for even if they are not a direct source for *Pom.*, they may witness intellectual currents that were relevant for its contents.

Although our primary focus has been on the ancient contexts for *Pom.*, it may in conclusion be useful to say a word about the advice of *Pom.* in comparison with **modern practices** of arboriculture: G. presents a remarkable synthesis of ancient knowledge, but how well can his recommendations be understood in light of modern agronomy? Even a modest attempt to answer this question yields interesting and often important insights into the details and purpose of G.'s instructions. In one case, the opinion of the modern authorities even enforces an emendation of the text.[113] In the commentary, I have attempted to compare G.'s instructions with a sampling of agricultural advice from both the late nineteenth and the late twentieth/early twenty-first centuries; for these purposes I have considered especially the practices in climates comparable to those of Italian (and other Mediterranean) fruit-growing regions, such as Spain or California.[114] Although G.'s advice does not admit

[112] See, respectively, 3.5.3n. (digging around the almond); 3.7.3n. (pig manure); 2.7.1n. and esp. 2.7.1n. *contra . . . alimentum* (peach nutrition).

[113] See 2.7.3n.

[114] These comparisons can only be approximate, however, and one should note that G. himself often calibrates his advice to planting conditions based on the climate of the region where the tree is to be planted: for a

of illuminating analysis in these terms in every case (especially his agricultural magic), in general these comparisons confirm the basic soundness of much of his advice: his remarks on, for example, irrigation, pruning, culling and preservation can all be paralleled and corroborated by modern practices,[115] and without attempting to vindicate G.'s theories with modern science (a tricky enterprise), it is still possible in some cases to speculate on their efficacy.[116] The study of *Pom.* from the perspective of the history of agriculture and technology is thus valuable not only for establishing its place in that history, but also for the ways in which it allows us to appreciate G. as a sensible and well-informed authority in his own time.

V LANGUAGE AND STYLE

One of the most intensely interesting features of *Pom.* is its language: as a technical work with literary ambitions, written in a period of scarcely attested but significant linguistic change, its language is a valuable object of study from a number of perspectives, not only for artistic aspects of its composition and its evidence for scientific and technical varieties of Latin, but also for the light it sheds on questions of historical change and social variation. In the commentary, in addition to attempting to explain the vocabulary and syntax of G.'s Latin, I have sought to determine the proper linguistic contexts for understanding what is, or is not (as the case may be), distinctive about his writing. In the following sections, I present summary remarks on the language of *Pom.* from four perspectives: its **literary** features (§v(a)); its **technical** and didactic features (§v(b)); its evidence for change and continuity in the **historical development** of

programmatic statement, cf. 2.3.1n. *Interest autem cuiusmodi statum caeli condicio regionis, calidum frigidumne, sortita sit.*

[115] See the respective references to notes in the commentary above, n. 105, and *passim.*

[116] See e.g. 1.1.10n. *cum amurca.*

Latin (§v(c)); and its relation to the question of **social variation** in Latin, especially that of 'Vulgar' Latin (§v(d)). Each of these sections will present a selection of themes and examples that are illustrative of the language of *Pom.* in respect of the perspective that is under discussion; these themes and examples will suffice to characterize a set of research interests that are pursued more extensively over the course of the commentary.

(a) Literary Features

The literary aspirations of *Pom.* are best judged not by a subjective reaction to its style, but by an index of G.'s attempts to enliven his work with elements drawn from the literary and rhetorical conventions of his time.[117] By this standard, G.'s literary ambitions are very clear: he quotes Virgil (4.1.1); attends to the rhythm of his cola and clausulae; uses metaphorical language and language drawn from a poetic register; creates multi-part sound effects with word choice and order; establishes symmetry between word groups and clauses; freshens his instructions with extensive and creative use of *uariatio*; and so on.[118] The literary quality of *Pom.* should not come as a surprise, for there was never a clean break between 'literary' and 'technical' varieties of writing as such,[119] and authors on technical subjects were

[117] Mazzini (1988) 44–6 and Maire (2002*a*) xxxi–xxxii collect impressions of the language of *Pom.* voiced by ancient and modern readers.

[118] Maire (1997), (1999), esp. 438, (2000), esp. 163–4, (2002*a*) xxxii–xxxvii, l–li, lx–lxiv, (2002*b*), esp. 534–5, has insisted on the fundamentally literary orientation of *Med.*, and argues that all features of its language (technical, sociolinguistic, etc.) must be understood in this light.

[119] Indeed, it is problematic to even speak of a 'genre' or uniform text-type of technical writing at all: the sentiment is provocatively captured by the exclamation of Formisano (2006) 133 that 'none of the ancient "technical writers" is a technical writer!' For a summary view on the heterogeneity of ancient technical writings, with some reflection on their literary elements, see Roby (2016) 26–42. Fögen (2009) 9–66 shows that there are certain implicit and explicit features that distinguished 'Fachtexte' in antiquity from other kinds of writing, although he also observes (65) that 'die Grenzen

often as keen to adapt their writings to their literary and intellectual milieux as were those from other traditions. Agriculture, in fact, was a paradigm case in this respect.[120] G.'s writings thus respond not only to the literary trends of his own period, but also to those of the agricultural tradition in which they were issued. I offer here closer examination of three of the literary features mentioned above; the remainder are discussed in the commentary.[121]

G.'s attention to **prose rhythm**, both in clausulae and colon ends, is indicative of the literary quality of *Pom.*[122] The rhythmical quality of G.'s prose may be concisely demonstrated by Table 3, in which are recorded the absolute number of occurrences and percentage in terms of total clausulae for three metrical patterns together accounting for approximately 80 per cent of all G.'s clausulae: (1) cretic-trochee or cretic-spondee (and resolved sub-types), (2) ditrochee or trochee-spondee and (3) dicretic.[123]

zwischen "schöner Literatur" und Fachtexten in der Antike fließend sind'; see also Fögen (2003), (2011), esp. 445–9, Adams (2016) 654–5.

120 For Varro, see e.g. Diederich (2007) 172–209, Kronenberg (2009) 76–131, Nelsestuen (2015), (2016); for Columella, Gowers (2000), Henderson (2002), Diederich (2007) 209–58, Doody (2007), Dumont (2008), Reitz (2013), (2017); for Palladius, Formisano (2005), Diederich (2007) 258–70. See also Henderson (2004) on Columella, Pliny and Palladius. 'Literary' from a different perspective is Cato's *De agri cultura*, which has been argued to present a somewhat distorting lens of the Roman farm with a view to aristocratic 'self-fashioning': see Reay (2005), Terrenato (2012).

121 For G.'s use of metaphor in *Med.*, see also Maire (2002*a*) l–li.

122 For concise introductions to metrical prose rhythm and the difficulties associated with its analysis, see e.g. Nisbet (1961) xvi–xx, Wilkinson (1963) 135–64, Nisbet (1990), Berry (1996*a*) 49–54, Berry (1996*b*), Whitton (2013) 28–32; for a survey of modern specialist approaches to the subject, Oberhelman (2003).

123 These figures assume that semi-colons, colons and periods in the modern editions of the Latin text (the present one for *Pom.* and Maire's edition for *Med.*) may be used as an index for the location of clausulae. Editorial punctuation is an imperfect guide to the identification of clausulae, to be sure, and misses sentence-internal cola, but the results are serviceable for the present purpose.

Table 3: Clausulae

Clausula type	Percentage of clausulae and absolute number of occurrences		
	Pom.	*Med.*	Combined
Type 1 Cretic-trochee or cretic-spondee, unresolved: — ∪ — — ×	31.0% (54)	35.7% (247)	34.8% (301)
Type 1a Cretic-trochee or cretic-spondee, cretic resolved: ∪∪∪ — — × or — ∪∪∪ — ×	6.3% (11)	7.0% (48)	6.8% (59)
Type 1b Cretic-trochee or cretic-spondee, trochee/spondee resolved: — ∪ — ∪∪ ×	7.0% (12)	4.3% (30)	4.9% (42)
Type 2 Ditrochee or trochee-spondee: — ∪ — ×[124]	17.2% (30)	19% (131)	18.6% (161)
Type 3 Dicretic: — ∪ — — ∪ ×[125]	16.1% (28)	14.2% (98)	14.6% (126)
Other clausulae or unclausulated	22.4% (39)	19.8% (137)	20.3% (176)
Total	100% (174)	100% (691)	100% (865)

G. shows concern not only for metre,[126] however, but also for accent. Accentual rhythms were not absent in prose of earlier

[124] It may be doubted whether the ditrochee/trochee-spondee by itself can really be called a clausula: for discussion of some related complexities, see Winterbottom (2017). 53 per cent of ditrochee/trochee-spondee patterns in *Pom.* (16x) and 28 per cent in *Med.* (37x) are preceded by a cretic.

[125] The statistics for the dicretic include forms with resolution in the first cretic. The first-paeon resolution (— ∪∪∪ — ∪ ×) occurs 4x in *Pom.* and 2x in *Med.* The fourth-paeon resolution (∪∪∪ — — ∪ ×) occurs 2x in *Pom.* and 7x in *Med.*

[126] I note here previous work on G.'s metrical prose rhythm. Mazzini (1988) 60–82 undertakes a detailed analysis of *Pom.* and provides statistics; the results are valuable but not entirely trustworthy, owing among other reasons to errors in scansion and in the identification of certain clausulae and cola (cf. Winterbottom (1980) 143; errors are still present in the second edition, as Önnerfors (1993) 272 notes). Önnerfors (1993) 270–4 analyses the prose rhythm of *Med.* and provides statistics. In our terminology, for *Pom.* Mazzini (1988) 72 counts type 1 = 33.7%, type 1a = 7.3%, type 1b = 6.7%, type 2 = 18%, type 3 = 11.8%. For *Med.*, Önnerfors (1993) 272 counts together type 1 and the resolved first-paeon (part of our type 1a) = 39%,

periods, but it is generally held that G.'s time (third century CE) saw the beginning of the transition from quantitative patterns to what would eventually be the 'purely' accentual cadences of the medieval period (the *cursus*).[127] The four principal *cursus* would evolve from and incorporate the metrical patterns above: the *cursus planus* types 1 and 2 (cf., respectively, *Pom.* 1.1.6 *ésse mit-ténda*; *Med.* 29.6 *praéstat meátus*), *tardus* types 1b and 3 (respectively *Med.* 45.10 *compositióne confícitur*, *Pom.* 2.2.1 *óssa custódiunt*), *velox* types 1 and 2 (respectively *Med.* 25.1 *éxprimit potestátem*; *Pom.* 3.6.4 *amýgdalis subueníri*), and *trispondaicus* type 1a (first-paeon resolved form *Med.* 3.5 *míxta subigúntur*, fourth-paeon resolved form *Pom.* 2.5.5 *léuiter adrórant*). Especially in the early centuries of the transitional period, many literary authors sought a coincidence of both metrical and accentual patterns (the so-called *cursus mixtus*), and G.'s practice may be interpreted in this context.[128] But while there is indeed an overwhelming agreement between clausulae of types 1–3 and the *cursus* in G.'s prose, he is not a purist: there are numerous metrically clausulated endings that cannot accept accentual patterns (cf. e.g. *Med.* 7.9 *efficaciūs pŭtāt sēmĕn ēius*; *Pom.* 3.3.2 *altitudinēm nŭcēs mērgunt*).[129] It is furthermore significant that

type 2 = 11.2% and type 3 = 13.7%. Mazzini's statistics for *Pom.* and Önnerfors' statistics for *Med.* agree with those presented here so far as establishing the precedence of the tabulated patterns, although differences in methods (definition, identification and resolution of clausulae) lead to statistical discrepancies.

127 I qualify 'purely' with quotation marks for the reasons voiced by Winterbottom quoted in the following n.

128 For a more granular treatment of the issue, and reservations, see Winterbottom (2011), (2017). Observing that '[n]o Latin prose could fail to have some proportion of accentual clausulae, and some proportion of clausulae that can be taken metrically' (2011 (267)), he cautions against the 'polarization' of metre and accent and doubts the general utility of the *cursus mixtus* ('we have no reason to suppose that there [was] a smooth and continuous historical development'). The proportions of metrical and accentual rhythms we find depend on the analytical criteria we adopt (cf. Winterbottom (2017) 234–5) and must be considered in view of an individual author's tendencies.

129 Önnerfors (taking aim at Mazzini) strongly rejects the attempt to find an intentional coincidence between metrical and accentual rhythms in G.'s

in sentence ends from which are absent the metrical clausulae stipulated above, *cursus* appear in only about a quarter of such places.[130]

Besides noting the general stylistic elevation that prose rhythm affords *Pom.* I should like to call attention to a few of the ways in which it has a more direct effect on G.'s compositional habits. First, and the 'subjective and fragile' nature of attempts to match rhythm to meaning notwithstanding,[131] there are a number of places in *Pom.* where G. clearly employs marked rhythmical parallelism in order to reinforce linguistic or conceptual symmetry. One example can be seen at 2.1.4 *Tenet quosdam hodieque uana persuasio ut putent* (sc. *persici*) *ossa quae dente contacta sunt aut prorsus extingui aut infeliciter nasci*.[132] We should observe not only the syntactic parallelism (adverb + infinitive) within the *aut*-phrases, which offer a rhetorical amplification of the harms posed to the seed by human or animal bite,[133] but also the rhythmical convergence of the cola (*prōrsŭs ēxtīngui / infēlīcĭtēr nāsci*, cretic-spondee *cursus planus*), which lays additional stress on these dangers. A second specific influence of prose rhythm on the text can be seen in G.'s verb forms. Thus, at 2.12.2 *Plerumque temptarunt* it is probable that the syncopated third-person perfect form *temptarunt* is motivated

writing, maintaining that we should identify accentual rhythms only where we cannot identify an established metrical pattern. Although Önnerfors reasonably criticizes Mazzini for his attempt to establish metrical and accentual cadences in every instance, which forces Mazzini to introduce singular patterns in his analysis, Önnerfors goes too far in rejecting the possibility of any intentional coincidence of accent and metre. Of course, it is true, as Önnerfors emphasizes, that the coincidence is in part a result of the natural relationship between metre and accent, but his method is too severe in its denial of any intentionality on the author's part.

130 That is, in 45 of the 176 places (25.6 per cent) identified as 'other clausulae or unclausulated' in Table 3 (10x in *Pom.*, 35x in *Med.*).

131 In the words of Whitton (2013) 30.

132 Other examples are discussed below, in connection with G.'s preference for symmetrical constructions.

133 For such amplification as characteristic of G.'s style, see esp. Maire (1999) 435–6.

metri causa (*plerūmquĕ tēmptārunt*, cretic-spondee *cursus planus*);[134] it is found several times in Imperial prose, and in many of those places rhythm may also be suspected as a motivation.[135] A third and final example of the effect of prose rhythm on G.'s composition may be seen in the verbal structures that he employs. At 4.3.4 *quae* (sc. *nuces*) *placere meruerunt*, the verb *mereo* is used in a weakened sense, scarcely making a semantic contribution to the verb *placere*.[136] We can appreciate a reason why G. might have used *mereo* here if we suppose a metrical purpose: *placērĕ mĕrŭērunt* is simply a resolved form of the cretic-trochee (here *cursus trispondaicus*) so common in his writing.

Another striking feature of the literary style of *Pom.* is G.'s fondness for imposing a formal **symmetry** on statements where a natural parallel or antithesis suggests itself. Also remarkable are his techniques for enlivening these symmetries with quantitative and accentual rhythms (noted above) and multi-part sound effects, and by careful application of *uariatio* within an overarching symmetrical structure in order to avoid staleness (discussed in greater detail below, under *uariatio*). Consider 1.1.12 *sed* (sc. *mala*) <u>*aspera*</u> *sunt* <u>*sapore*</u>, *quamuis et* <u>*aspectui speciosa*</u> *procedant*. Against the backdrop of the finely balanced antithesis should be noted the rhythmical play, sibilance, consonance of (*s*)*p*, alliteration, chiasmus (adj. + noun / noun + adj.) and slight *uariatio* (adj. + abl. / dat. + adj.). Another good example is 2.12.1 *cutis exilitas et carnis squalor consentanee prodibunt ad parsimoniam uetustatis*.

134 For examples of other perfect verb forms syncopated for metrical reasons in G.'s writing, cf. *Med.* 21.15 *uitia sanarunt*; 33.6 *ante gustarint*; 38.4 *saeuire consuerunt*; 43.3 *ciere consuerunt*.

135 Cf. Plin. *HN* 2.83 *indagare temptarunt*; Plin. *Ep.* 3.4.3 *eximere temptarunt*; [Quint.] *Decl.* 9.5 *retinere temptarunt*; *SHA Claud.* 12.1 *uastare temptarunt*.

136 For the weakening of some verbs into such auxiliaries in late Latin, see 2.1.3n. *Graecis . . . siquidem*. The *TLL* (VIII 806.43–807.10 (Bulhart)) documents the weakening of *mereo* in particular from the meaning 'favore vel casu donari, accipere, nancisci' into certain diffuse applications ('abit in notionem posse . . . licet, contingit'); see also Löfstedt (1911) 211. At *Med.* 23.3 the verb *mereo* is used with its full force (*nobilissimis antidotis meruit ascribi*; of the herb *nepeta*).

In this wordy sentence, the parallel expressions *cutis exilitas* and *carnis squalor* are integrated into alliterative sequences of *c* and *p*/*b* building to a conclusion in the rhythmical *parsimoniām uĕtūstātis* (*cursus velox*).[137] Still another strongly enforced symmetry is found at 4.2.4 *alterum spondet de incolumitate perseuerandi, alterum prodest de uirtute generandi*, where the natural antithesis of *alterum . . . alterum* is reinforced both by rhythm (cretic-spondee) and by syntactic parallelism (verb + *de* + noun + gerund). Larger-scale symmetrical antitheses are also evident in *Pom.*, as at 4.4.4 *Inuenti tamen qui de nucibus et hoc genere plantarum audeant aliquod facere discrimen et nuces credant caeduis arboribus aptiores, plantas uero nucibus procreandis esse utiliores, si tamen fas est ut admittere in animum debeamus quod lignum in poma proficiat, fructus in ligna desistat.* Here, G. presents a group (*inuenti qui*) who maintain that slips are best grown for fruit, whereas nuts are best for timber; the reason (as they maintain) is that wood grows into fruit, whereas fruit reaches its end in wood. The corresponding claims and explanations are organized in a chiastic pattern (abba), that is, *nuces . . . caeduis arboribus aptiores* (the first claim) corresponds to *fructus in ligna desistat* (the second explanation), and *plantas . . . nucibus procreandis esse utiliores* (the second claim) is explained by *lignum in poma proficiat* (the first explanation). Stylistic parallelism is enforced in the claims (i.e. within phrases ab) and the explanations (i.e. within ba), with the slight variation in word order of the central elements of the first clauses (*caeduis arboribus* / *nucibus procreandis*) motivated by rhythm.

The most important stylistic principle in *Pom.* is ***uariatio***, which by freshening material that might otherwise grow dull, and by surprising or delighting G.'s readers, induces them to continue reading.[138] His careful techniques for achieving *uariatio*

[137] For the densely clustered abstract nouns in the line, see the remarks on nominalization below, §v(b).

[138] For *uariatio* in *Med.*, see esp. Maire (1997), (2002*a*) xxxii–xxxvi. Maire (1997) 317 aptly notes that '[e]n utilisant une multiplicité d'expressions pour exprimer un même concept, Gargilius privilégie la composante littéraire au détriment de la technicité pure'.

also contribute to a more specific appreciation of the literary agenda of his work: they demonstrate to the audience G.'s ingenuity in the appropriation and transformation of traditional sources, establishing his contribution not only as an agricultural author but also as an artful re-fashioner of previous advice. For the most part, G.'s use of *uariatio* manifests itself as an inconspicuous shift in verbs and nouns, often technical terms, of the same semantic field, as when *demergo* (1.1.9) follows *obruo* (1.1.8); *corium* (3.8.3) is used for *cortex* (3.8.2);[139] and *iudico* (4.3.4), *tempto* (4.3.4) and *probo* (4.4.4) are exchanged in close succession. There are a number of cases, however, where we are especially apt to notice his striving for *uariatio*. First, this may happen when the search for variety seems to lead G. to use an unusual word (*hiasco* at 3.1.1), or a word used in a singular sense (such as *cingo* for engrafting, 2.10.6).[140] The reader is apt to be arrested at these places, at least momentarily, but context prevents the words from becoming an impediment to understanding: this is the showmanship of *uariatio* that carefully toes the line between creative expression and obscurity. Second, we notice G.'s *uariatio* when a whole thought is strikingly repatterned for use elsewhere, as is the case with 2.5.3 *Infigenda . . . terrae pars acutior, unde radicis exordium est* / 3.1.4 *Acuta pars nucis . . . prima defigitur: radices exitum moliuntur*. G.'s impressive means of introducing *uariatio* are on full display in his reworking of the line at 3.1.4. We note (1) variation in word order *pars acutior* / *acuta pars*; (2) variation of comparative and positive forms *acutior*/*acuta*;[141] (3) variation in verb choice *infigo*/*defigo*; (4) variation in predicates *terrae* and *prima*; (5) exchange of the imperatival third-person passive *defigitur* for the gerundive *infigenda*; (6) replacement of the hypotactic connection through *unde* with causal asyndeton;[142] and (7) variation in structure of

[139] See 3.8.2–3n. *cortex . . . corio*.

[140] See, respectively, 3.1.1n. *hiascere coeperunt*; 2.10.6n. *in ligno fortiore cinguntur*.

[141] For this variation in the comparative and positive with no semantic difference, see 1.1.15n. *Diligentioribus*.

[142] For the asyndeton, see 3.1.4n. *radices exitum moliuntur*.

the final expressions *radicis exordium est* / *radices exitum moliuntur*, employing *exordium* and *exitus* to the same effect.

Finally, G.'s *uariatio* demands our attention when it is embedded in broader and stylistically marked symmetries. An excellent example is found at 3.1.4, where G. reports the varying opinions of authorities about how best to orient the almond seed in its planting hole:[143] *aciem lateris in Aquilonem Mago inuertit, Celsus ad Fauonium dirigit, Columella cum dicat, anceps nucis ad Fauonium spectet,* [*et*] *sine dubio et ipse unum latus opponit Aquiloni.* The sentence presents sophisticated chiasmus (*Aquilonem . . . Fauonium . . . Fauonium . . . Aquiloni*), with the final verb (*opponit*) displaced from the unmarked sentence-final position typical in *Pom.* in order to shape and emphasize the figure. Variation is introduced through verb choice, word order, preposition and case, but in such a way that the variation reinforces rather than destroys the chiasmus. Consider the equivalences that the sentence presents in its four members as given by Table 4.

Table 4: An Example of Stylistic *uariatio* (*Pom.* 3.1.4)

Elements in *uariatio*	Clause 1: Mago	Clause 2: Celsus	Clause 3: Columella	Clause 4: *et ipse* = G.'s interpretation of Columella
Verb	*inuertit*	*dirigit*	*spectet*	*opponit*
Preposition and/or case	*in* + acc.	*ad* + acc.	*ad* + acc.	dat.
Word order	prep. phrase + subject + verb	subject + prep. phrase + verb	subject + prep. phrase + verb	subject + verb + dative (in lieu of prep. phrase)

[143] For an attempt to explain this advice (and the concerns that may motivate it), see 3.1.4n. *aciem . . . Aquiloni.*

The similarity in preposition, case and word order in members (2) and (3) reinforces their central position in the chiasmus (a**bb**a), and the varied word order in members (1) and (4) allows the words *Aquilonem . . . Aquiloni* to bracket the entire sentence (**a**bb**a**). Variation is particularly strong at the boundaries between members (1) | (2) and (3) | (4).

Another good example of variation-in-symmetry is presented by 2.4.4 *ne siccitas anni ruat et admodum exigua poma constringat ac frequens umor enormi uastitate et inani sapore distendat.* The sentence is carefully constructed to present the equal dangers to the peach of an immoderately dry or wet climate. *ac* is the fulcrum for two eight-word clauses sharing rhythmical endings (*pōmă cōnstrīngat* / *sapōrĕ dīstēndat*, cretic-trochee *cursus planus*).[144] There are some conspicuously balanced antitheses, notably *siccitas anni* / *frequens umor*, *admodum exigua* / *enormi uastitate* and *constringat* / *distendant*, but the clauses are also structurally distinguished by *uariatio*: *et* joins two verbs in the first, two nouns in the second; *poma* must be resupplied in the second clause from the first; and the proleptic *exigua . . . constringat* is replaced by the complementary ablatives *uastitate* and *sapore.*

(b) Technical Features

As a work self-consciously composed in the written tradition of agriculture,[145] *Pom.* shares elements with many other Greek and Latin writings, agricultural and otherwise, which purport to instruct an audience in a τέχνη or *ars*. The conventional use of the term 'technical' to describe the full range of such works overstates the coherence and stability of this ill-defined body of ancient texts;[146] it is perhaps advisable to

[144] For the isocolon, see Lausberg (1998) 321–3.

[145] The description 'self-conscious' is warranted by G.'s thick citation of and comment on previous authorities.

[146] Langslow (2005) and Fögen (2011), drawing on studies of modern technical languages, highlight the wide range of possible technical *Textsorten* (for which concept see Roelcke (2005) 32–49) in antiquity. Cf. also above, n. 119.

employ the label simply in a tautological sense, that is, in reference to those works about any of the branches of specialized knowledge which could be called τέχναι or *artes*. The great diversity of ancient technical writing notwithstanding, much of it – including *Pom.* – can be understood in terms of a properly extra-textual goal of communicating a body of information clearly, completely and accurately enough to allow the audience to achieve a desired competency, whether of knowledge or praxis.[147] This extra-textual goal is realized not only in a set of conceptual, didactic tools that the author employs, but also, importantly, in specific linguistic phenomena. It is the latter fact that gives a sense to the notion of 'technical Latin'. Although there remain difficulties in determining which features of technical Latin are truly distinctive in isolation from their practical context, there are nevertheless well-defined aspects of the language that serve the extra-textual goal described above. I discuss four of the most salient aspects here.

The most distinctive feature of technical Latin is the development of a rich **technical terminology**.[148] *Pom.* is replete with technical agricultural terms, which range from words with wider currency but employed by the agronomists in a specific sense, such as *pono*, 'to plant', or *transfero*, 'to transplant', to words that never appear outside of the agronomists, such as *adobruo*, 'to heap earth up around', 'to earth up'.[149] There

147 It bears emphasizing that the description given here is intended to be rigorous only for a subset of ancient technical writings; cf. the previous n.

148 For the identification of technical terminology, I adopt the broad definition formulated by Langslow (2000) 25, namely, a word or expression 'which is recognized and used in a standard conventional way by the relevant community of specialists and which unambiguously (and often uniquely) names an object or concept of the discipline, and therefore, because of this attachment, lends itself to absolute synonymy and total translation'. Cf. also the definition of Schiefsky (2005) 253.

149 For *pono*, see below, §v(d); for *transfero*, see 2.5.6n. *transferant*; for *adobruo*, see 3.2.3n. *adobruunt*.

are also found two ἅπαξ technical terms in *Pom.*: *punctorium* (2.5.8), a kind of dibble, and *ferularis* (3.3.5), an adjective denoting relation to the fennel.[150] Both are of regular formation and would have been very unlikely to pose any interpretative difficulties to a reader.

G.'s use of technical agricultural terminology is documented fully in the commentary, but the subclass of these words comprising abstract nouns (e.g. 1.1.15 *austeritas*, 3.1.2 *admixtio* or 3.7.4 *amaritudo*), that is, deadjectival or deverbal **nominalizations**,[151] deserves further discussion here, insofar as it has wider repercussions for our understanding of technical Latin and of the style of *Pom.* Langslow (2000) 377–430 has made an important study of the effect of these nominalizations in medical Latin, but his terms of analysis and conclusions are more broadly applicable to other varieties of technical prose. It is worth rehearsing them here before examining G.'s practice with nominalizations more closely.

Langslow (2000) 378 establishes two stylistic poles useful in assessing technical (Latin) prose:

> One pole may be characterized as being 'unscientific', in making appeal only to ordinary linguistic, as opposed to scientific knowledge on the part of the reader/hearer. In formal terms, it is wordy, descriptive, unconventional, and uncompressed, using a range of synonymous phrasal expressions based on verbs and adjectives, which take part in a richly varied syntax.

[150] For discussion of their formation and meaning, see 2.5.8n. *punctorio*; 3.3.5n. *ferularem*. For discussion of another, non-technical ἅπαξ in *Pom.*, *amaritosus*, see below, §v(d).

[151] Langslow (2000) 412, discussing medical writing, voices some reservations about calling such nouns 'abstract': 'One's instinct is, indeed, to call this all "abstract": deverbal and de-adjectival nominalizations are, after all, what are standardly called in the grammar books abstract formations. But I wonder if "concrete" is not more appropriate. Nominalizations create the illusion, at any rate, of objects. They imply that the symptom (*contractio neruorum, siccitas linguae*) resulting from a process (*contrahuntur* (*nerui*)) or a state (*sicca est* (*lingua*)) is somehow graspable and treatable as much as the spot or the swelling.'

Langslow calls this the 'diffuse' or 'verb-based' style (*ibid.*). Next he presents the 'compact' or 'noun-based' style (378–9):

> The opposite extreme . . . is typified by a syntax that is much less varied in construction, to the point of being seriously impoverished, thanks to a more-or-less normalized terminology based on nouns and their adjectival and verbal derivatives, and that shows the results of both formal and semantic compression: semantic compression is seen in the selection of particular features of the referend to serve as the basis of its name, morphological compression, in the incorporation of these features in complex monolexematic structures [i.e. technical words] of conventional form and meaning. The interpretation of prose written in the compact style may call for more specialized knowledge of the linguistic coding of the technical discipline in question than is usually commanded by the average native speaker.

As Langslow's introduction of these two stylistic poles already suggests, an important consequence of this distinction is to show that the *prima facie* lexical question of abstract nouns is in fact a question of syntax as well, for the compact style eschews diffuse phrasal expressions (which introduce syntactic variation) in favour of informationally dense nominalizations (see 383–408 for Langslow's development of the contrast): in effect, nominalizations supplant their semantically equivalent but syntactically diverse verbal or adjectival alternatives. They should therefore be understood as 'lexical-cum-syntactic' choices that influence the broader shape and style of the text as much as they do our account of an author's diction (Langslow (2000) 408–9; cf. 377–8).[152] The compact, noun-based style that relies on

[152] This conclusion is important, for it poses a counterargument to the statement of André (1986) 9 that 'les langues techniques latines sont des langues réduites au lexique'. Indeed, this is one of the positions that Langslow seeks to critique (cf. (2000) 377), although, to be sure, others had already implicitly or explicitly indicated how such a view could prove inadequate; see e.g. Callebat (1990) 45, Langslow (2005) (with further references).

nominalizations can obviously be useful for authors of technical works who assume a knowledgeable audience, because it allows for brevity and precision in referring to the special 'objects' that would ordinarily require periphrastic or otherwise diffuse description – in medical Latin these are typically symptoms or conditions, but in agricultural prose the qualities of plants, produce, soil and climate, as well as various cultivation techniques, are all candidates.

For the reason that nominalizations encode information, so to speak, about an author's stylistic decisions, they can also be valuable for establishing the nature and register of a piece of prose. They must be used for that purpose carefully, however. Nominalizations have sometimes been said to belong to Vulgar or colloquial Latin (e.g. Löfstedt (1911) 147, Svennung (1935) 517–21, H–Sz Stilistik §25), less often to literary Latin (e.g. Callebat (1982) 704–5). As Langslow (2000) 412–13 has argued, the picture is more complicated: '[p]robably, the question is misplaced and it is quite misguided to attempt to characterize nominalization as in and of itself either literary or colloquial'. It can nevertheless be suggested – tentatively (see the cautions of Langslow (2000) 412–15), but very plausibly – that such nominalizations and the noun-based style are 'more technical than vulgar' (Langslow (2000) 415, referring to H–Sz Stilistik §22; see further 415–18). It is plain why the compression and precision that the nominal style affords would commend it to technical writers; but whether these gains could be justified at the cost of other stylistic sacrifices and a reduced intelligibility for the lay reader would depend on an author's ambitions for a given work.

We find nominalizations in *Pom.*, but neither an excessively large amount nor especially abstruse cases; it is hard to imagine that even a non-specialist reader would have found most of them a challenge. This is partly a consequence of the fact that agricultural writing in general requires fewer unfamiliar terms than, for example, medicine, and partly that G. exhibits a great concern for *uariatio* and general stylistic polish.[153] Indeed, the nominalizations themselves can work in tandem with literary aims, whether by introducing variety, achieving specific rhetorical effects (e.g. derivational play at 3.7.2–5 *ex amaris . . . amaritudinem amygdali . . . amaritosa*) or producing vivid metaphors (e.g. 2.6.1 *medicina continuae runcationis*). The possibilities for stylization through such abstract nouns are on full display in a wordy and alliterative sentence such as 2.12.1 *cutis exilitas et carnis squalor consentanee prodibunt ad parsimoniam uetustatis* (for stylistic analysis, see above, §v(a)). These few examples suffice to illustrate Langslow's caution that it is not safe to oppose a 'literary' style to the nominal, both for the reason that nominalizations can accomplish a variety of goals, not only or even compression of information (one could hardly say that 2.6.1 *medicina . . . runcationis* is more precise or brief than *runcatio* alone), and because stylistic norms are not static but subject to change and influence from many sources (so Langslow (2000) 412–14; see on the same point Haverling (2012), (2014)).

A partial but representative conspectus of nominalizations in *Pom.* pertaining to agricultural usage is given here (in order of first occurrence): 1.1.5 *sapore . . . mellis* (*sapor* also 1.1.12, 2.1.3, 2.4.4, 2.5.6, 3.5.2); 1.1.15 *austeritatem eius* (sc. *mali*); 2.1.3 *culturae adminiculis* (*cultura* also 3.3.2, 3.5.1, 4.1.2, 4.1.8); 2.4.1 *remotione loci*; 2.4.4 *siccitas anni*; 2.5.1 *Ordo sationis*; 2.5.2 *trium mensura palmorum* (*mensura* also 4.7.3, 4.7.4); 2.5.3 *radicis exordium*; 2.5.8 *punctorio*; 2.5.9 *inertia ligni*; 2.6.1 *densitas contegentium*; 2.7.1 *substantia radicum*; 2.10.8 *infirmitatem*; 2.12.1 *cutis exilitas et carnis squalor*; 2.12.1 *parsimoniam uetustatis*; 3.1.2 *stercoris admixtione*; 3.2.2 *pars aliqua fossurae*; 3.4.1 *Et natura et mediocritas arborum*; 3.5.2 *incursus pecorum*; 3.5.3 *fossio*; 3.7.4 *amaritudinem amygdali*; 3.8.1 *signum maturitatis* (*maturitas* also 4.2.2); 4.2.4 *incolumitate perseuerandi*; 4.6.1 *caeli qualitate*; 4.6.1 *uarietas opinionis*; 4.6.6 *uiciniam fluminis*.

[153] For technical medical terms in *Med.*, see Maire (2002*a*) xli–l, (2003*a*). As Maire observes, G.'s medical language is not overly obscure, which suggests a lay audience (cf. above, n. 23).

See further the respective nn. The nominalizations are both deadjectival and deverbal (cf. Langslow (2000) 383–96); for some general features of nominal derivation, see Rosén (1983). Among the above tally could also be included the more vivid and metaphorical cases of the type 2.3.5 *subsidium teporis et umoris* or 3.1.5 *alimenta umoris*, but these are better treated in a different place.[154]

Besides the technical terminology and the syntactic effects of nominalization that are the most conspicuous features of technical Latin, there is also found in *Pom.* a tendency towards what we might call **over-specification**. That is, G. shows an occasional inclination to use pronouns, adjectives and other words that appear strictly redundant but function to remove any possible source for misunderstanding by 'over-specifying' the matter under discussion. This phenomenon is exemplified, for example, by 4.3.2–3 *supra eas ita <ut> uniuersas obtegat funditur harena fluuialis, quae et pinguis et dulcis est. Post trigesimum diem eadem harena remouetur*, etc. The demonstrative *eadem* is used anaphorically (like *ea*) and in an apparently redundant way to refer to *harena fluuialis* mentioned in the previous sentence: for such use of *idem*, see Hofmann (1926) 108, Svennung (1935) 300–7, Josephson (1940) 229–30, *TLL* VII 1.205.28–206.12 (Hofmann), Adams (2013) 494–8, (2016) 291–2. Adams (2013) 495 has stressed that a redundant usage of *idem* is conventional in technical writing, a function of the 'traditional obsession of technical writers with making things clear by specification'[155]. Another example of over-specification is 2.1.3 *ideoque plerique non ossa sed poma ipsa matura integra existimant obruenda, ut ea res et ossibus uicem stercoris praebeat et pomis futuris illibatum saporem . . . acquirat.* The participle *futuris* here is attached to *pomis* purely in order to eliminate a possible

[154] See 2.3.5n. *subsidium teporis et umoris.*

[155] Nor should we, therefore, automatically infer that *idem* is always weakened in such cases. For other cases of 'over-specifying' *idem* in *Pom.*, cf. 1.2.3, 2.3.5, 2.4.1, 2.4.6, 4.1.4, 4.3.5. Maire (1999) 437, (2002a) xl, (2003b) 362 observes the occurrence of a similar usage of *idem* in *Med.*, and states that it reflects a Vulgar influence on the text: on the interpretation offered here I would resist that conclusion (and see further below, §v(d)).

confusion with *poma ipsa* of the previous clause, despite the fact that this confusion is only a remote possibility in the context.[156] Finally, this same tendency may explain what in places may appear to us to be awkward repetition, as with *mense Ianuario* at 2.1.2 *Curtius Iustus et mense Ianuario calidioribus locis putat posse disponi; Quintilii armenia sola consentiunt Ianuario mense ponenda* (but note the variation in word order).

A final feature of the technical language of *Pom.* that deserves mention is the wide variety of **directive expressions** that G. deploys.[157] Conspicuous is the absence of second-person imperative forms (the direct quotation from Mago at 4.1.2 excepted), and indeed, second-person verb forms are essentially restricted to generalizing conditional statements (2.11.2 *ponas*; 3.7.6 *auferas*; 3.8.2 *resolues*). But the avoidance of the second person is no barrier whatsoever to G. in conveying to the audience the desirability or necessity of following a given course of action. In the didactic context of *Pom.*, virtually any statement can take on a directive force, even if only obliquely, and G. employs a great diversity of expressions for these purposes, including auxiliary verbs;[158] impersonal expressions;[159] adjectival and adverbial complements, strengthened by litotes and other figures of speech;[160] generalizing conditions; gerundives; third-person

[156] A comparable phenomenon is G.'s frequent use of *eiusmodi* in situations where it is not required for clarification: see 2.7.3n. *eiusmodi folia sortita.*

[157] For a comprehensive study of Latin directive expressions, see Risselada (1993); for directives in technical contexts, with emphasis on the wide range of possible forms these can take, see Adams (1995*a*) 460–8, Gibson (1997), Sharrock (1997), Grocock and Grainger (2006) 98–102, Hine (2011), Adams (2016) 369–70.

[158] *debeo* is especially common (8x); for remarks on some possible cases of weakening with *debeo*, see 3.6.4n. *debeant germinare.*

[159] Cf. 2.8.1 *ablaqueare . . . oportet*; 3.7.3 *utile est si effundatur stercus*, etc.; and especially *necesse est* (1.1.15, 2.5.1, 4.3.5, 4.6.6).

[160] For adjectives, cf. 2.4.4 *Demissa temperies necessaria eis*; and especially *iustus* (5x; see on 2.11.2n. *mole iusta*). For adverbs, cf. 3.4.3 *Bene Diophanes ait*, etc.; 4.1.8 *Iulio Frontico . . . non fuit . . . norma quando rectissime castaneae sererentur*; and especially *melius* (6x: 1.1.2, 1.1.7, 2.1.1, 2.10.1, 2.10.7, 4.3.1). For litotes, see 1.1.7n. *non leui pondere.*

jussive subjunctives; predicate statements of the type *moris est*; third-person imperatival indicatives in both the active and passive;[161] verbs of teaching, ordering and persuading;[162] verbs of thinking, speaking, regarding, agreeing and esteeming;[163] and so on. This broad set of directive expressions, in many cases inherited from earlier agricultural authors,[164] affords G. a means both of achieving rhetorical *uariatio* (see above, §v(a)) and of shifting between a number of different presentational styles or voices in accordance with the descriptive and prescriptive didactic aims of *Pom.* discussed above (§III(b)).

G.'s pervasive use of direct and indirect directive expressions can at times be correlated in interesting ways with other stylistic features of *Pom.*, such as the regular occurrence of a future or future perfect verb in a temporal clause, regardless of the tense of the main verb, where the temporal clause specifies an action that must be fulfilled before that of the main verb. Consider the following examples: 2.3.3 *denique tunc inseruire arboribus debent, cum extra hiare coeperint* (in spite of *denique tunc*); 3.2.1 *Vbi primum . . . ostenderint . . . palus affigitur*; 3.6.4 *cumque iam tempus uidebitur . . . fodere atque remouere . . . praecipiunt*; and in a clause not strictly temporal but with the same force, 3.2.3 *tantumque radicibus cumuli largiuntur quantum altitudinis truncus adiecerit*.[165] The apparent looseness of tenses in G.'s clauses – pairing future or future perfect with present indicative main verbs – is to be traced back to an

161 These are indicatives of the type 'they do *x* [and so should you]' (active) or '*x* is done [and should be done by you]' (passive): see concisely Gibson (1997) 71; also, Adams (1995*a*) 466–7, 489, (2016) 370. (For remarks on the difficulties of distinguishing prescriptive/descriptive indicatives, see Adams (1995*a*) 466, Gibson (1997) 75–7, Sharrock (1997) 101–2.)

162 E.g. *suadeo* (3.3.4), *praecipio* (3.6.4) and *iubeo* (3.6.4).

163 A very wide range of verbs is employed, from e.g. *existimo* (8x) and *puto* (7x) to *affirmo* (4x), *asseuero* (2x), *arbitror* (2x), *consentio* (3x), *polliceor* (2x) and *promitto* (1x). We also find *existimo* + infinitive used once with modal force: see 1.1.10n. *illiniri existimant*.

164 See esp. Hine (2011).

165 For some additional examples of this pattern, see 1.1.7n. *ubi tempus erit . . . pollicentur*.

extension of a pattern very common in technical writing, in which the future verb of a temporal clause is paired with an imperative or other expression of necessity in the main clause, which is forward-looking in sense (cf. K–S II §203.3.a, *NLS* §217.6 with §239.3). The basic pattern seems to have been codified at an early time in technical writing, which naturally has many such injunctions as 'do *x* [only] when *y*'. In agricultural writing the pattern is already regular in Cato (e.g. *Agr.* 65.1 *postridie aut post tertium quam lecta erit facito*) and Varro (e.g. *Rust.* 1.63.1 *aliquanto post promere quam aperueris oportet*). It gradually became more flexible, allowing greater choice in temporal conjunction and main verb (e.g. Col. 11.3.23 *brassica cum sex foliorum erit, transferri debet*; 5.11.3 = *Arb.* 26.2 *cum primum germina tumebunt . . . eos legito*). Although the tense and mood of the main verb are quite free even among early authors, since nearly any expression can carry imperatival force in the didactic context of a technical treatise, we see a highly developed state of this phenomenon in *Pom.*, where in each case the present indicative verb (2.3.3 *debent*; 3.2.1 *affigitur*; 3.2.3 *largiuntur*; 3.6.4 *praecipiunt*) carries an obliquely imperatival force and thus passes as a future-equivalent expression.

(c) The Place of *De arboribus pomiferis* in the History of Latin

Pom. holds much of value for understanding the historical development of Latin. Written not long after the conclusion of a period of Latin that is relatively well described (the period extending approximately to the second century CE),[166] it tentatively points the way towards developments that would proliferate only much later. But in no sense does *Pom.* present a clean

[166] Sometimes prejudicially well described, owing to more or less arbitrary periodizations of Latin that prioritize writings before this time: see the references in the following n., and note also Goodyear's criticism (1983) of the *Oxford Latin Dictionary*, which stops at 200 CE.

break with the conventions of educated Latin in the centuries preceding its composition, and for this reason it can be difficult to periodize closely even those features of its language that are relatively distinctive. It may be said to contain genuinely 'late' features, if 'late' is taken to mean that such features would become more common sometime in the period roughly 200–600 CE,[167] but the greatest share of linguistic phenomena in *Pom.* have roots, often deep ones, in earlier periods. The linguistic change demonstrated by G.'s writing can thus only be understood against a background of linguistic continuity. Periodizing the language of *Pom.* is further complicated by long-standing confusion between the notions of late Latin and, variously, technical, Vulgar and colloquial Latin (see further below, §v(d)), and by the fact that G.'s literary ambitions (above, §v(a)) often lead him to bring into *Pom.* higher-register elements that were no longer a part of the living language (i.e. 'early' features). This is all to say that we should resist slotting *Pom.* into a simple narrative of linguistic change; it is not in every case inappropriate to apply the label 'late', but the word should not by itself be taken to have explanatory value. I offer here two case studies illustrative of the complications attached to periodizing the language of *Pom.*

The relative frequency and usage of **connectors** such as *atque/ac*, *-que* and *et* provide a clear example of linguistic change and continuity in G.'s writings. They also touch on questions concerning the stylistic level of his prose and illustrate the need to understand G.'s evidence for the historical development of Latin in close connection with the literary and rhetorical conventions that he adopts. Table 5 shows the distribution of three conjunctions meaning 'and' in *Pom.* and *Med.*[168]

[167] The cutoff is arbitrary, and the temporal range too broad to provide a rigorous understanding of the linguistic change in the period. For trenchant criticism, but also constructive reflection, on the notion of 'late Latin', see Adams (2011), esp. 257–8, (2016) 643–4.

[168] For the sake of these figures, I exclude 18 instances of *-que* found in the expression *itemque*, where it is probable that *-que* is to some degree weak-

Table 5: Three Conjunctions: *atque/ac, et, -que*

	Number of occurrences			Relative frequency by percentage of total		
Conjunction	*Pom.*	*Med.*	Combined	*Pom.*	*Med.*	Combined
atque/ac	12	34	46	9.1%	7.9%	8.2%
et	104	356	460	79.4%	82%	81.4%
-que	15	44	59	11.5%	10.1%	10.4%
Total	131	434	565	100%	100%	100%

-que was the regular coordinator in ancient Latin texts (K–S II §152.2, H–Sz §254, Penney (2005) 41, Torrego (2009) 457). It is the most frequent copulative coordinator, for example, in the Twelve Tables and Cato's writings, and it is noteworthy that it is especially common in the 'archaizing' style of Cicero's *De legibus*, where it alternates with asyndeton (Albrecht (2003) 40–1). As is well known, *-que* retreated before *et* in the history of Latin, eventually disappearing altogether from the living language: see Löfstedt (1911) 87, Svennung (1935) 489–91, *Syntactica* II 340–2, H–Sz §254, Rosén (2009) 395–6, Torrego (2009) 460. A variety of evidence can be adduced to this end from texts that have been held to cleave most closely to spoken Latin (i.e. many traditional sources for 'Vulgar' Latin), including portions of Petronius (Löfstedt (1911) 87, *Syntactica* II 341), the Latin curse tablets (Jeanneret (1922) 20–1), Pompeian graffiti and so forth.

The retreat of *-que* occurring in spoken language manifested itself more slowly in (formal) written Latin, however, where

ened (see 4.5.2n. *itemque*); I also exclude three instances of *-que* found in the expression *hodieque* (but not because it is weakened; see 2.1.4n. *hodieque*). I include, however, three instances of *-que* in *ideoque*, where there is not decisive evidence of weakening (see 2.1.3n. *ideoque*). It goes without saying that the conjunctions tabulated here do not exhaust the range of coordinators that G. uses: consider the remarks also, for example, at 2.2.2n. *neque sol neque uentus*; 2.9.3n. *truncis ramisue*. Examination of unusual cases of the conjunctions discussed below will be found in the respective nn. in the commentary.

it continued to appear more or less frequently depending on the usage of a given author (Svennung (1935) 489, H–Sz §254). The stylistic norms of the literary language were not as quick to change as those of the spoken, so it is no surprise that *-que* retained currency in certain registers long after it was fading from the living language. Eventually *-que* came to be associated with an elevated, literary style (Linderbauer (1922) 95–6, *Syntactica* II 341, 342 n. 2, Petersmann (1977) 240, *TLL* II 1050.10–24 (Klotz) on *atque*, Adams (2016) 501), owing in part to this gap between the quicker evolution of spoken idiom, where *-que* was felt to be increasingly archaic, and the slower development of the written idiom, where *-que* had a long and august pedigree.[169] The statistics for G. given in Table 5 show that all the copulatives, including *-que*, were ceding to *et* in his writing, an outcome that is in line with our picture of language change in this period; but while a rate of occurrence of 10–11 per cent for *-que* is not high in absolute terms, it is nevertheless significant for G.'s time, and offers a positive indication of the literary nature of *Pom.* (even if we do not go further and claim that it must suggest an elevated or lofty tone).

The usage and convention of *-que* is reviewed at K–S II §152. In *Pom.*, G. uses it to connect (1) nouns: 2.3.5 *frigus imberque*; 4.5.2 *carbunculus itemque tofus*; 4.5.3 *glarea cretaque*; (2) adjectives: 3.1.2 *altos latosque*; 4.3.2 *angustum locum siccumque*; and, most often, (3) clauses: 1.1.10 *subigunt soleque . . . existimant*; 2.2.2 *componuntur, idque . . . infodiunt*; 2.8.1 *congerere quantaque maiore****; 2.10.5 *fuerit, uixque proueniet*; 2.10.6 *inseritur . . . magisque durabilis arbor efficitur*; 2.12.1 *potest uixque . . . peruenit*; 3.2.3 *adobruunt . . . tantumque . . . largiuntur*; 3.3.3 *obruunt parique . . . transferunt*; 3.3.4 *seramus eamque . . . compleamus*; 3.6.4 *aggregant, cumque . . . uidebitur*. Examination of the instances where G. uses *-que* as a clausal connector shows them to be, on the whole, in agreement with the remarks of K–S II

[169] The story is more complicated, to be sure, and includes other factors such as the continued use of the word in poetry for metrical purposes and in artful prose for 'Klauselbildung' (cf. H–Sz §254; also §271 on *ue*).

§152.4: 'Sätze knüpft *-que* häufig in der Weise an, daß es einem aus dem vorhergehenden sich ergebenden Gedanken eine Folge anschließt.' Furthermore, in joining nouns and adjectives *-que* continues to indicate a close relation, as, for example, 2.3.5 *frigus imberque*; 3.1.2 *altos latosque*.[170]

What has been said above about the history of the distribution and register of *-que* applies to *atque/ac* as well: like *-que*, *atque/ac* declined in favour of *et*, this change becoming pronounced in the Imperial period (Löfstedt (1911) 85–7, H–Sz §255, Rosén (2009) 380, Torrego (2009) 460); like *-que*, *atque/ac* became associated with higher-register language (see Klotz's *TLL* article cited above, Löfstedt (1911) 86, Petersmann (1977) 239–40, Calboli (2003) 274–5, Rosén (2009) 402, Galdi (2014) 74, Adams (2016) 11); and, as a consequence, like *-que*, the appearance of *atque/ac* in a late Latin text can serve as an indication of its literary nature. For the use of *atque/ac*, see generally K–S II §153. In most cases in *Pom.*, *atque/ac* retains the function of drawing attention to the second element that distinguishes it from an unmarked coordinator such as *et* (Torrego (2009) 459; 'steigernd' in the terminology of K–S): cf. 2.2.2 *tectum atque oblitum*; 2.2.1 = 4.3.1 *anxie atque sollicite*;[171] 2.10.3 *fragile ac putre*; 2.10.4 *Decembri ac deinde Ianuario*; 3.6.4 *fodere atque remouere*.[172] It may be said in conclusion that G. shows good control over the conjunctions

[170] As regards *-que* in *Med.*, G. uses it in the same ways as in *Pom.* plus in some additional situations, such as to coordinate a participle and relative clause (18.8 *ulcera . . . manantia quaeque in uesicam tumescent*) or two relative clauses (29.10 *umor capitis qui . . . comminatur quique . . . praestruit*). That it is a marked conjunction semantically distinguishable from *et* is shown by G.'s fondness for using it to connect participles indicating sequential action: cf. *Med.* 18.17 *elixum illitumque*; 22.8 *tritum haustumque*; 28.13 *feruefactum admotumque*; 30.36 *radicem plantae auulsam colloque suspensam*; 36.5 *maceratum coxaeque feruens superpositum*; 37.7 *tritum collectumque*.

[171] So cf. Sen. *Ep.* 43.5 (sc. *conscientia*) *anxia atque sollicita*; Apul. *Met.* 4.23 (sc. *latrones*) *anxii atque solliciti*. Note also the reverse order: Cic. *Att.* 2.24.1 (sc. *amor*) *sollicitus atque anxius*; Sen. *Ben.* 7.14.5 (sc. *te*) *sollicitum atque anxium*.

[172] The same is true for *Med.*: cf. 7.1 *umidae atque frigidae*; 21.10 *contritus atque illitus*; 23.6 *dissoluit ac purgat*; 30.10 *insomnia ac uigilias*; 39.2 *tempori ac fronti*; 42.10 *tumores atque liuores*.

atque/ac and *-que*; this control represents an element of linguistic continuity with earlier periods of Latin that stands out against the predominance of *et* in the living language of his day and indicates the relatively formal nature of his prose.

Another good example of linguistic change-in-continuity is offered by G.'s use of ***quod*-clauses after *uerba sentiendi et dicendi*** (3x in *Pom.*: 1.1.7 *affirmamus quod*; 2.5.6 *affirment quod*; 3.3.4 *asseuerare quod*). It is a well-known feature of later Latin that, in reported speech, there are occasionally found a conjunction and finite verb where in an earlier author accusative–infinitive syntax might have been expected.[173] The finite construction is increasingly well attested from the first century CE, at first and most often with *quod* and *quia*, but also eventually with other conjunctions such as *quoniam*, *quomodo*, etc. Two facts about *quod*-clauses are of particular note in connection with *Pom.* First, *quod*-clauses in reported speech belong well through the time of G. not to colloquial or Vulgar Latin, as is sometimes assumed, but rather to literary Latin (or at least Latin with literary pretensions): '[i]t is a remarkable fact that in non-literary Latin there is not so far a single example attested of a dependent *quod*-clause functioning like the accusative + infinitive . . . The fact of the matter is that in the later Republic and early Empire (of about the first three centuries A.D.) the construction is entirely restricted to literary sources' (Adams (2005) 195, with n. 6; see also Adams (1977) 61–2, Halla-Aho (2009) 76, Adams (2013) 743–4). Although absence of evidence for *quod*-clauses in non-literary texts is not proof that they could not occur in lower-register writing, it seems prudent to take G.'s usage of such clauses more as an indication of the literary character of his text than of Vulgar or late influence *per se*, especially when

[173] The best recent synopsis of the matter in English is Adams (2013) 743–6, with bibliography. See also concisely Herman (2000) 87–90; and further, Svennung (1935) 499–505, Herman (1963) 32–51, H–Sz §312, Coleman (1975) 119–22, Herman (1989), Baños Baños (2009*b*) 550–6, Adams (2016) index s.v. 'accusative + infinitive and substitutes'.

his *usus scribendi* is considered (see further below). The second noteworthy observation is the relatively low frequency of the construction in the Latin of G.'s time: until after the fourth century, the use of a conjunction and finite verb following *uerba dicendi et sentiendi* is found, on average, no more than once for every ten instances of accusative–infinitive syntax (Herman (1989) 134–5, (2000) 89; see Calboli (2012) for some refinements in connection with biblical texts). The statistics for *Pom.* agree with this figure, with 3 *quod*-clauses against approximately 30 cases of accusative–infinitive syntax. If the statistics for *Med.* are integrated, G.'s proportional usage will fall even lower, for there are no entirely unambiguous examples of the dependent-clause construction in about 100 instances of reported speech.[174]

There are further observations that we can make about G.'s use of dependent *quod*-clauses. First, *quod*-clauses in place of accusative–infinitive syntax in *Pom.* are found only after the semantically similar verbs *affirmo* (1.1.7, 2.5.6; used also with accusative and infinitive at 4.1.4) and *asseuero* (3.3.4; used also with accusative and infinitive at 2.9.1). Second, these *quod*-clauses always immediately follow their verbs (a universal tendency of such clauses, observed by Herman (1989) 137–40; cf. Adams (2013) 745). When taken together, these facts cause G.'s *quod*-clauses to appear strikingly formulaic, suggesting that their use in place of accusative–infinitive syntax owes more to self-conscious literary choice than to colloquial or sub-literary influence creeping into the text. This intuition may be corroborated by the evidence

[174] There occur some linguistically related 'look-alike' dependent clauses, which, however, admit of alternative interpretations: cf. e.g. *Med.* 41.1 *de punici arbore quod uim stypticam obtinet omnes fere medici una opinione consentiunt* (*quod* need not be analysed as the object of *consentiunt*); 41.14 *Dioscorides simplicius in hunc modum tradit ut cum primum cytini erumpere incipiunt, tres numero additi <in os> sine contactu dentium transuorentur* (*ut* is jussive, communicating a medical prescription, as it also is in the following sentence: 41.15 *ceteri, in quibus Plinius, efficacius putant ut*, etc.); 51.4 *Galenus de siliquis pessime sentit quod uix degerantur et uentrem restringant et umores noxios nutriant* (it should rather be construed, 'Galen has a very negative opinion about carob fruit, *because*, etc.').

of *Med.* Recall that the historical origin of the dependent *quod*-clause is in the deletion of the pronominal object in the pattern *uerba dicendi et sentiendi* + pronominal neuter object + explanatory *quod*-clause. (This fact probably also explains the absence of dependent *quod*-clauses from non-literary texts, for the pattern originally belonged to educated Latin.[175]) It is therefore striking that in *Med.* G. writes the following sentence: 47.1 *mespili arborem Dioscorides contentus agrestem nominasse praeteriit, hoc tantum asseuerans quod poma eius et folia uirtute non discrepant.* The occurrence of this 'proto-*quod*-clause' (note that *hoc* and not *quod* is the object of *asseuerans*) in G.'s writing helps us to locate the *quod*-clauses of *Pom.* in their proper linguistic-historical context, for it suggests that they are a part of a broader development in the literary language of his and earlier periods, not a purely late or colloquial departure from the infinitive construction.

Finally, it is worth mentioning briefly the problem of the mood of verbs found in *quod*-clauses. H–Sz §312.a.β notes that the distribution of moods is obscure: 'Der Modusgebrauch bei *quod, quia,* und *quoniam* nach den Verben sentiendi und dicendi unterliegt im Spätlatein starken Schwankungen und ist zusammenfassend immer noch nicht untersucht.' It can in some cases be shown that authors preserved a distinction between indicative and subjunctive, but in other cases the mood and tense were simply assimilated to those of the main verb (Adams (1976) 94–6, Burton (2000) 189–90). At 1.1.7 and 3.3.4 the indicative is used in the *quod*-clause, matching the mood of the main verb at 1.1.7 (*affirmamus*) and at 3.3.4, although the governing verb is an infinitive, matching the mood of the finite verb upon which it depends (*nec desinunt asseuerare quod*). At 2.5.6 *affirment quod . . . tribuantur,* the mood of *tribuantur* could be influenced by that of the main verb *affirment* or it could represent an attributed reason or belief. As regards the latter explanation, however, if G. had consistently discriminated between the use of the indicative for

175 See Adams (2005), esp. 196–7. On the historical development of *quod*-clauses, see Cuzzolin (1994) 83–134.

facts and the subjunctive for attributed reason in subordinate clauses, we might have expected that in reporting the opinions of Mago and Diophanes on the fennel method (3.3.4 *nec desinunt asseuerare quod . . . prouenit*) we would find *proueniat* rather than *prouenit*. Our sample size is ultimately too small to discern whether there is a clear pattern to G.'s usage of moods in *quod*-clauses. In conclusion, we should emphasize that G.'s *affirmo/asseuero* + *quod*-clauses represent an evolving 'intermediate' stage in the development of the use of dependent clauses in reported speech. They do not directly herald the demise of accusative–infinitive syntax, which would come only centuries later, and although they are in some respects novel when compared to the usage of the formal Latin of earlier periods (for example, the late Republic), they also have roots in the literary conventions of those times.

(d) Style and Social Level

In this final section on the language of *Pom.* I wish to address primarily the supposition that the Latin of *Pom.* is 'Vulgar' or, more cautiously, that it evinces extensive 'Vulgar' influence, as has been argued by Mazzini,[176] the most recent editor and commentator of the text and the only scholar to devote an extensive discussion to its language. I have reserved this section for the end because I am sceptical of the value of the notion of 'Vulgar Latin' for elucidating the language of *Pom.*: as I try to show in the few examples below and more extensively

[176] Mazzini (1988) 46 promises to 'individuare e descrivere quegli elementi linguistici di diverso livello che possono contribuire ad un giudizio globale', before going on (46–58) to catalogue numerous features that are presented as 'elementi volgari'. Maire (1999), (2000) 160–1, (2002*a*) xxxvii–xli, (2003*b*) has argued that many Vulgarisms can be found in *Med.*, which she takes in part as evidence for the influence of oral instruction on the work. She argues more cautiously than Mazzini, emphasizing the need to distinguish technical and literary features from Vulgar; but while she plausibly identifies some colloquial aspects of *Med.*, I would nevertheless resist on methodological grounds the labelling of these as 'Vulgar': see below.

in the commentary, those features of G.'s language that have been argued to be Vulgar, or which could be so argued, are in most cases more satisfactorily accounted for in terms of one or more of the diverse linguistic influences sketched above (§§v.(a)–(c)).[177] If the analyses under that set of rubrics are cogent, it is not clear what is to be gained by bringing together features that spring from quite distinct sources under one and the same heading of 'Vulgar'. In such conditions, what explanatory value would the concept have? But the vagueness attached to the notion of 'Vulgar Latin' is old.[178] Let us adopt the more concrete definition proposed by Herman (2000) 7, who prefers to understand Vulgar Latin as being 'the usage, particularly but not exclusively spoken, of the Latin-speaking population who were little or not at all influenced by school education and by literary models'.[179] Under this definition, would it be fair to call *Pom.* 'Vulgar'? Here, too, I have found no unambiguous

177 Mazzini's argument for the Vulgar nature of the *Pom.* depends not only on commonplace features about whose significance we disagree (e.g. G.'s usage of composite verbs; see below, this section), but also on several singular features of the transmitted text, such as a possible use of *ad* + dative (see 2.4.5n. *hortensia . . . destinanda*) or the ἅπαξ word *roncinia* (see 3.3.5n. *Tarentinas*), which Mazzini (1988) 56 attributes to the '*sermo provincialis* o *rusticus*'. In these and a number of other places where Mazzini defends the text of **N** (cf. also e.g. 2.1.4n. *Tenet quosdam . . . persuasio . . . ut putent*), I am not convinced that the manuscript readings are plausible on independent grounds, and it seems to me circular to argue at the same time that they are evidence for the Vulgar nature of *Pom.* and that they should be admitted into the text because *Pom.* is Vulgar.

178 See concisely Adams (2013) 3–13.

179 Cf. also Coleman (1993) 2: 'By *Vulgar Latin* is meant primarily that form of the language which was used by the illiterate majority of the Latin-speaking population.' This is not to say that even these reasonable and circumscribed definitions avoid all the difficulties attached to the term. The statement cautiously ventured by Adams (1995*b*) 131, that '[i]t is perhaps best to think of Latin as a single language which embraced the usual types of sociolinguistic and dialectal variations', sc. rather than in terms of a sharp Vulgar/educated or Vulgar/literate divide, has been borne out by research in recent decades (not least Adams' own). See also Clackson and Horrocks (2007) 231, Halla-aho (2012), Adams (2013) 3–13.

cases of the influence of such Vulgar Latin on the text.[180] In denying the allegedly Vulgar character of *Pom.*, I do not mean to deny the presence of any features of colloquial origin whatsoever in G.'s language; this would be strange, for developments in written Latin, including G.'s, are obviously driven in large part by developments in spoken Latin. Rather, what we should doubt is that the influence of spoken Latin on *Pom.* – which has not in any case been extensively demonstrated – can be shown to belong exclusively to the low-register language of those 'who were little or not at all influenced by school education and by literary models'. The claim is indeed *per se* implausible when seen from the perspective of G.'s skilful handling of many high-register features of educated and literary Latin (see above, §v(a), §v(c)). Below, I offer three examples of allegedly Vulgar elements of *Pom.* (in either the broad or narrow sense of 'Vulgar' identified above) and argue for their reinterpretation on other bases; other suspicions of Vulgar or low-register features of *Pom.* are addressed in the commentary.

One phenomenon that has been used to argue for the Vulgar character of *Pom.* is G.'s purportedly extensive usage of **compound for simple verbs with no difference in meaning**, that is, *compositum pro simplici*. One might hope to substantiate this claim (as Mazzini does) with the example of *pono*, the *uox propria* for planting (Svennung (1935) 597, *TLL* x 1.2636.35–63 (Reineke-Hillen), *OLD* s.v. *pono* 4). From the simplex verb were developed a number of prepositional compounds with various nuances. In *Pom.* we find:

1. *depono* 2.5.2, of planting down into the earth (*TLL* v 1.576.74–577.1 (Jachmann), *OLD* s.v. 5a);
2. *dispono* 2.1.2, 2.5.6, 3.1.3, of planting distributively (*TLL* v 1.1422.15–29 (Hey), *OLD* s.v. 1a);

[180] It would confuse the situation to insist that features which are potentially colloquial or low-register, but which were also firmly established in technical language, are evidence for the Vulgar character of the text; see the discussion below, this section, on the usage of composites for simples, and cf. Callebat (1992).

3. *expono* 2.5.5, 3.1.5, 4.2.4, of planting 'out' (but rarer among the agronomists; see *TLL* v 2.1758.30–9 (Hiltbrunner)).

The uncompounded *pono* appears more frequently than any one of its compounds (8x in total: 2.1.2, 2.4.1, 2.5.4, 2.7.4, 2.11.2, 3.4.3, 4.6.6, 4.7.1).

Mazzini (1988) 51–2 argues that the variety of compounds that G. employs for *pono* is precisely an example of the 'semantic banalism' which he takes to be evidence of Vulgar influence in G.'s writing: 'più evidenti esempi di banalismi semantici nel nostro frammento sono vari composti verbali con lo stesso significato dei corrispondenti verbi semplici: *dispono* . . . *depono* . . . *expono* . . . = *pono* . . .'. The usage of compounds for simples is indeed often alleged to be characteristic of Vulgar Latin (e.g. Cooper (1895) 246–50, Svennung (1935) 546–7, 550, H–Sz §166.d.β), in comparison with the poetic or archaizing *simplex pro composito* (e.g. Wackernagel II 188–9, Svennung (1935) 546, H–Sz §166.d.α). As one might expect, the reality is often more complicated. Although the phenomenon of *compositum pro simplici* is certainly not a fiction, it is sometimes applied without sufficient consideration of alternative possibilities in context. In *Pom.*, for example, it is demonstrably untrue that prepositional compounds carry no additional semantic nuance: on closer scrutiny they are found in almost every case to carry subtle but meaningful specifications for the actions that they describe. Thus at 2.5.2, the prefix *de-* in *depono* retains a local force, responding to the earlier portion of the sentence in which there is described the opening of a trench in the earth (*in terra sulcus . . . aperitur*) in which (*in quo*) the stones are 'set/planted down' (*deponi*). At 2.5.6, the *dis-* of *dispono* quite clearly retains its distributive force: the context makes it clear that the seeds are being planted in several vessels (*gastra*) or baskets (*quala*). So too at 3.1.3 *disponunt*, the point is that the almonds must be 'planted apart' in a triangle (*triangulari lege*) with a fixed measure of space retained between the seeds. Again, at 3.1.5 *seminibus expositis*, the prefix *ex-* picks out the fact that the seeds have been planted 'out' beforehand into the nursery plots (so also 2.5.5 *exponatur* of planting in the nursery). This principle holds for other compounds, too: at 3.2.3 *adobruunt*, *ad-* distinguishes the technical operation of 'earthing up' (*adobruere*) from the action of 'burying' or 'planting' that is ordinarily described by *obruo*.[181] *Pace* Mazzini (1988) 52, it is thus not an example of *compositum pro simplice* for *obruo*. Examples could be multiplied (and are observed *passim* in the commentary), but these will suffice to illustrate the point.[182]

But even were we to set aside the fact that many alleged cases of *compositum pro simplici* in *Pom.* are mistaken, we would still have to take into consideration the precedents of earlier agricultural writers when assessing G.'s diction. Suppose, for a moment, that the prefix *dis-* in one of the occurrences of *dispono* did not have any special force, and that we could make a case for a genuine *compositum pro simplici*. From this fact we could still not arrive at the conclusion that the influence of Vulgar Latin stands behind G.'s choice of the verb, both for the reason that *dispono* had long been fixed as a standard element of the agricultural lexicon (cf. Col. 3.13.4, 5.12.3, 11.3.40, *Arb.* 1.4, 19.1, Plin. *HN* 17.78) and for the reason that literary *uariatio* could play a role in the verb choice. These motivations would play an important role in explaining G.'s use of the verb, whatever other explanations we might attach to it notwithstanding. In general, we may say therefore that it is fallacious to infer from the frequency and

[181] See further 3.2.3n. *adobruunt*; cf. also 2.1.3n. *obruenda*.

[182] Although we are here concerned primarily with the usage of compounds for simples, it merits notice that the standard claims about the register and periodization of the usage of simples for composites have also been subject to qualification and revision. Thus, while the phenomenon has been variously called early, archaizing or poetic, it is in fact not uncommon in late Latin texts of a wide variety of stylistic levels. The interpretation of some of these simples can be difficult (e.g. on *plico/applico* see Löfstedt (1959) 45–6, Väänänen (1990), Adams (2007) 349–50, *TLL* x 1.2441.30–8 (Schrickx)), but it is in general not hard to find examples of the phenomenon in sub-literary or late texts: consider the attestations in the *TLL* for *pareo* under the first heading 'fere i. q. appareo' (x 1.373.4–5 (Breimeier)), and more generally H–Sz §166.d.α, with references.

variety alone of compounded verbs in *Pom.* (or in other authors' writings) that such verbs are used indifferently or that they have lost the force of their prepositions and are thus evidence of the Vulgar character of the text. The frequency of such compounds in G.'s writing can be explained on the basis of a special concern for semantic precision, a concern that is natural enough in technical works such as *Med.* and *Pom.*[183]

A second example that could be adduced as a purportedly Vulgar aspect of the language of *Pom.* is G.'s use of the ἅπαξ adjective *amaritosus* (3.7.5). The adjective is formed regularly, by the denominative adjectival suffix *-osus* meaning 'provided with, rich in' or 'resembling' (Leumann §305, Adams (2013) 574, 577, (2016) 68). Although *amaritosus* has a full and appropriate meaning in the context, we must reckon with various claims that have been made about the register of adjectives formed by the ***-osus* suffix**. The most recent and comprehensive treatment of such claims is that of Adams (2013) 571–8, who assesses the different attempts to assign *-osus* adjectives to a single stylistic level. Adams considers four registers with which such adjectives have been connected: (1) Vulgar (Olcott (1898) 205, Baehrens (1922) 119), (2) 'plebeian' (Cooper (1895) 122, Ross (1969) 59), (3) rustic (Cooper (1895) 122–3, Knox (1986) 92–3) or (4) unpoetic (Axelson (1945) 60–1). As Adams argues in a careful review of the evidence, however, the alleged connection between the *-osus* suffix and various low stylistic levels has largely to do with the things which such adjectives are used to describe (and thus the context in which they appear): the reasoning goes that, since *-osus* adjectives describe objects that are socially 'low' – objects of medical or agricultural attention, for example – , then the *-osus* adjectives *themselves* must be socially 'low'. But, with Adams (2013) 578, we should 'resist the idea that, if a suffix often

[183] G.'s usage of compounds may also reflect to an extent now difficult to appreciate the stylistic norms of formal Latin of his day: see the remarks on stylistic change in connection with nominalizations, above, §v(b); cf. also Coleman (1999) 346, 348 on the 'diachronic perspective' on change in register.

appears in terms referring to things that may be mundane, unpleasant or despised, that makes the suffix itself despised, vulgar or substandard'. Even more relevant for the interpretation of *amaritosus* in *Pom.* is the observation of Adams (2013) 574 that *-osus* adjectives are especially common in technical treatises because of the semantic force of the suffix, which was 'capable of expressing concrete features of phenomena' (cf. 577).[184] The evidence for Vulgar Latin here is therefore scant.

Finally, Mazzini (1988) 49 purports to find in a number of places in *Pom.* the **unmotivated use of the subjunctive** in a mixed condition, an 'elemento caratteristico del volgare', as he maintains. He points to 3.7.3 *utile est si effundatur stercus*, etc., as an example. But there is another way to construe the grammar. It is preferable to take the *si*-clause as a noun clause dependent on *utile est*: although *utile est* typically takes an infinitive, there is substantial overlap between substantive *si*-clauses and infinitive clauses (Nutting (1922) 131–2), and substantive *si*-clauses are indeed regular at all periods of Latin (Nutting (1922), (1925) 78–80, 86, H–Sz §366, *OLD* s.v. 12). The subjunctive would not be unmotivated in that context. Mazzini adverts to 3.7.6 *Si uiridibus auferas tenuiora, quaeque liqueris meliora sunt* as another example of the same phenomenon. Here, too, the Vulgar character of the phenomenon is not proven, for it is regular in the educated Latin of earlier periods to put a (present) generalizing second-person singular verb in the protasis into the subjunctive even where the apodosis contains an indicative (Nutting (1923) 151, (1925) 82–5, *NLS* §195, Pinkster 659). For an early example, cf. Cato *apud* Gell. *NA* 11.2.6 *si* (sc. *ferrum*) *exerceas, conteritur; si non exerceas, tamen robigo interficit*. It may not be irrelevant to note that as a general rule the subjunctive is preferred by agricultural authors in these indefinite second-person clauses (Hine (2011) 637–8).[185]

[184] Before we peg *amaritosus* to a technical register in *Pom.*, however, we should remark that the word seems to be used in part for the purpose of rhetorical variation after *amarus* and *amaritudo* earlier in the same section (as noted above, §v(b); see also 3.7.5n.).

[185] Exceptional are Cato and *Arb.*: for discussion, see Hine (2011) 638–41.

VI RECEPTION

The reception of *Pom.* can now be traced largely in the use that Palladius made of it in his fourth-century agricultural manual *Opus agriculturae.*[186] Palladius mentions G. by name 13 times,[187] but comparison of chapters in *Pom.* with their corresponding treatments in Palladius shows that his use was probably more extensive, for G. often goes uncredited in his advice.[188] Although the tenth-century Byzantine agricultural compilation called the *Geoponica* does not mention G. as a source, it seems probable in view of the significant parallels in their advice that *Pom.*'s influence is felt, directly or indirectly, in parts of it.[189] Especially Palladius, but also the *Geoponica*, is useful in reconstructing the sense and sometimes the text of *Pom.*[190] Some have traced the influence of *Pom.* all the way to the twelfth-century Arab agriculturalist Ibn al-ʿĀwwām, who mentions a Marsiyāl, but it has been denied that this figure is to be identified with G.[191] The paucity of secure allusions to his work in later times notwithstanding, it seems likely that the reception of G.'s agricultural writings was greater than can now be easily appreciated: this supposition is indirectly corroborated by the later success of *Med.* (despite its detachment from *Pom.*), by the appreciative remarks of Servius and Cassiodorus,[192] and by the existence of a later codex (for

[186] On Palladius, see esp. Bartoldus (2012); also, concisely, Fitch (2013) 11–27.

[187] Pallad. 2.15.10, 2.15.19, 4.9.9, 4.10.5, 4.10.16, 4.10.34, 5.3.4, 6.6, 7.5.2, 11.12.5 (*bis*), 11.12.7, 13.4.1.

[188] Wellmann (1908) 2 argues that 'Palladius ist . . . im Grunde weiter nichts als ein verwässerter Gargilius Martialis'; Svennung (1927) 131–54, without denying borrowings from G., shows that Palladius used other sources too (e.g. Columella).

[189] For the composition of the *Geoponica* and its agricultural sources, see Teall (1971) 39–44, Georgoudi (1990) 15–89, Decker (2007), Lelli (2010) I vii–lxxxv.

[190] For Palladius' role here cf. e.g. 3.6.4n. [*aut*] *pruina*; 3.7.2n.

[191] The extracts (in French translation) of Ibn al-ʿĀwwām mentioning a Marsiyāl are printed by Condorelli (1978) 46–52, Maire (2002*a*) cx–cxiii. As Fischer (1997*a*) 271 n. 9 notes, Ullmann (1972) 444 n. 15 rejects the identification of Marsiyāl with G. ('[e]s handelt sich jedenfalls nicht um Gargilius Martialis'), although Ullmann gives no explanation for doing so.

[192] Quoted above, n. 3.

which see below, §vii) that bound G. in a *quadrigae* along with the towering agronomists Cato, Varro and Columella.

VII HISTORY OF THE TEXT

As was remarked above (§i(b)), *Pom.* represents in all probability only one part of a much larger work on the cultivation and therapeutic use of garden produce.[193] The widely circulated excerpts from this work that became *Pom.*'s 'sister' treatise, *Med.*, are entirely pharmacological; that is, they contain no horticultural advice. This circumstance makes the palimpsest **N** (Naples iv.A.8),[194] which transmits *Pom.*, extraordinarily valuable: it is the sole trace not only of G.'s agricultural writings, but also of the composite original work, for, as mentioned above, **N** conjoins a part of the quince pharmacology of *Med.* (43.1–8) to the preceding horticultural advice of *Pom.* (1.1).[195] We need not conclude from the sentences that are common to **N** and the manuscripts of *Med.* that **N** is itself the manuscript from which the excerpts constituting *Med.* were prepared, but there is also nothing to exclude the possibility.[196]

The facts surrounding the discovery, editing and publication of *Pom.* have been closely investigated by previous editors: see Condorelli (1978) xxxii–xlii, Mazzini (1988) 11–16. The interest in

[193] This section draws in parts on Zainaldin (forthcoming 3).

[194] For photographs of the manuscript, see Cipolla (1907) tabb. i, x (*Pom.*), xi, xxxvi, xlii, Lowe (1934–71) iii nos. 392, 400–4 (*Pom.* 404). It is described also by Collura (1943) 50–2. Scotti includes with his edition 16 pages of the copper plates made of the primary script for *Pom.* (see below), which were reproduced and attached as an appendix to the edition of Condorelli. See also for appraisals of the manuscript Mai (1828) 387–90, Scotti (1833) 119–40.

[195] Add that it allows for the identification of G. as the author of *Med.* The portion of shared text is reproduced as Appendix 1.

[196] The only good readings of consequence that the manuscripts of *Med.* have against **N** are 43.3 *ciere* (*acere* **N**) and 43.5 *itaque et* (*et* om. **N**). A careful scribe could have corrected either of these; regarding the former (a more difficult correction), it would have helped that *urina* and *ciere* appear three times together elsewhere in *Med.* (25.13, 29.4, 38.5).

these questions is stimulated not least by the fact that *Pom.* came to light as a result of the famous palaeographic investigations of Cardinal Angelo Mai in the early nineteenth century, and further by the uncertainty around the question to whom the *editio princeps* is to be attributed: is it to Mai himself, whose earliest edition is dated 1828, or to A. A. Scotti, notwithstanding the apparently later date of the publication of his edition, 1833? Without tracing Condorelli's and Mazzini's detailed reconstructions of the individual and collaborative efforts of Mai and Scotti in the years 1826–8, the salient conclusions can be given in brief here. Mai discovered *Pom.* in 1826. At the time of its discovery, he undertook a hasty initial transcription of the text, prepared in a matter of days, which would serve as the basis for his edition. Between 1826 and 1828, Scotti undertook a much more careful examination of the manuscript; the edition based on this examination was accepted for publication in 1828 in the *Memorie della regale accademia ercolanese di archeologia* (II), which would not however see a public printing until 1833. Consulting Scotti's edition, Mai improved his own and published it immediately, placing it in his volume containing also fragments of Cicero (*De re publica*), Sallust and Archimedes. Here too the date of publication is misleading, for although the volume is stamped 1828, Condorelli has argued that it did not appear until 1829. In sum, although the discovery of *Pom.* is Mai's, the credit for the *editio princeps* must be given to Scotti. So much for the discovery of *Pom.*: now a word on the manuscript.

N contains the fragments of *Pom.* in a sixth-century uncial that was written over with the *Liber pontificalis* at Bobbio in the seventh/eighth century; the palimpsest is illustrated in Figure 1. The primary scripts of **N** further include *Paralipomena* (half-uncial, sixth century), Lucan (rustic capital, fourth century) and the *Iustiniani digesta* (sixth century), written over by Charisius, *Ars grammatica* and Servius, *De centum metris* (both thirteenth century).[197] The manuscript remained at Bobbio until

[197] See the references above, n. 194.

the second half of the fifteenth century (it is still in the catalogue of 1461), whence it travelled to Rome and then in the sixteenth century to Naples, where it remains today, its leaves taken apart and kept between glass.[198] The manuscript has degraded to the point of almost complete illegibility as a result of the early attempts of Scotti and Mai to improve the text with the aid of chemical reagents.[199] Thus, the text of *Pom.* is now known to us almost entirely in a mediated form, through the engraved copper plates of the primary script (called the *tabulae Neapolitanae*) which Scotti had produced and attached to his edition (Figure 2). Nonetheless, Condorelli was able

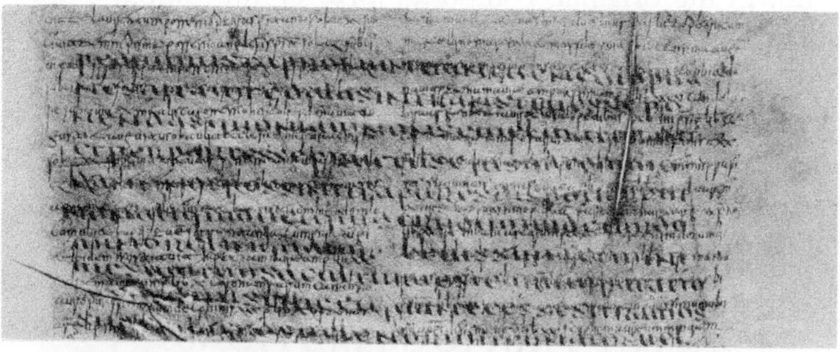

Figure 1: The palimpsest from Bobbio (Naples IV.A.8), showing *Pom.* 3.4.2–3.5.1

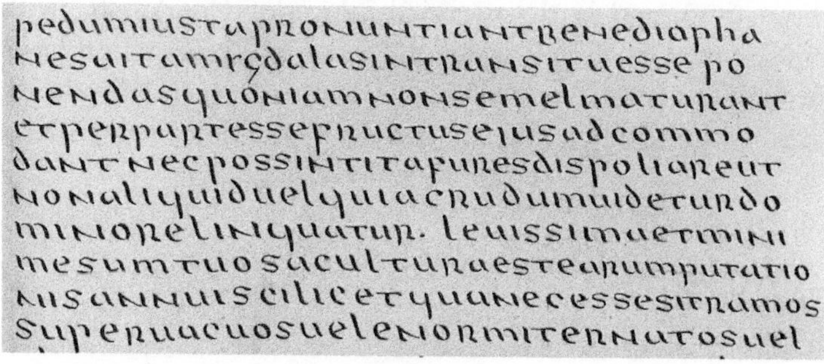

Figure 2: The *tabulae Neapolitanae*, showing *Pom.* 3.4.2–3.5.1

198 On the travels of **N**, see Mazzini (1988) 19–21.
199 See Condorelli (1978) xvii–xviii, Mazzini (1988) 22–4.

upon repeated inspection to extract some readings (= **N**^{c}) from the remains of the manuscript that differ from those of the *tabulae*.[200] In fact, Scotti and Mai had already attested to readings of the codex that may have differed from those preserved in the *tabulae*: Mai's divergent readings (= **N**^{m}) are presumably the result of his own examination mentioned above, whereas the reasons for Scotti's discrepancies are unknown.[201]

We hear of two other manuscripts of *Pom.*, neither surviving: the codex of Gargilius that Cassiodorus mentions depositing in his *Viuarium* at Scylacium (Squillace in Calabria), which we may surmise from his words contained both *Pom.* and *Med.* (no doubt in the original, composite form of the work);[202] and a codex from S. Marco collated by Politian in 1482 and seen by Petrus Victorius in 1541, whose content included the agricultural works of Cato, Varro, Columella and Gargilius Martialis (identified as 'Claudius Gargilius' in the codex, according to Victorius).[203] The *codex Marcianus* was already missing Columella and Gargilius when Politian collated it.[204]

VIII PREVIOUS EDITIONS

There have been four major editions of *Pom.* since its discovery: in the nineteenth century those of Scotti and Mai (see further above, §VII), and in the twentieth, those of Condorelli (1978) and Mazzini (1978 first edn, 1988 second). Mai's edition went through numerous re-printings, which saw very few alterations to the text; besides the first Rome edition (dated 1828),

200 For his remarks on these efforts, see Condorelli (1978) xliii–xliv.

201 For Mai's recollections of his own (hastier) study, see Condorelli (1978) xl–xli; and for Scotti's (unexplained) discrepancies from the *tabulae*, Condorelli (1978) xli–xlii. I have not made use of Scotti's discrepancies in establishing the text (and hence have not assigned them a siglum).

202 For the quotation (Cassiod. *Inst.* 1.28.5), see above, n. 3.

203 Victorius (1542) 3–4. See also Condorelli (1978) xv–xvi, Mazzini (1988) 21.

204 As well as the final portion of Varro: so Politian *ad* Varro, *Rust.* 3.17.4, 'hactenus cum uetustissimo codice emendaui: nam reliquum in illo quidem deest'.

I have made use of the edition of 1832 (= **Mai**²), published in Lüneburg, Germany,[205] and the edition of 1846 (= **Mai**³), published again in Rome, for a handful of improved readings that they present. The editions of Scotti and Mai are augmented by prefaces and footnote-style commentaries, which contain text-critical notes and occasional exegesis of language and content and exhibit select *loci similes* from the Roman agricultural authors and others. In 1848, there was published under the auspices of Joseph Antonelli the first translation of *Pom.* into a modern language (Italian), printed in columns facing the Latin text of Mai. Although the text of Mai is not altered in the printing,[206] the 17 pages (34 columns) of text-critical and exegetical notes appended to the translation offer numerous suggestions for the construal or improvement of the text. The very translation of *Pom.* also constitutes an important step forward in its interpretation, and in numerous places establishes the basis for the efforts of Mazzini, who was the next scholar to translate the text.

We must wait until the twentieth century for further critical and exegetical work on *Pom.* In 1915, Marx extracted several sentences from *Pom.* for his collection of fragments pertaining to Celsus' *Artes*. In doing so, he also offered a few proposals for improving the text, in one place (4.1.3) with a happy result.[207] The 1978 editions of Condorelli and Mazzini (the latter publishing a second edition in 1988, see below) represent the next significant advance in the study of *Pom.* since its discovery and initial publication. Both offer new texts of *Pom.* (Mazzini also an Italian translation) based on the *tabulae Neapolitanae* and the

205 This edition and the other German edition published two years earlier (1830, in Hanover) share a peculiar set of errors and misprints which have no textual authority: see Condorelli (1978) xxxvii–xxxix. For a case in which the German edition improves the text – whether fortuitously or by design – , see 2.3.5n. *cum iam* [*iam*].

206 Excepting a few places 'ove la scrittura del codice ci pareva troppo discordante dalle nostre regole comuni' (Antonelli (1848) coll. 1613–14).

207 See 4.1.3n. *sed . . . reiectat.*

discrepant readings of Scotti and Mai and improved by the conjectures of Scotti, Mai, Antonelli and Marx, to which they add their own proposals; Condorelli's edition further incorporates the results of his inspection of the remains of **N** (=**N**^{c}), which gave a few readings different from those in the *tabulae*,[208] as well as the occasional conjectures found in the marginal notes to a copy of Scotti's edition of *Pom.* at the Archiepiscopal seminary at Padua (= **adn. Pat.**).[209] Both Condorelli and Mazzini provide a general overview of *Pom.* and extensive discussions of the history and constitution of the text. Condorelli further prints extracts from Palladius and Ibn al-'Āwwām relevant to the lost agricultural sections of *Pom.*;[210] he also reproduces the *tabulae Neapolitanae* in an appendix to his volume. Mazzini's introduction, along with the footnote-style comments to his text and translation, presented by far the most substantial analysis of the sources and language of *Pom.* to date. In 1980, there appeared a two-page review of Mazzini's edition by M. Winterbottom which contains numerous suggestions for improving the text. In response to the edition of Condorelli and the review of Winterbottom, as well as to the advent of further work on *Med.* (Tapper (1980), Riddle (1984)), Mazzini released a second edition in 1988 augmented by 21 pages (127–47) of critical reflection on *Pom.* and reactions to the editorial decisions of Condorelli and the criticisms of Winterbottom. Although several important publications on G. have appeared in the years since then, and Önnerfors (1993) 264 n. 8, 270–4 has in passing drawn attention to certain shortcomings in Mazzini's treatment

[208] In two places these readings improve on those of the *tabulae*. The first is 2.3.2 *In calido merito sub ipsum tempus autumni ossa merguntur*, where the *tabulae* transmit the subjunctive *mergantur*, unmotivated in context. The second is 3.6.2 *alii in locum taedae cuneum qui esset ex robore*. Here, the *tabulae* have *illi*, but Condorelli saw *uli* in the fragments: not the true reading, but closer to *ali(i)*, the reading Mai recollected from his own examination.

[209] Condorelli (1978) 1 believes that the annotator may be Joseph Furlanetto.

[210] But for doubts about the pertinence of the references in Ibn al-'Āwwām, see above, n. 191.

of G.'s language, no work devoted solely to *Pom.* has appeared in the last three decades.[211]

IX CONVENTIONS ADOPTED IN THIS EDITION, TRANSLATION AND COMMENTARY

The reedition offered here is based on the readings of **N** as they are reproduced in the *tabulae Neapolitanae*, plus those readings of the manuscript reported by Mai (**N**[m]) and Condorelli (**N**[c]) that differ from the *tabulae.* This text differs in 294 places from the readings of **N** preserved in the *tabulae Neapolitanae*, orthography and trivial spelling errors representing the largest share of these.[212] For the most part I report in the apparatus only those errors which cannot be corrected mechanically, or which present a choice in their correction that would affect the interpretation of the text; virtually all of these are discussed in greater detail in the commentary. The full register of spelling errors and orthographic differences in **N**, both those reported in the apparatus and those not, may be found in Appendix III. In preparing the text, I have as a general rule attempted to retain the readings of **N**, even where they are surprising or difficult, if it is possible to make a reasoned defence on the basis of our knowledge of the Latin of the period and of G.'s compositional

[211] The most important of the scholarship on G. appearing after the edition of Mazzini is Maire's 2002 Budé edition of *Med.* (2002*a*); in the same year she published a concordance to G.'s works (see *CG*). Maire has also published numerous articles on *Med.*, focusing chiefly on aspects of its language (1997), (1999), (2003*a*), (2003*b*), but also providing a specialist overview of the work (2000), discussing its sources (2002*b*), and providing an introduction to *Med.* in the context of ancient medicine (2010). Other work devoted to *Med.* includes Condorelli (1995–8) (relationship of *Med.* to Pliny's writing; a general discussion of the original form of *Med.* in light of the evidence of *Pom.* is found on pp. 241–52) and Ferraces Rodríguez (2006) (review article of Maire's edition). Stok (1993) reflects on G.'s status as an intellectual, focusing mostly on the evidence of *Med.*

[212] Characteristic errors are studied and classified by Mazzini (1988) 24–34.

habits in *Pom.* and *Med.* There are a number of places where a relatively idiomatic transposition or other small alteration would produce one of those clausulae found most commonly in G. (on which see above, §v(a)),[213] but I have preferred not to alter the text in these places unless I found independent grounds for doubting it.

I note here editorial decisions concerning the ordering and numbering of chapters and sections in *Pom.* Mai assembled the leaves of **N**, which gave an uncertain picture of the proper order of the chapters,[214] as follows: (1) quince, (2) peach, (3) almond, (4) chestnut. Scotti, on the other hand, placed the peach at the end: (1) quince, (2) almond, (3) chestnut, (4) peach. (This is accordingly the order of the *tabulae.*) I have adopted the ordering of Mai, which is preferable for the reason that it brings the chapters of *Pom.* into alignment with the order of the corresponding chapters of *Med.* (see above, §1(b)). A more pragmatic reason is that Mai's order has become the reference standard, especially since it was adopted by Mazzini. As regards the sections of *Pom.*, I also follow Mai in printing the horticultural part of *cydon.* in one section, not two, as Scotti does when he begins a new section before 1.1.2 *infundunt.* For *persic.*, however, I follow the numbering of Mazzini in 12 sections, not Scotti (11 sections) or Mai (13 sections).[215]

213 These are reported *passim* in the app. crit., to which add the following (due to Reeve): 1.1.5 *affixi mellis* (*mēllĭs āffīxi*, cretic-spondee); 2.3.4 *teneantur aquae innantia* (*aquae teneāntŭr īnnāntĭa*, dicretic); 2.5.7 *pluuia ei subuenerit* (*plŭuiă sūbuēnĕrit*, resolved dicretic, by deleting or transposing *ei*); 2.11.1 *quibus seponitur sucus pinguis* (*pinguis sepōnĭtūr sūcus* or *sucus sepōnĭtūr pīnguis*, cretic-trochee); 2.11.2 *harenoso ponas* (*ponās hărēnōso*, cretic-spondee); 2.12.4 *saepe in breue tempus flaccescunt* (*in breue tempus saepĕ flāccēscunt*, cretic-spondee); 3.7.4 *dulcior sucus hauriatur* (*hauriatur dūlcĭōr sūcus*, cretic-trochee); 4.3.4 *integrae debent credi* (*debent īntĕgrae crēdi*, cretic-trochee (scanning *-egr* -short)).

214 See Mai (1828) 389.

215 Scotti does not recognize a lacuna between 2.8–9 and therefore runs the sections together as 2.8, giving 11 sections in total in the chapter; Mai on the other hand introduces a tenth 'section' in the lacuna between 2.8–9 consisting of material drawn from Palladius, thus giving 13 chapters rather than 12.

Two further remarks on the Latin text are in order. First, I have introduced a novelty in numbering the sentences of the sections in order to make consultation of the commentary and the insertion of cross-references easier. In order to avoid confusion, I have not attempted to force the numbering in the Latin onto the English translation. Second, I have not printed in my edition the portion of **N** which corresponds to the first seven sentences and a partial eighth sentence of *Med.* 43 (= 43.1–8). That text is edited by Maire in her Budé edition with the aid of the manuscripts of *Med.* A slightly modified version of her text with a selective apparatus is given as Appendix 1. For notes and commentary on the passage, see Tapper (1980) 194–205, Maire (2002*a*) 146–7.

Cross-references in the introduction, commentary and appendices that give the chapter, section and sentence number without Latin words, e.g. '2.1.1n.', indicate the first and principal note to the respective Latin sentence of *De arboribus pomiferis*. Cross-references that give only the chapter and section number, e.g. '1.1n.', indicate the introductory note to a section or, when given as a range, e.g. '4.1–7n.', to a group of sections. Where further specificity is required in cross-referencing notes, the words of the Latin lemma may be added. Thus distinguish e.g. '3.1.1n. *cum hiascere coeperunt*' from '3.1.1n. *hiascere coeperunt*'. In cases where such a reference is to a note whose lemma contains more than five words, the reference to that lemma is abbreviated by the first and last word of the lemma with an ellipsis intervening. For example, instead of '2.7.1n. *contra aliarum arborum legem de superioribus partibus mutuatur alimentum*', the cross-reference reads '2.7.1n. *contra . . . alimentum*'.

A final note on the translation. G.'s Latin poses a challenge for the translator as much for its technical terminology and alternately abrupt or repetitive patterning of language and thought as for its careful *uariatio* and elaborate symmetries. I have sought a middle course between an excessively literal (and thus sometimes obscure) and excessively free (and thus sometimes inaccurate) translation and have tried to represent G.'s rhetorical figures with English diction and figurative language appropriate to the effect of the Latin, although sometimes employing different means to that end. Above all, it is intended that the translation will serve as a first helper and guide to the interpretation of the Latin, although it must be admitted that it will not always be completely adequate to the purpose; in such cases, the commentary will sometimes, in addition to full explanation of the text, provide an awkward but more literal construal.

SIGLA

N	codex Neapolitanus IV.A.8 fragmenta *de arboribus pomiferis* continens, cuius lectiones in tabulis Neapolitanis seruantur
N^{m}	eiusdem codicis lectiones cum tabulis Neapolitanis discrepantes quas A. Mai testatur
N^{c}	eiusdem codicis lectiones cum tabulis Neapolitanis discrepantes quas S. Condorelli testatur

Editiones uel commentationes quae ad constitutionem textus pertinent:

Mai	A. Mai, Romae 1828
Mai^{2}	A. Mai, Lunaeburgi 1832
Scotti	A. A. Scotti, Neapoli 1833 (editio princeps)
edd.	consensus Mai 1828 et Scotti 1833
Mai^{3}	A. Mai, Romae 1846
Antonelli	coniecturae quas in adnotationibus proposuit I. Antonelli, Venetiis 1848
adn. Pat.	adnotationes Patauinae, de quibus uid. Condorelli pag. l
Marx	F. Marx, Berolini & Lipsiae 1915
Condorelli	S. Condorelli, Romae 1978
Winterbottom	M. Winterbottom, *CR* 30 (1980) 144–5
Mazzini	I. Mazzini, Bononiae 1988 (editio altera)
Reeve	coniecturae quas mihi per litteras proposuit M. D. Reeve

< >	addenda uel inserenda
[ ]	delenda
	typis Italicis (e.g. 3.1.4) indicantur ipsissima uerba auctoris a Gargilio laudati

TEXT, CRITICAL APPARATUS AND TRANSLATION

DE ARBORIBVS POMIFERIS

<I DE CYDONEIS>

1.1,2 ***uerum de fructibus pauca nobis dicenda sunt. Sane ad mon-
tium deuexa melius proueniunt, non in planis locis . . . urasin
. . . ci . . . sun . . . de raro nunc inuenies . . . me . . . infundunt
3 et na . . . nt fa . . . ut putant, ampliora quae Nouember affert,
et matura cydonea legenda sunt . . . iuncta con . . . subinde . . .
interd . . . aut ex . . . rere . . . quo . . . mi . . . capessere . . . omne
4,5 con . . . condiri etiam per cydoneum posse qu . . . affirmant ali
6 . . . traxisse aliquid olim de sapore affixi mellis. Aliqui putant et
in dolium uini olla noua conclusa, ut in eo obtecta natent, esse
7 mittenda. At nos non leui pondere affirmamus quod si melle
linantur, in eo melius perseuerant et, ubi tempus erit, speci-
8 osa exire pollicentur. Quidam patina noua sicco gypso sepa-
9 rata obruunt. Eadem alii milio siue palea separata demergunt.
10 Sunt qui figularem cretam cum amurca subigunt soleque sic-
catis cydoneis illiniri existimant, si in loco sicco et frigido repo-
11 nantur. Multi uel solam cretam sufficere crediderunt, si cum
ramulis suis mala detracta et oblita in sole siccentur ac, si quid
12 extra hiauerit, luto contegatur. Seruantur et gypso illita, sed
aspera sunt sapore, quamuis et aspectui speciosa procedant.
13 Consentiunt aliqui posse et scrobibus et in doleis eodem modo
14 quo punica supra uel alia m<ala> seruari. Quintilii cydonea
15 in scobe seruata diuturna promittunt. Diligentioribus rectissime
uisum est praemonere ne cum ceteris pomis cydoneum miscea-
tur: sibimet censent reponendum, austeritatem eius timentes,
quae corrumpere solet, ut necesse est, quidquid de proximo
16 adfuerit. Cydoneorum pondus adeo iumentis praeter cetera
infestum est ut remedia quoque contra haec nullas uires habere
compertum sit.

1.2 ura sin *edd., sed diuisio litterarum incerta* 3 *post* putant *lac. stat. edd.* 4 condiri] condire *Scotti* 8 separata obruunt *Scotti*: separatas obruunt *uel* separata sobruunt *N* (*unde* separata subruunt *Mai*) 10 illiniri] *an* illini? 13 alia mala *Scotti in adn.*: aliam *N*: alia *Mai* 15 solet ut] solent N^c *dub.*

ON FRUIT TREES

I On the Quince

1 But a few things ought to be said about fruits. To be sure, they come forth better on the slopes of mountains, not in level places . . . you will now seldom find . . . they pour in and . . . according to them, those which November produces are larger, and the ripe quinces ought to be picked . . . (sc. it?) can also be flavoured through the quince . . . (sc. some?) assert . . . they have acquired at some point a taste of the honey which was applied. Some think that they also should be enclosed in a new jar and put into a vat of wine, so that the bottled fruit float in it. But we maintain with no little weight that if they are smeared with honey, they keep better in it, and, when the time comes, they promise to emerge with a fine appearance. Some bury them in a new pan, separated by dry gypsum powder. Others sink them separated by millet or chaff. There are some who knead potter's clay with olive lees, and are of the opinion that this mixture should be smeared onto the quinces once they have been dried by the sun, if they are to be stored in a cool and dry place. Many have expressed the belief that even clay alone suffices for the purpose, if the fruits that have been picked along with their twigs and coated are dried in the sun and, if any of them splits open, patched with mud. They are preserved smeared also with gypsum, but these have a harsh flavour, although they also come forth with a fine appearance. Some agree that they can be preserved in both holes and vats in the same way as for pomegranates mentioned above or other fruits. The Quintilii promise that quinces preserved in sawdust will last for a long time. Those who are more diligent have, very rightly, thought it correct to advise against mixing the quince with other fruits: they recommend that it should be stored by itself, since they fear its harshness, which, as is inevitable, often spoils whatever is close by. The weight of quinces is so much more aggravating to pack animals than that of other fruits that it has been established that remedies indeed have no efficacy against them.

<II DE PERSICIS>

1.1,2 ***ossa sererentur. Curtius Iustus et mense Ianuario calidiori-
bus locis putat posse disponi; Quintilii armenia sola consentiunt
3 Ianuario mense ponenda. Graecis omnibus ferme autumnalis
satio ratione naturali uidetur electa, siquidem eo tempore cum
pomis caducis ossa delapsa in nouas plantas saepe uideamus
sine culturae adminiculis sua sponte fruticare; ideoque plerique
non ossa sed poma ipsa matura integra existimant obruenda,
ut ea res et ossibus uicem stercoris praebeat et pomis futuris
4 illibatum saporem, sicuti arbitrantur, acquirat. Tenet quosdam
hodieque uana persuasio ut putent ossa quae dente contacta
5 sunt aut prorsus extingui aut infeliciter nasci. Sed frustra id cre-
didisse tunc sentiunt, cum coeperint experimenta consulere.

2 Sunt qui ab exortu Fauoni persicum serere melius existiment;
2 medio tamen tempore anxie atque sollicite ossa custodiunt. Et
quidem primo recentia paucis diebus in cinere mergi patiun-
tur ut quidquid umidi continent illo siccante deponant; post in
umbra explicantur et biduo uentilata cum terra soluta in qualo
uel in fictili uase componuntur, idque tectum atque oblitum
sulco ita infodiunt ut accessum neque sol neque uentus inueniat.

3 Interest autem cuiusmodi statum caeli condicio regionis,
2 calidum frigidumne, sortita sit. In calido merito sub ipsum
tempus autumni ossa merguntur; metus enim non est gelum:
immo si qua corrupta depereant, frigida relapsa hieme, cessa-
bunt cum ossibus exterius tempore resolutis totam germinis sani
3 uitam nuculeus experitur. Mala denique tunc inseruire arbo-
4 ribus debent, cum extra hiare coeperint. Huic sationi potest
prodesse quod generaliter Celsus agi scribit, ut ossa priusquam
terrae committantur, diebus quinque teneantur aquae innantia.

1.4 putent *edd.*: putant *N*   **2**.1 anxie *Scotti in adn.*: anxius *N*   2 quidem *Scotti*: quidam *N*   idque *Winterbottom*: adque *N*: atque *Scotti*   **3**.2 merguntur *N*[c]: mergantur *N*   3 Mala *Scotti in adn.*: Mali *N*: Ossa *uel sim. Reeve, fort. recte*

II On the Peach

1 . . . that the stones be sown. Curtius Iustus thinks that they may be planted out also in January in warmer places; the Quintilii accept that it is the apricot alone that should be planted in January. It seems to be by reasoning from nature that nearly all of the Greeks have chosen an autumn sowing, since at that time we often see the stones that have dropped with the fallen fruits sprouting into new saplings of their own accord, without the support of cultivation. And for this reason, most of them are of the opinion that not the stones, but the ripe fruits themselves should be buried whole, so that that material may both serve in place of manure for the stones and, according to them, provide the future fruits with an undiminished flavour. Still today a vain belief grips some people that stones which have been touched by a tooth are either entirely destroyed or grow unfruitfully. They see that their belief was groundless, however, as soon as they begin to consult experience.

2 There are some who are of the opinion that it is better to sow the peach from the onset of the west wind; in the intervening period, they store the stones with painstaking care. This is how: they first have them set down into ash for a few days while they are fresh so that they may give up whatever water they contain as the ash dries them out; afterwards they are spread out in the shade and, once they have been aired for two days, are put along with loose earth into a basket or a clay vessel. They bury this, covered and sealed, in a trench, so that neither sun nor wind can find ingress.

3 It makes a difference, however, what type of climate the nature of the land possesses, whether warm or cool. In a warm climate the stones are set down shortly before autumn with good reason, for frost is not a danger: on the contrary, if any should be damaged and begin to perish once chill winter has returned, they will cease to do so when in due course the outer covering of the stone is shed and the seed experiences the unvitiated life of the healthy bud. The stones should be used for growing trees at the time when they have begun to split open. This method of sowing can be benefited by what Celsus writes is done generally, namely, that

5 Vt asseruentur hiemalibus brumis, subsidium teporis et umoris
habemus in ceterum autumni; idem facere superuacuum est,
cum iam [iam] gelidae hiemis aduentu frigus imberque ueniat.

4 Quolibet caelo et quamuis macro solo persicorum arborem
poni Celsus existimat; idem tamen laetissimas fieri confidit si
2 remotione loci blandimentum ceperunt. Curtius Iustus irriguis
et frigidis locis mallet †aquam† et, adiungit, doleo aliquo supra
posito quod ab Aquilone defendat: sic enim fieri ut semina
3 pecorum stercus ita ament ut alia germina. Quintilii calidum
4 aerem persico et umidum solum addunt. Demissa temperies
necessaria eis ne siccitas anni ruat et admodum exigua poma
constringat ac frequens umor enormi uastitate et inani sapore
5 distendat. Graecorum celeberrimi auctores irrigua et hortensia
huic arbori loca destinanda senserunt, ubi umor manu temper-
6 atus nocens ita auerti potest ut et possit induci. Idem persico
solum pingue contrarium [tum] crediderunt: calculosum et
harenae et sabulo proximum conueniet, sicuti ceteris omnibus
quae gummis glutina excludant.

5.1,2 Ordo sationis in persico talis est. Crebrior in terra sulcus
duobus intermissis pedibus aperitur, in quo ossa deponi siccata
erecta non altius debent quam quod amplectitur trium men-
sura palmorum: obruta ulterius germine infirmo ad superna
3 non scandent. Infigenda autem terrae pars acutior, unde radicis
4 exordium est. Commodissime quattuor palmis inuicem separata
5 ponuntur. Multi et fimo ossa circumdant, aqua leuiter adrorant;
robustior tamen, etsi tardior, planta procedit ex osse quod sim-
6 pliciter exponatur. Sunt qui in gastris uel qualis cribrata terra

5 teporis *Reeve*: temporis *N* iam *Mai*[2]: iam iam *N* **4**.1 arborem] arbores *Reeve* 2 *cruces apposui*: aptari *uel potius* apponi *pro* aquam *Scotti in adn.*: mallet, aquam et (*sc.* etiam) *dist. Antonelli* 4 eis] *fort.* est 5 hortensia *Antonelli*: hortensiad *N*: hortensia ad *edd.* senserunt] consenserunt *edd.* auerti *Scotti*: uerti *N* ut et *Winterbottom*: et ut *N* 6 tum *del. Antonelli* harenae *Mai in adn.*: harena *N*: *fort.* harenosum proximum *edd.*: proximata *N*: proxima terra *Condorelli* gummis *ego*: gummi *N* excludant] ex(s)udant *Reeve* **5**.5 procedit *Scotti in adn.*: praecepit *N*

before the stones are entrusted to the earth they be kept floating in water for five days. In order that they be kept safe during the cold of winter, we have the aid of mild weather and moisture for the remainder of autumn. There is no need to do the same once cold and stormy weather comes at the arrival of freezing winter.

4 Celsus is of the opinion that the peach tree may be planted in any climate and even in meagre soil; yet the same author is confident that they become most fruitful if they receive relief from a removed location. Curtius Iustus would prefer them to be planted in places that are well watered and cool and, he adds, with a vat placed over them for protection against the north wind, for in this way it results that the seeds become as fond of the manure of livestock as are other buds. The Quintilii provide a hot climate and moist soil for the peach. A moderate climate is necessary for them, so that seasonal drought not attack and stunt the fruit, so that they become excessively small, nor frequent water cause them to swell to a great size and bland flavour. The most famous Greek authors hold the belief that places well watered and suitable for a garden should be devoted to the peach tree, where the water, controlled by hand, can be channelled away when it is harmful in such a way that it can also be channelled in again. The same group has expressed the belief that rich soil is bad for the peach: soil that is pebbly, and very close to a fine or coarse sand, will suit them, just as it suits all other trees which put out sticky secretions of gum.

5 The procedure for sowing in the case of the peach is as follows: fairly close-set trenches are opened in the earth at two-foot intervals, in which the dried-out stones are planted upright at a depth no greater than what is embraced by a measure of three palms; those buried any further will not climb to the surface owing to the weakness of the bud. The pointed end of the nut, where the origin of the root is, should be set into the earth. They are best planted at a distance of four palms from one another. Many also surround the stones with dung and sprinkle them lightly with water, but a stronger albeit later plant grows from the stone that is planted in a simple manner. There are some who plant the stones in large-bellied vessels or

repletis ossa disponant, unde mox plantas adultas in seminar-
ium cum cespite suo transferant, et affirment quod si pertranse-
7 ant, gratioris saporis et suci amplioris poma tribuantur. Planta
uero quae de codice auellitur autumnale tempus expostulat:
huic namque arescendi periculum incumbit nisi largus umor et
8 frequens pluuia ei subuenerit. Moris est fimo oblitam et inferi-
oribus foliis deputatam, subacto solo et praecedente punctorio,
9 sicuti olera demergere. Nam calcata plus iusto et terra densante
constricta inertia ligni maturum sentit occasum.

6 Sed utrisque frequens sarculi opera et desiderata medicina
continuae runcationis adhibenda est, ne tenero germini noceant
flatus herbarum neue densitas contegentium solis atque aurae
2 nutrimenta praecludat. Danda etiam calamo siue uirgulto pro-
pius affixo adminicula, quibus se tueri contra uiolenta uento-
3 rum flabra consuescant. Quae sollicitudo non amplius quam
unius anni spatio complectitur; ceterum bima translata iam suis
uiribus nulli non repugnat iniuriae.

7 Scrobes altos arbor persici non amat, quippe cum modica
radicum concessa substantia esset; <et> tamen cito bibit: contra
aliarum arborum legem de superioribus partibus mutuatur ali-
2 mentum. Hanc maxime existimant causam propter quam sola
3 maturius consenescat. Nec interualla maiora desiderat eiusmodi
folia sortita, quorum tenuis umbra opacitatis incommodum
4 non parat [quam scilicet binos pedes]. Sic Curtius optime pos-
5 itas arbitratur. Quin immo inuenti qui densius consitas [tutas]
existiment tutiores dum se inuicem protegunt a solis ardoribus,
quibus difficulter hoc genus arboris creditur repugnare.

6 disponant *Mazzini*: dispont *N*: disponunt *edd.* affirment] affirmant *edd.* 9 inertia *Mai*: inertiam *N*: <ob> inertiam *Scotti* **6**.1 noceant] noceat *Reeve* contegentium *Scotti in adn.*: condegentium *N* 2 propius *edd.*: proprius *N* uiolenta *Reeve*: uiolentia *N* 3 spatio] spatia *Reeve* **7**.1 esset; et (*fort.* sit; et) tamen cito bibit *dub. temptaui*: essetamnecito uiuit *N*: est et tam nec cito bibit *Condorelli* (*dub.*, *in praef.* xx): esset, tamen cito uiuit *Mazzini*: esset ei, nec cito uiuat *Mai*: sit, et ea nec cito uiuit *Scotti* 2 consenescat *Mai*: consenescant *N* 3 quam scilicet binos pedes *del. Winterbottom* 5 tutas *del. Winterbottom* protegunt *Antonelli*: pretegunt *N*: praetegunt *edd.*

baskets filled with earth that has been passed through a sieve, soon transplanting the grown plants from here into the nursery along with their soil, and maintain that if they pull through, they acquire fruit with a more pleasant taste and more abundant juice. The slip which been torn from a tree stock, however, requires an autumn planting time, for the danger of parching threatens unless copious water and frequent rain come to its aid. It is customary to coat it with manure and strip it of its lower leaves and, after the dibble is applied to the well-worked soil, to sink them in the same way as vegetables. For when it is packed down more than is appropriate and constricted by the earth pressing closely on it, it experiences an early demise owing to the feebleness of its wood.

6 To each of them, however, must be applied frequent hoe-work and the welcome medicine of continual weeding, so that the exhalations of the weeds do not harm the tender bud, nor the density of those covering it shut out the nourishments of sun and air. Props should also be provided, with a reed or cut brushwood fixed close to the plant, whereby they can become accustomed to protecting themselves against violent blasts of wind. This care is encompassed by no more than the space of one year; apart from that, once transplanted at two years old there is no assault that it does not withstand by its own strength.

7 The peach tree does not take well to deep holes, insofar as the substance of the roots with which it is endowed is slight. Nonetheless, it drinks rapidly: contrary to the manner of other trees, it takes in nourishment from the upper parts. (It is believed that this is the chief reason why it alone grows old earlier.) Nor does it want intervals that are too large, because the leaves it possesses are such that their slight shade does not produce the drawback of shutting out light. Curtius reckons that they are best planted this way. In fact, there are found some who are of the opinion that those sown more closely together are safer because they protect one another from the heat of the sun, which this kind of tree is thought to have difficulty withstanding.

8 Ablaqueare omnibus annis oportet nouissimo autumno et inhalantibus fibris de radice purgatis folia sua in uicem stercoris cum terra molliore congerere quantaque maiore***

9 ***et magna fieri asseuerant si, cum arbor in flore sit, lactis
2 caprini per triduum ternis sextariis irrigetur. Omnibus persicis, maxime tamen praecoquis et armeniis, quotiens abundantibus pomis arbor onerata est, decerpenda sunt aliqua: sic enim rel-
3 iquis celerior maturitas imperatur. Sparto ligandam aduersum plurima incommoda, ut amphora, existimant, et hodieque nonnulli iumentorum solias in itinere derelictas pro remedio truncis ramisue suspendunt.

10 Inseri persicus utroque modo potest, uel in cortice uel in ligno
2 praecedente fissura. Sed quantum propius ad terram recisa sit,
3 eo melius inseretur: fortior est enim pars eius ima. Quidquid in summo est, quoniam suci minus retinet, ut fragile ac putre
4 damnatur. Inserendi tempus aptissimum est nouissimi autumni, si loca non sint adeo pruinosa, uel inter brumam et exortum Fauoni, id est extremo Decembri ac deinde Ianuario, ita exigente praecoquo germine, quod saepe et ante ueris aduen-
5 tum florere festinat. Meliores ad inserendum surculi habentur aliquatenus pleniores et de ea parte sublati quae propinqua arbori fuerit, uixque proueniet aut non diu durat pars summa
6 ramorum. Persicus et in semet ipsam inseritur et in amygdalum et prunum, magisque durabilis arbor efficitur cum surculi eius
7 in ligno fortiore cinguntur. Sed armenia uel praecoqua peculiariter prunis insita floridiora sunt; cetera, id est duracina, sic
8 insita in amygdalo melius assuescunt. At haec ipsa prunorum atque amygdali germina difficilius <recipiunt> et cito recurrunt in senectutem; quotiensque surculo alterius generis insita sint infirmitatem suam confessa moriuntur.

8.1 inhalantibus] inhaerentibus *Winterbottom*: inalentibus (*sc. concrescentibus*) *Antonelli* **9**.2 sunt *Antonelli*: sit *N*: sint *edd.* imperatur] impetratur *Reeve* 3 ligandam *Mazzini*: ligata *N*: liganda *Scotti* ut] aut *Scotti* pro remedio *N*^{m}: prae medio *N* **10**.6 cinguntur] panguntur *Reeve* 7 floridiora *Scotti*: floriora *N* 8 At *Winterbottom*: Ad *N* recipiunt *addidi* et *Antonelli*: ed *N*: sed *edd.* sint infirmitatem *edd.*: sinfirmitatem *N*

8 One ought to expose the roots every year at the end of autumn and, once the odorous fibres are removed from the root, in place of manure heap their leaves together along with a softer soil, and by how much greater . . .

9 . . . and they maintain that (sc. the fruit) become large if, when the tree is in flower, it is watered for three days with three sextarii of goat's milk. With all peaches, but especially early-ripening and common apricots, whenever the tree is loaded with abundant fruit some should be picked, for this way the rest are forced to a swifter ripening. People are of the opinion that the tree should be bound with esparto, just as an amphora is, against a great many kinds of harm, and still today some hang the shoes of pack animals left in the road from the trunks or branches as a remedy.

10 The peach can be grafted in both ways, either in the bark or in the wood after a fissure has been made. It will, however, be better grafted in proportion to how much closer to the ground it is cut back, for its lowest part is stronger. The highest parts, since they hold less sap, are disapproved on the grounds that they are apt to break and fall apart. The most suitable time for grafting is at the end of autumn, if the place is not very prone to frost, or between the winter solstice and the commencement of the west wind, that is, at the end of December and then into January, as is required by the bud of the early-ripening apricot, which often hastens to bloom even before the arrival of spring. Those scions are considered better for grafting which are somewhat fuller and have been taken from a part close to the tree; those from the tips of the branches will thrive with difficulty, or even come to a swift end. The peach is grafted onto itself as well as onto the almond and plum, and a longer-lasting graft is produced when its scions are enclosed in a stronger wood. Apricots or early-ripening apricots, however, flower more when grafted onto the plum in particular; the rest, that is, clingstone peaches, adapt better when so grafted onto the almond. The last, however, take the buds of plums and almonds with more difficulty, and lapse quickly back

11 Sic quoque fiet emplastratio: olea, persicus, sicuti et nonnulla
ficus et huius generis cetera quibus seponitur sucus pinguis,
2 rite inoculantur prius trunco reciso. Aduerti deinde mole iusta
plerumque esse et diutius asseruari si non, ut Graeci, in glarea
3 argilla sed in solo harenoso ponas. Et quidem solito more pos-
sunt emplastrari, quod agere in fine prodest mensis Aprilis, sed
initio Maii satius proueniet.

12 Persici pomum, nisi conditum muria et oxymelle, asseruari
diu non potest uixque hoc modo peruenit in alteram aestatem:
cutis exilitas et carnis squalor consentanee prodibunt ad par-
2 simoniam uetustatis. Plerumque temptarunt ossibus detractis
more ficorum in sole siccare atque in frigido seruanda sus-
3 pendere. Vix in his aliquid suci inest, cuius admonere inane
4 uidetur. Ex ueteribus multi Suessatibus ollulis impicato singula
poma clausere umbilico atque in sapam purissimam demissa
merserunt, sed hoc modo saepe in breue tempus flaccescunt.
5 Graecorum deprehendi***

<III DE AMYGDALIS>

1 ***<in> fimum bubulum transferunt, nec ex eo nuces pro-
2 ferunt ad serendum nisi cum hiascere coeperunt. Mago et
Celsus scrobes altos latosque trinis pedibus effundunt; in eos
terram congerunt quam minutissimam cum stercoris admix-
3 tione eatenus ut pars dimidia compleatur. Ibi tria amygdala in
medio triangulari lege disponunt quo inter se non palmo, ut
Columellae uidetur, sed pedis siue dodrantis spatio separen-
4 tur. Acuta pars nucis ab utrisque prima defigitur: radices exi-
tum moliuntur; sed in posteriores dissentiunt, siquidem aciem

11.1 olea *edd.*: oleum *N* 2 *fort.* glarea <atque> argilla: glarea <aut> argilla *Scotti* **12**.2 frigido *Antonelli*: rigido *N* 3 uidetur] *an* uideatur? **1**.1 in *suppleui* hiascere *edd.*: hiacere *N* 2 effundunt] effodiunt *Marx* terram . . . minutissimam *Antonelli*: tercorum (st- *N*[c] *dub.*) . . . minutissima *N*: terrae . . . minutissima *Mai*: stercorum . . . minutissima *Scotti* 3 triangulari *Mai*[3] *in adn.*: trina angulari *N* 4 inde *post* defigitur *add. Marx* posteriores] posteriora *Antonelli*: posterioribus *Mai* aciem *Antonelli*: faciem *N*

into old age – whenever they are grafted with scions of another kind, they reveal their weakness and perish.

11 Patch-budding will be accomplished too, in the following way: the olive and the peach, just like some figs and others of this kind from which is secreted a thick sap, are properly budded after the trunk is first cut back. I have further observed that they tend to be the right size and last longer if you plant them not in gravelly white clay, as the Greeks do, but in sandy soil. We add that they may be patch-budded in the customary manner, which it is good to do at the end of April but will go better at the beginning of May.

12 The peach fruit cannot be long preserved unless it is pickled in brine and honey-vinegar, and by this means it lasts to the next summer, but with difficulty: the thinness of the skin and roughness (?) of the flesh will consistently result in a scant old age. It has often been tried in the manner of figs to dry them in the sun with their stones removed and hang them up for preservation in a cool place. There is hardly any juice in them, a fact which it seems unnecessary to mention. Many of the ancients pitched over the stem, enclosed the fruits singly in little Suessian jars, and set them down submerged into very pure grape syrup; but in this way they often begin to wither in a short time. I (?) have discovered (sc. some of?) the Greeks . . .

III On the Almond

1 . . . (sc. they) put into cow dung, and do not bring the nuts out for planting until they have begun to split open. Mago and Celsus empty out holes three feet deep by three feet wide. Into these they heap extremely fine soil with which manure has been mixed, enough to fill them halfway. There they arrange three almonds in the middle in a triangular manner, so that each is separated from the others not, as seems best to Columella, by a palm, but by the space of a foot or nine inches. They both set the pointed end of the nut down first: the roots force an exit

lateris in Aquilonem Mago inuertit, Celsus ad Fauonium dirigit,
Columella cum dicat, *anceps* nucis *ad Fauonium spectet,* [et] sine
5 dubio et ipse unum latus opponit Aquiloni. Supra nuces terra
sesquipedis ingeritur et, si nec ullus imber fuerit, ter in mense
seminibus expositis umoris alimenta sumministrentur, et quid-
quid in summo scrobe induruit leui sarculo relaxetur.

2 Vbi primum se germina ostenderint, quoniam tenuis illis
stilus et flexuosus est, iuxta uastior palus affigitur quo rami
adminiculo et in ualido requiescant, et scandere in rectum
2 †conpletum† docent consuescere. Triennio exacto unum ger-
men in loco permanet, reliqua eodem mense quo sata sunt ad
scrobes transferuntur, ut pars aliqua fossurae conuexa relin-
3 quatur. Hanc paulatim adobruunt singula arboris incrementa
numerantes, tantumque radicibus cumuli largiuntur quantum
altitudinis truncus adiecerit.

3 Quintilii omnem agrum in seminario amygdalis destinatum
2 ad sesquipedis altitudinem fodiunt. Hunc in areas diuidunt
quo canalibus umor infusus diutius immoretur, dein per areas
intermisso pedis spatio modicis scrobibus excitatis ad digitorum
3 quattuor altitudinem nuces mergunt. Stercora in terra obruunt
parique qua ceteri cultura lene nutritas in scrobes post triennium
4 transferunt. Mago primus et post eum Diophanes nouam <nor-
mam> amygdali suadent, ut ferulam primo seramus eamque
post annum recisam et in media parte diffissam nuce amygdali
qua medullae sucus mollis est compleamus; nec desinunt asseuer-
are quod arbor eiusmodi et robore et fructu pulcriore prouenit.
5 Sic et Columella describit ferularem; alio modo scribit, nam hoc
tantum praestare amygdalis ferulas ut illas faciant Tarentinas.

4 et *del. Antonelli* 5 nec ullus *edd.*: nucullus *N*, *unde* nullus *edd. in adn.* **2**.1 et in ualido] et *fort. delendum* *obelis notaui*: conplexu N^{m} docent] *fort.* docentur, *quod post* consuescere *malim transferre* **3**.3 scrobes *edd.*: scrobe *N* 4 normam *add. Condorelli*: sationem *suppl. Mazzini* (*sed post* amygdali): nouum *Scotti* amygdali] amygdalis *Condorelli* diffissam *edd.*: diffusam *N* nuce *Antonelli*: nuci *N* pulcriore] pulcrior *Antonelli, fort. recte* 5 Sic et *Antonelli*: Sicut *N* at *ante* alio *add. Antonelli* Tarentinas *Mai* (*ex* Ter- N^{m}): roncinias *N*

there. But they disagree on the following parts: Mago turns the sharp edge of the side to the north; Celsus directs it to the west; Columella, when he says 'let the double-edge of the nut look west', no doubt himself also makes one side face north. A foot and a half of soil is heaped over the nuts and, if it does not rain, the seeds that have been planted out should be aided with the nourishment of water three times a month, and whatever has grown hard on the surface of the hole should be loosened with a light hoe.

2 As soon as the buds appear, since their stems are thin and flexible, a fairly stout stake is set beside them so that, with its support, the branches may rest on something strong; and they teach them to become accustomed to climbing upwards. After the third year, one of the buds remains in place; the rest are transplanted into holes in the same month in which they were planted, so that some part of the pit remains unfilled. They gradually fill this portion while reckoning up the incremental growth of the tree and add a measure of soil to the roots in proportion to the increase in the height of the trunk.

3 The Quintilii dig the whole field in the nursery that has been designated for almond trees to a depth of one and a half feet. They divide this field into plots so that water that has been poured in through channels will remain longer. Next, they excavate small holes throughout the plots at one-foot intervals and set the nuts down to a depth of four fingers. They mix manure into the earth and, providing them with the same gentle cultivation as others recommend, after the third year transfer them into holes. Mago first, and after him Diophanes, recommends a novel method for planting the almond: first, to sow giant fennel, and then, after a year, to lop off the upper part, split the fennel through the middle, and fill the part of the pith where the sweet sap is with the almond seed. They continually assert that a tree planted in this way comes forth with finer wood and fruit. Columella likewise discusses the almond that is grown in the fennel; he writes of it in a different way, for he says that the giant fennels only render this benefit to the almonds, that they make them 'Tarentine'.

4 Et natura et mediocritas arborum suggerit ne altissimo scrobe
2 recondant. Illa interualla etiam simili ratione moderanda sunt:
uiginti et quinque nonnulli, quidam et uiginti pedum iusta pro-
3 nuntiant. Bene Diophanes ait amygdalas in transitu esse ponen-
das, quoniam non semel maturant et per partes se fructus eius
accommodant, nec possint ita fures dispoliare ut non aliquid—
uel quia crudum uidetur—domino relinquatur.

5 Leuissima et minime sumptuosa cultura est earum, puta-
tionis annuae scilicet, qua necesse sit ramos superuacuos uel
enormiter natos uel aliqua necessitate praefractos falce purgare.
2 Exigunt sane propter incursus pecorum custodiam diligentem,
quorum dente contacta in amarum saporem ex dulcibus transe-
3 unt. Fossio contraria illis habetur, nam florem statim amittunt;
quod malum illis uel a nobis non debet inferri, cum et alias
periclitentur in flore, nubilo prius marcidae deinde Austris inse-
quentibus concrematae.

6.1,2 Plurimum generant. Si ferax non erit, terebrari in radicibus
debet et taxillus e taeda in foramen aptari; alii in locum tae-
3 dae cuneum qui esset ex robore. Quidam radicibus perforatis
silicem adiciunt et ita arboribus librum patiuntur inolescere.
4 In locis frigidis, in quibus metus est ne in flore deprensas [aut]
pruina comburat, hoc modo tradunt amygdalis subueniri:
antequam floreant nudare radices iubent et albi generis lapi-
des quam minutissimos cum harenae cumulo aggregant, cum-
que iam tempus uidebitur ut debeant germinare, fodere atque
remouere quae fuerant infusa praecipiunt.

7 Vt teneras nuces procreent, utile erit ablaqueatis radicibus,
2 antequam floreant, aqua calida per aliquos dies irrigata. Fiunt
dulcia ex amaris si circumfosso stipite tribus a radice digitis latior

4.1 recondant. Illa interualla] recondant illas. Interualla *Mai in adn.* **6**.1 generant *edd.*: generent *N*: *lac. ante* Plurimum *stat. Mazzini* 2 alii *scripsi pro* ali *N*[m]: illi *N*: uel *edd.*: uli *N*[c] cuneum] cuneus *edd.* 4 aut *del. Mai*: <nix> aut pruina *Scotti* aggregant (adg- *N*)] *melius* (ad)aggerant? **7**.1 utile] utilis *Mazzini*

4 Both the nature and the small size of the trees advise against setting them down in a very deep hole. The intervals between them should also be regulated by a similar calculation: some declare that a distance of twenty-five feet is appropriate, others say also twenty. Diophanes rightly says that almonds should be planted along a crosspath, since their fruit do not ripen all at once and become ready for picking bit by bit, and this way thieves would not be able to plunder them so that there remains no fruit for the master – or if any, only because it appears unripe.

5 Their care is very simple and not at all costly: one pruning a year, in which it is necessary to remove with a billhook the superfluous branches, those that have grown irregularly, or those that have been broken off short through some difficulty. They require diligent guarding against the forays of animals, since the almonds touched by their teeth pass from a sweet to a bitter flavour. Digging is considered to be harmful for them, for they immediately lose their flowers. They should not be exposed to that injury, at least not by us, because they undergo dangers at other times too while in flower, first wilting under cloudy weather, then withered by the south winds that follow.

6 They are extremely fruitful. If the tree is not productive, a perforation should be made in the roots and a small die of pine fitted into the hole; others use a wedge of oak in place of pine. Some perforate the roots and insert a pebble and then have the bark of the trees grow over it. In cool places where there is a fear that frost will catch the trees in flower and burn them, they recommend that aid be brought to the almonds in the following way: before they flower, they give instructions to lay bare the roots and to pile on them white stones, the smallest possible, along with a heap of sand. When it comes time for them to bud, they enjoin us to dig up and remove the mixture that had been poured over them.

7 In order for the trees to produce tender nuts, it will be useful to expose the roots of the trees before they flower and to irrigate them with warm water for a few days. They become sweet from bitter if the soil around the trunk is dug and a fairly broad

cauerna ponitur qua pituita decurrat; post media trunci parte
terebrata cuneum qui esset ex arboris quidem radicibus simplici
3 delibutum melle fixerunt. Nec minus utile est si effundatur ster-
cus suillum, humano lotio resolutum et radicibus superfusum.
4 Aristoteles in georgicis amaritudinem amygdali fraglanti soli
5 assignat, quum crebrius exustae dulcior sucus hauriatur. Illico
[praesto] uidetur ex morsu pecorum omnia haec arbori amar-
itosa contingere, cum inopinata in medullas saliua consederit.
6 Si uiridibus auferas tenuiora, quaeque liqueris meliora sunt.
8 A Kal. Aprilis, ut putat Celsus, incipiunt esse matura, ac si
asseruanda deligi placeat, signum maturitatis ostendunt cum
2 corticem remiserunt. Sunt quibus cortex difficulter abscedat,
3 sed si paleis obruuntur, calore solari illico resolues. Tum corio
liberatae, si quis aqua marina***

<IV DE CASTANEIS>

1 ***corripiat quales illae fuerunt de quibus pastor ita prae-
2 cinit, *mea quas Amaryllis amabat.* Mago breuiter, ut Poenus et cui
peregrinae eiusmodi arboris minus nota cultura sit, *castaneam*
3 *in umido sicut Graecam nucem serito.* Cornelium tamen Celsum,
Italicae disciplinae peritissimum, decuit aliquatenus edocere
quidquid Magonis ignorantiam fugerat; sed et ipse castaneas
4 amygdalorum exemplo reiectat. Columella, cui curae fuit dil-
igentius et de castanea conscribere, etiam ipse indifferenter
et uarie sententiam promit; fluxe enim ista scribit: affirmat a
Nouembri mense per totam hiemem seri posse, nunciat Kal.
Martias in Idus easdem ne quid intersit, et plantat ut <uix>

2 pituita *scripsi*: pruna *N*: pruina *edd.* quidem] eiusdem *Reeve* 4 fraglanti soli *Mai*: fraglanti solis *N*: flagranti soli *Condorelli*: flagrantiae solis *Scotti* 5 praesto *del. Reeve* uidetur] *fort.* uidentur haec] *fort.* huic amaritosa] calamitosa *Reeve* 6 quaeque liqueris] quae reliqueris *Reeve* sunt] fiunt *Reeve* **1**.2 sit] ait *Scotti* 3 amygdalorum *Marx in adn.*: amygdalo eorum *N* 4 fuit *Scotti*: fit *N* uix *add. Reeve*: non *post* scire *add. Scotti*

channel is set three fingers from the roots for the foul sap to run down into. Afterwards, they bore a hole into the middle of the trunk and fix in a wedge from the tree's roots that has been smeared with unalloyed honey. No less useful is it to pour out swine manure, softening it with human urine and spreading it over the roots. In his *Georgics*, Aristotle attributes bitterness in the almond to the blazing sun, since the sweeter part of the sap of a tree that has been scorched too often is consumed. It seems to occur immediately as a result of the bite of beasts that all this bitterness comes to the tree, whenever the saliva falls upon the pith unawares. If you cull the weaker of the green almonds, all those left behind are improved.

8 They begin to be ripe from the first of April, according to Celsus, and, if you intend to have them harvested for preservation, they show a sign of their ripeness once they shed their hulls. There are some nuts whose husks it is difficult to remove, but if they are buried in chaff, you will separate them straightaway through the heat of the sun. Then, freed from their shell, if anyone with sea water . . .

IV On the Chestnut

1 . . . would pluck (sc. chestnuts) like those that the shepherd sings about, 'which my Amaryllis used to love'. Mago is brief, as might be expected from a Carthaginian and one to whom the cultivation of this kind of foreign tree is not well known: 'Sow the chestnut in a humid place, like the almond.' Yet it was fitting that Cornelius Celsus, a man very experienced in the Italian practice, should teach in some measure whatever had escaped Mago in his ignorance; but he too refers chestnuts to the example of almonds. Columella, who was at pains to write diligently about the chestnut as well, nonetheless also expresses his opinion in an undiscriminating and inconsistent manner, for he writes carelessly about the subject: he asserts that they can be sown from the month of November through the whole winter and reports the first of March to the fifteenth of the same

5 quando quaeque sine culpa manus ueniat scire possimus. Etiam
Gaius Plinius, quamquam ipse sationem caeduam assentiatur,
6 hiemali tamen primum tribuit. Iulius Atticus ex Nouembri in
7 Februarium mensem serere decreuit. Curtius Iustus sationem
8 nucis primam Decembri, secundam Februario assignat. Iulio
Frontico, cum et soli de genere et alia ratione culturae nonnulla
tractaret, non fuit [culturae alia] norma quando rectissime cas-
9 taneae sererentur. Quintiliis in satione castaneae uergiliarum
non omittendus occasus.

2 Sed et quibusdam Graecis placet ab Arcturi [et] exortu serere
2 castaneam. Nec immerito naturam secuti quibuscumque tem-
pus hoc placuit: tunc enim seminis, id est nucis, iusta maturitas
creditur adeo ut hiante tegumento, si legent, tum manus condat
3 in terra, cum caduca labuntur. Sed ad hoc accedit quod nux
4 reseruata difficile sit contra putredinem uindicare. Sana igitur
et matura debet exponi, quorum alterum spondet de incolumi-
tate perseuerandi, alterum prodest de uirtute generandi.

3 Contendunt plerique Graecorum castaneam melius ex
uerna satione procedere, si tamen nux anxie atque sollicite con-
2 dita seruetur expansa. Et tunc in angustum locum siccumque
translatae eriguntur in cumulum: supra eas ita <ut> uniuersas
obtegat funditur harena fluuialis, quae et pinguis et dulcis est.
3 Post trigesimum diem eadem harena remouetur et [in] nucibus
in aquam frigidam missis sanas a corruptis sic licet segregare:
4 nam corruptae supernatant, sanae residunt. Illae igitur quae
placere meruerunt rursus in harenam sicut antea reponuntur
et post alterum mensem, si quae forsan integrae debent credi,

4 quando quaeque *Antonelli*: quandoque quae *N*: quando quaequae *Scotti* 6 decreuit *Scotti* (inde de- *in adn.*): indecreuit *N* 8 Frontico] *fort.* Frontino *uel* Frontoni, *uide intro. cap. II* soli de genere] *fort.* de genere soli: soli *del. Winterbottom*: solide <de> genere *edd.* culturae alia *del. Winterbottom* castaneae sererentur *Mai*[3] *in adn.* (sererentur *iam Mai*[2]): castaneas serere rentur *N* 9 uidetur *post* uergiliarum *add. Winterbottom* **2**.1 et *del. edd.* 2 legent] *fort.* legentur cum *Winterbottom*: ut *N* labuntur *Scotti*: labantur *N* **3**.2 ut *add. edd.* 3 in *del. adn. Pat.* 4 non *post* forsan *add. Mazzini*

month without any distinction at all – his plantings are such that that we are hardly given to know when any shall grow without manual error. Gaius Pliny too, although he would agree that it is a sowing for timber, nevertheless assigned first place to winter. Iulius Atticus prescribed sowing from November to the month of February. Curtius Iustus assigns the best planting of the nut to December, the second best to February. Iulius Fronticus, although he treated some matters concerning soil type and other aspects of cultivation, had no rule for when the chestnuts might be sown most properly. For the Quintilii the setting of the Pleiades must not be neglected in sowing the chestnut.

2 There are furthermore some Greeks who think it right to sow the chestnut from the rising of Arcturus. Not without cause do they follow nature, whoever have decided on this time, for the seed, that is, the nut, is believed to be ripe enough at that time that if one gathers the chestnuts, their covers splitting open, one's hand may cause them to be buried in the earth when they drop, prone to fall as they are. Besides, there is also the fact that the nut which is saved up is hard to protect against rot. It is the healthy and ripe chestnut therefore that ought to be planted out: the one of these qualities vouches for its resilience, the other is beneficial for its productivity.

3 Most of the Greeks contend that the chestnut comes forth better from a spring sowing, if, at least, the nuts, stored up with painstaking care, are spread out for preservation. Afterwards they are transferred into a narrow and dry place and raised into a mound; river sand, which is both rich and sweet, is poured over them so that it covers them entirely. After the thirtieth day the aforementioned sand is removed and the nuts are placed in cold water, which makes it possible to separate the healthy from the rotten ones, for the rotten float while the healthy settle. Consequently, those which have happened to pass the test are placed back into the sand as before and if, after the second month, any ought perhaps to be believed to be sound, an

5 aqua iudicante temptatur. Hoc idem fit tertio mense donec uer-
num tempus adueniat, quo satis nucibus nec includi iam necesse
sit nec probari diligentioribus.

4 Non in pauimento sed uasibus seruanda, harena ut similiter
2 influat. Quidam in plantis iuxta arborem sponte nascentibus
3 castaneas conserendas esse dixerunt. Sed eiusmodi planta tard-
issime prouenit et continuo biennio languida est, uix ut demum
uel illam sucus recreet qui consequitur, quem non habuerat,
4 cum <nuce> dilata non esset. Inuenti tamen qui de nucibus
et hoc genere plantarum audeant aliquod facere discrimen et
nuces credant caeduis arboribus aptiores, plantas uero nucibus
procreandis esse utiliores, si tamen fas est ut admittere in ani-
mum debeamus quod lignum in poma proficiat, fructus in ligna
desistat.

5 Solum castanea desiderat molle et subactum, non tamen
2 harenosum; patitur et sabulum, si umor adfuerit. Pulla terra uel
maxime prodest sed et carbunculus itemque tofus optime seruit
3 diligenter infractus. Spissum rubricosum locum non uult, uix
prouenit in argilla, in glarea cretaque nec nascitur.

6 De caeli qualitate nonnulla inter auctores uarietas opinionis
2 exoritur. Celsus, ante oculos habens quid pomis expediret, cal-
3 idas apricas regiones castaneae aptissimas credit. Id illo argu-
mento probat, quod Neapolis, Campaniae ciuitas, eo fructu
maxime glorietur; prudentissime tamen postea adiunxit quod
et frigidis locis conseri posset, quae fere omnes anteponenda
4 senserunt. Iulius Atticus opacis et septentrionem spectanti-
5 bus cliuis laetissimum fieri castaneae pollicetur. Columella et
6 Plinius in eadem fere uerba consentiunt. Quintilii montana sine
dubio et umida et frigida principaliter probauerunt; si tamen in

4 iudicante *Antonelli*: iudicanto *N*: iudicando *edd.* 5 diligentioribus] curis *uel sim. add. Reeve*: diligentius *Scotti* **4**.2 in *del. Antonelli* conserendas *Antonelli*: conseruandas *N* 3 recreet *edd.*: recreat *N* nuce *addidi* 4 et hoc *Mazzini*: ex hoc *N* ligna] ligno *Scotti* **5**.3 in argilla, in glarea *edd.*: in argillam glarea *N*: in argilla. Glarea *Mazzini* **6**.2 apricas] Africas *edd.*: Africae *Marx* 3 posset *Condorelli*: possae *N*: possit *edd.*: posse *Mazzini*

appraisal is made by using water to judge. The same thing is done in the third month, until springtime arrives, when, after the nuts are sown, they need neither laying up nor (for those who are more diligent) testing.

4 They should be preserved not on the floor, but in vessels such that sand is poured in in the same way. Some have said that chestnuts should be sown from the shoots that grow spontaneously around the tree. This sort of plant grows very slowly, however, and is weak for an entire two-year period, so that in the end it is barely restored by the attainment of the nourishment which it had lacked because it had not been sown from seed. Yet there are found those who presume to make some distinction between nuts and shoots of that sort and believe that the nuts are more suited for trees for timber but that the shoots are more fit for producing nuts, if, at least, it is right that we should allow that wood grows into fruits and fruit reaches its end in wood.

5 The chestnut wants soil that is soft and worked over, but not a fine sand. It tolerates coarse sand, too, if water is present. Dark soil is especially beneficial, but both sandstone and tufa serve very well if diligently crushed. It rejects a dense place of *terra rossa*, hardly thrives in white clay, and does not even sprout in gravel and potter's clay.

6 Regarding the nature of the climate, there exists no little variation of opinion among the authors. Celsus, keeping before his eyes what might benefit the fruit, believes that warm, sunny lands are most suited to the chestnut. He proves it on the evidence that Naples (the city in Campania) especially prides itself on that fruit; yet he most prudently added afterwards that it may be sown in cool places, too, which nearly all have believed should be preferred. Iulius Atticus promises that it is most fruitful for chestnuts on slopes that are shady and look towards the north. Columella and Pliny agree in nearly the same words. The Quintilii in the first instance approved of mountainous locations, doubtless both wet and cool; but if it is necessary to plant them on a level, they are of the opinion that a lean place,

campo necesse est ponere, tenuem locum, et hunc harenosum, existimant deligendum, aut iuxta uiciniam fluminis, si tamen
7 limus ex eo stare non soleat. Sunt qui putent et in decliuibus locis castaneas optume prouenire.

7 Maiori parti placet in pastinatos omnino ponere ita ut terra
2 confracta in duorum pedum altitudinem subleuetur. Columella et Atticus addendum adhuc altitudini semissem putauerunt.
3 Sunt aliquot ex Graecis quibus mensura non satis iusta sit nisi
4 quae tribus pedibus excreuerit. Et fortasse singulis satio consistat, nam qui dipondi breuitate contenti sunt seminarium cogitant, et potest qualiscumque mensura nucibus exceptis alere plantas post biennium recessuras; sed qui semina mansura condebant non immerito ampliorem radicibus parauerunt uti arbor haberet quantum***

6 hunc *edd.*: tunc *N* **7**.1 pastinatos] pastinatis *Mai* 4 haberet *Condorelli*: haberit *N*: habuerit *edd.*

and that a sandy one, should be chosen, or near the vicinity of a river, provided that mud from it is not given to lying stagnant. There are some who think that chestnuts come forth very well on slopes, too.

7 Most think it right to plant them in thoroughly trenched ground in such a way that the broken earth is raised under them to a height of two feet. Columella and Atticus thought that a further half foot should be added to this height. There are some Greeks for whom the depth does not suffice unless it has come up to three feet. And perhaps a sowing could be based on each of these, since those who are satisfied with two feet have a nursery in mind, and after the nuts have been taken up by the earth any depth is capable of nourishing plants that will be removed in two years. But it was not without cause that those who planted the seeds intending for them to remain provided a greater depth for the roots, so that the tree might have as much . . .

COMMENTARY

COMMENTARY

<I DE CYDONEIS>: Fr. 5 Condorelli. Ch. 1 Mai. Ch. 1 Scotti. I follow the numbering of Mai, not Scotti (Introduction §IX), who counts two sections because he establishes a new section immediately before 1.1.2 *infundunt*.

1.1 The first sentence of this section establishes that G. has turned from another topic, probably the cultivation of vegetables or herbs (see Introduction §I(b)), to a discussion of fruits (*de fructibus*). The subject of discussion is the quince (Plate 1), but we know from a remark later in this section (1.1.13 *eodem modo quo punica supra uel alia m<ala> seruari*) that this fruit was not the first species that G. discussed (cf. Rose (1864–70) II 126, Condorelli xxvi); beyond that, we are limited in what we can say with certainty about the character of his prefatory remarks and about the number of fruits that were discussed before the quince.[1] At all events far more has been lost between *uerum de fructibus*, etc., and where the text can be read from *traxisse* than is suggested by the blank spaces left in **N**, which are enough to allow at most several sentences. The lacunose state of the opening part presents further specific difficulties of interpretation. It is uncertain, for example, whether G.'s claim that 'they' (cf. 1.1.2 *proueniunt*) are more productive on slopes than on plains is intended to apply to fruit trees in general, or whether it is stated with particular reference to the quince or another fruit. If, however, it is right to connect this statement with the quince (cf. 1.1.2n.), then it is likely that the remaining fragments of 1.1 pertain to it as well.

When the text can first be read continuously at 1.1.5 *traxisse* the subject is the preservation of the quince. Comparison with other chapters of *Pom.* (see Introduction §III(a)) shows that much is missing from G.'s treatment of the fruit, but we catch among the fragments that precede only a glimpse of other topics, such as the proper time for picking the fruit (*ampliora quae Nouember affert, et matura cydonea legenda sunt*). The gist of the missing information can be supplied from ancient and modern accounts of quince culture: see Pallad. 3.25.20–6, *Geop.* 10.26–8, Meech (1888), Wickson (1891) 362–3; and concisely, but in several respects less applicable, *EFN* 637–42. The quince was often painted in Roman antiquity (Plate 2).

With regard to preserving the quince, G. has a great variety of methods to report, and his meticulousness in laying out the techniques propounded by different authorities might hint at the shape of the truncated sections on

[1] For a discussion of a possible missing preface or prefaces, see Introduction §III(a); for a hypothesis about the original structure of the work that places the surviving chapters of *Pom.* in context, see Introduction §I(b).

preserving peaches (2.12), possibly also almonds (3.8) and chestnuts (where the discussion has been wholly lost). The techniques for the quince are a more reliable guide for fleshy fruits than nuts, however, since preserving is a simpler matter for the latter (3.8n.). Although he mentions only the Quintilii by name, G. clearly draws much of his material from the robust tradition of instructions for preserving quinces that is represented by Col. 12.47.1–4. After G., most of the advice for the quince is recycled by Palladius (3.25.25–6) and the *Geoponica* (10.28). G.'s specific borrowings, and those of later agronomists, are observed *passim* below.

The fullness and variety of G.'s instructions for preserving the quince are also instructive in that they reflect more generally the significant attention paid by the Roman agronomists to the preservation, packing and storage of fruits and other crops. From the earliest times, the agronomists show concern for bringing to market unblemished fruit with good flavour (Plate 3; cf. Kron (2012) 14–16). Their techniques for preserving are diverse and, it may be added from our modern vantage, scientifically sound: see the overview of Thurmond (2006) 173–87. Archaeological discoveries throughout Roman territory of amphorae packed with fruit remains attest to the thriving trade in fruit and show that the methods of preservation suggested by the agronomists were actually put into practice (Callender (1965) 39 and *passim*, Parker (1992) 92–3, Cool (2006) *passim*, Sadori et al. (2009) 47). Among the Roman authors, the most thorough discussion of preserving fruits, vegetables and legumes is Col. 12,[2] but useful recommendations are scattered among all of the agronomists. Varro (*Rust.* 1.59) offers basic advice on the construction of the *oporotheca* ('fruit-room'; see De Angelis (1995) 50–1, and cf. 1.1.9n. *palea*), and simple but effective guidelines for storing fruits in a cool, dry location (Plate 4). Palladius, like G., often appends detailed instructions for preserving at the end of the discussion of each fruit tree.

Of the many techniques recommended by the ancient agronomists for preserving fruit, three of the most important are dehydration, conserving and pickling. G. mentions conserving (cf. 1.1.7 *melle linantur*) in connection with the quince, but we must wait until 2.12 (n.) for dehydration and pickling,

[2] In keeping with his intention to provide comprehensive instructions for farm management, Columella also provides preliminary advice regarding e.g. the delegation of tasks (it is the *uilica* who is in charge of preserving), the layout of storage facilities, the preparation of pickling liquid and storage containers and so on. As often in *Pom.*, it is not certain whether G. provided this sort of introductory material in a section that is now lost or simply assumed that his audience knew the basic principles and were seeking advice specific to the fruits that they hoped to store more successfully: on the question, see Introduction §III(a).

which G. vigorously recommends for the peach. Widely found among the agronomists, the many other techniques that G. mentions in 1.1 – from burying the fruit in containers to coating them with various substances to sinking them in desiccants – all share the fundamental aim of creating a stable environment for the fruit through the provision of 'natural refrigeration' and a 'CO_2 blanket' (Thurmond (2006) 183; cf. 1.1.6n., *EFN* 640). By preventing the accumulation of moisture and trapping stale air against the fruit, the desiccants, hard coating or sealed containers would all act similarly to inhibit their degradation.

1.1.2 Sane ad montium deuexa melius proueniunt, non in planis locis: A number of trees are said to enjoy hilly or mountainous areas (e.g. *Pom.* 4.6.6–7 chestnuts, Pallad. 2.15.14 walnut, 11.12.4 cherry; for the olive, Verg. *G.* 2.179, Col. 5.8.5, *Arb.* 17.1), and it is sometimes implied that the seedlings of fruit trees in general should be started out on slopes before they are transplanted to flat places (cf. Col. 5.10.7, Pallad. 3.19.3). Nevertheless, the fact that advice very similar to G.'s in both substance and language is found in Palladius' discussion of the quince suggests that G. is speaking of the quince here: cf. Pallad. 3.25.21 (sc. *cydoneae*) *et in planis et in decliuibus proueniunt, magis tamen inclinata et deuexa desiderant.* (This is the only occurrence of the substantive use of *deuexum* in Palladius: it likely comes from G.) Palladius does not explicitly state that he is relying on G. in his treatment of the quince, but similarities of language and content here and elsewhere in his discussion of preserving (3.25.25–6) establish that G. is one of his sources (see *passim* below).

Sane: Here *sane* is a sentence adverb that adds force to the entire statement (*OLD* s.v. 3a, Chahoud (2011) 378; see also Risselada (1994) 333–6, (1998) 228, 232–4); at 3.5.2 it may have the same function, or it may simply be a connective particle (like δέ/*autem*; see Svennung (1935) 405, Adams (1995*a*) 629–30).

ad montium deuexa: *deuexum* in its substantive usage is found once in Caesar (*BGall.* 7.88.1) and thereafter commonly in Latin of the first century CE and later: see *TLL* v 1.858.5–29 (Lommatzsch).

melius . . . non: Semantically like *melius . . . quam*, for which *melius . . . non* could be said to work out the 'implicación pragmática negativa' of the comparison (Torrego (2002) 262): see further Bertocchi and Orlandini (1996), esp. 223 ('la conjunction *quam* véhicule un sense négatif ("*et non*")'), Torrego (2002) 261–2, Cuzzolin (2011) 580–3, Tarriño (2011) 378; also Panagl (1975).

melius proueniunt: *prouenio* is not found in Cato, *Agr.* or Varro, *Rust.*, but in the agricultural lexicon of the Empire it is used frequently to describe the growth or, more positively, flourishing of plants (*TLL* x 2.2310.21–39 (Galdi-Spoth), *OLD* s.v. 3a). This meaning originates in the notion of growth attached to physical 'coming forth', an idea that is conspicuous in the earlier attestations of its application to plants: cf. e.g. Caes. *BGall.* 5.24.1 *frumentum in Gallia propter siccitates angustius prouenerat.* The simplex may also be so used, for which cf. Boldrer on Col. 10.106 *Megaris ueniant . . . semina bulbi.* The use of *prouenio* in agricultural contexts appears almost formulaic by G.'s time: besides the present instance it is applied to plants five times in *Pom.* (2.10.5, 3.3.4, 4.4.3, 4.5.3, 4.6.7),[3] and in all but one of these cases (3.3.4, where it is joined with descriptive ablatives), it is paired with an adverb (e.g. *uix, optime, tardissime*) to form a statement that concisely describes the suitability of a particular condition for the plant's growth. The verb *procedo* is used similarly (2.5.5n. *planta procedit*).

1.1.3 ampliora quae Nouember affert: Evidently in reference to the size of the quinces; the observation is not found in any of the other agronomists. The implied harvest time is plausible, as quinces ripen relatively late in comparison with other fleshy fruits; most cultivars will not be ready for picking until mid-to-late October at the earliest (*EFN* 637, 640).

affert: The word is not used by Cato in this application, and only once by Varro (*Rust.* 1.8.6 *uitis quae ostendit se afferre uuam*), but is regular among the Imperial agronomists: see *TLL* I 1199.35–57 (v. Mess), *OLD* s.v. 18.

matura cydonea legenda sunt: Cf. Pallad. 3.25.25 *legenda sunt matura cydonea.* The adjective *maturus* is the *uox propria* for describing ripe fruit: see *TLL* VIII 498.24–58 (Brandt), *OLD* s.v. 1. For remarks on expressions designating ripeness, see also 3.4.3n. *maturant.*

cydonea: *cydoneum* (sc. *malum*) and its variants derive from Gk. κυδώνιον (sc. μῆλον), which in antiquity was traced to the Cretan city Κυδωνία, but in truth derives from an older Anatolian word (preserved in κοδύμαλον *apud* Alcman): see E–M 261, Chantraine 596, Beekes I 797; also Solmsen (1911), Nehring (1923). The quince was also known to the Romans as *cotoneum* (sc. *malum*), and Pliny identifies this as the Latin name of the fruit (*HN* 15.37): *mala quae uocamus cotonea et Graece cydonea.* Cato and Varro use only *cotoneum*, as do Celsus and Pliny, but Scribonius Largus, Petronius and Columella use *cydoneum*, which becomes the regular form in later Latin (Solmsen (1911) 242). *cotoneum* is usually held to be an independent loan via an Etruscan intermediary rather than a simple corruption of *cydoneum.* For a conspectus of forms and variation in spellings, see *TLL Onomasticon* II 786.17–18 (Reisch) and *passim* in the article. For Romance survivals, see *REW* no. 2436.

1.1.4–5 condiri etiam per cydoneum posse qu . . . affirmant ali . . . traxisse aliquid olim de sapore affixi mellis: The connection

[3] Possibly six times, if 2.11.3 *satius proueniet* is counted (n.).

between the words *condiri etiam per cydoneum posse* and *traxisse aliquid olim de sapore affixi mellis* is obviously obscured by the lacunae, but the two fragments can be tentatively related if we suppose that the subject of *condiri . . . posse* is *mel.* In this case, G. would be discussing the means of producing *melomeli* (μηλόμελι), the honey that has been flavoured by quince and has thereby obtained therapeutic properties. This interpretation is corroborated by the reference to honey at the end of the sentence (*affixi mellis*): with *traxisse*, etc., we should understand that G. has turned from the improvement of the honey to the reciprocal improvement of the quinces, which are preserved and sweetened by the liquid (cf. also 1.1.7 *si melle linantur, in eo melius perseuerant et, ubi tempus erit, speciosa exire pollicentur*).

The double benefits of immersing quinces in honey – i.e. not only the production of *melomeli* but also the preservation of the fruit – makes the method a favourite of Columella, who provides the fullest version of the technique: 12.47.2 *nihil tamen certius aut melius experti sumus quam ut cydonea maturissima integra sine macula et sereno caelo decrescente luna legantur et in lagona noua, quae sit patentissimi oris, detersa lanugine quae malis inest, componantur leuiter et laxe, ne collidi possint; deinde, cum ad fauces usque fuerint composita, uimineis surculis sic transuersis artentur ut modice mala comprimant nec patiantur ea, cum acceperint liquorem, subleuari; tum quam optimo et liquidissimo melle uas usque ad summum ita repleatur ut pomum summersum sit. haec ratio non solum ipsa mala custodit sed etiam liquorem mulsi saporis praebet, qui sine noxa possit inter cibum dari febricitantibus; isque uocatur melomeli.* The drug *melomeli* was an important enough therapeutic that Columella's contemporary Dioscorides also provides instructions for its preparation (5.21), claiming that the honey that has become προσηνής ('gentle'; cf. Col. *mulsi saporis*) through the quinces shares a host of medical benefits with quince-wine (κυδωνίτης, Dsc. 5.20).

Pliny has a terse mention of preserving the quince in honey but does not mention *melomeli* in the same place: *HN* 15.60 (sc. *cotonea*) *incoqui melle ea mergiue oportere.* Nor does Palladius mention it, although he offers two ways of preserving the quince in honey, one of which has the fruit first cut into quarters: 3.25.25–6 *alii canna uel ebore in quattuor partes diuisa, sublatis omnibus quae in medio sunt, in uase fictili melle obruunt. alii in melle sic integra demittunt, in quo genere condiendi satis matura deliguntur.*

1.1.4 condiri: On the interpretation offered in the previous note (which assumes *mel*, honey, as the subject), *condio* will mean simply 'flavoured' here; it is used of 'pickling' fruit at 2.12.1 *conditum.*

1.1.5 affixi: Lit. 'stuck on'. The verb is perhaps substituted for *(il)lino* (cf. 1.1.7 *melle linantur*) from a desire for *uariatio.* The more typical sense of *affigo* is found at 2.6.2, 3.2.1.

1.1.6 Aliqui putant et in dolium uini olla noua conclusa, ut in eo obtecta natent, esse mittenda: The recommendation, if rather involved, is based on sound principles, for the sealed *olla* floating in the wine would provide a cool and stable environment and prevent oxygen from permeating to the fruits and hastening their spoiling. The same principle is at work in G.'s more conventional advice to bury the quinces in holes or vats: cf. 1.1.13 *et scrobibus et in doleis . . . seruari*. A possible forerunner to G.'s advice here is found in the technique of 'potting', which is mentioned by the agronomists in connection with a number of fruits but used especially for grapes, so-called *uuae ollares*: cf. Varro, *Rust.* 1.54.2, Col. 12.45, Plin. *HN* 14.16, Brenni (1985) 25. The grapes are placed in pots (*ollae*) that are subsequently packed into a *dolium* filled with grape skins and other leavings (*uinaceae*). G.'s method for the quince is similar save for his use of *uinum* rather than *uinaceae* in the storage *dolium*. Furthermore, it is relevant that Varro (*ibid.*) reports as an alternative to potting the grapes placing them in a pitch-sealed amphora and sinking them in a pool (*alia quae in piscinam in amphoram picatam descendat*). If we suppose that Varro joins the methods because he assumes that the same principle is at work in both, then it will be the cool and stable conditions afforded by the surrounding substance, rather than the nature of the substance *per se*, that is the common thread in these techniques for preservation. Note thus that the same purpose is fulfilled by *sapa* at 2.12.4 *multi Suessatibus ollulis impicato singula poma clausere umbilico atque in sapam purissimam demissa merserunt*. This is by no means to rule out, however, the belief that the liquid around the pots could somehow (magically?) affect the produce, despite having no direct contact with it. Why else would G. include a detail (2.12.4) such as *purissimam*?

Quite different is Palladius' advice to immerse quinces in new wine (followed by *Geop.* 10.28.1): 3.25.26 *alii doliis musti mergunt atque ita clauduntur, quod odoratum reddit et uinum*. (A detailed recipe for quince-wine itself is actually given by G. at *Med.* 43.11–17.) The *Geoponica* when it reports the technique seems to conflate the recommendations of G. and Palladius, promising both that the quinces, sealed in a jar, will emerge fresh (νεαρά) and that the wine will be perfumed by them (εὐώδης): 10.28.2 καὶ εἰς χύτραν καινὴν ἐντεθέντα, καὶ εἰς πίθον οἰνηρὸν τῆς χύτρας ἐμβληθείσης, ὥστε πλεῖν ταύτην, τοῦ πίθου ἐπιχρισθέντος, ἔσται τὰ μὲν κυδώνια νεαρά, ὁ δε οἶνος εὐώδης.

dolium: The *dolium* (Gk. πίθος) was the largest Roman earthenware vessel, ranging from 2½ to 6½ feet tall (Hilgers (1969) 48, 171–6, White (1975) 146, Brenni (1985) 6–7, 29–31, Curtis (2016) 592). It served many purposes. As here, it was often used for the long-term storage of a variety of foodstuffs, not only fruits, vegetables and wheat but also wine, olive oil and other liquids (Brenni (1985) 18–27, Curtis (2016) 593); but it could also serve, for example, as a temporary holding vat for a fisherman's catch until it could be processed into *garum* (Manil. 5.678–9 with Curtis (2016) 592); a vessel for keeping dormice while they were being fattened for consumption (Varro, *Rust.* 3.15.1–2, Plin.

HN 8.224, Brenni (1985) 26); or a semi-permanent fixture on cargo ships for transporting foodstuffs (Brenni (1985) 53–6, Hesnard (1997), Curtis (2016) 593).

olla noua conclusa: With *conclusa* supply *cydonea*. The word *olla* (older form *aula* or *aulla*; see *TLL* II 1453.11–20 (Münscher), de Vaan 62) refers to a pot or jar (Gk. χύτρα) of varying size, used especially for cooking or boiling vegetables (so cf. *Med.* 41.12 *punicum in olla noua operculo illito in furno exustum*), but not uncommonly for preserving fruit (as here): see Hilgers (1969) 39–40, 112–16, White (1975) 176–9. It was used often enough for preserving that one could speak of potted grapes (*uuae ollares*): cf. Stat. *Silu.* 4.9.42, Mart. 7.20.9, and see also 1.1.6n.

obtecta: A rare word among the agronomists (Cato, *Agr.* 40.4, whence Plin. *HN* 17.112, and by conjecture at Plin. *HN* 18.252) but found once more at *Pom.* 4.3.2, where it refers to covering a heap of chestnuts with sand (4.3n.). Unless *obtego* is taken in both places as the equivalent of *tego* (cf. *TLL* IX 2.268.47–56 (Paschoud)), the sense that *ob-* will contribute is that of protection (i.e. provision of a protective cover): see *OLD* s.v. 1, and cf. *TLL* IX 2.269.26–37.

1.1.7 At nos non leui pondere affirmamus quod si melle linantur, in eo melius perseuerant et, ubi tempus erit, speciosa exire pollicentur: G. asserts, apparently from personal experience (*nos . . . affirmamus*), the superiority of coating the fruit with honey to floating them in a vat of wine: not only do they keep better (*melius perseuerant*), but they also promise to turn out more attractive (*speciosa exire*). It is difficult to know whether by *melle linantur* G. is merely referring to his previous discussion of honey (*affixi mellis*) or introducing a separate technique; it may be that there is intended a contrast between smearing or coating (*linantur*) the fruit with honey and immersing them, if that is in fact how G. described the earlier technique (cf. 1.1.4–5n.), but the lacuna makes it impossible to tell.

non leui pondere: For this use of *pondus* ('impetus sim., quo voces sim. proferuntur', *TLL* X 1.2626.70–2627.3 (Hajdú)), cf. Quint. *Inst.* 6.1.2 *quae . . . enumeranda uidentur, cum pondere aliquo dicenda sunt.* The expression is strengthened in the present instance by litotes (*negatio contrarii*), a stylistic figure which G. commonly uses to stake an evaluative position of various degrees of strength (cf. Hoffmann (1987) 42 and *passim*, (1989) 592) on a fact or piece of advice that he reports. G. employs litotes again at 3.3.4 (*nec desinunt asseuerare*) to emphasize the vehemence of an authority's support for a technique; in other contexts the figure indirectly signals G.'s approval of a piece of advice (4.2.2 *Nec immerito*; 4.7.4 *non immerito*), describes a quality of a plant (2.6.3 *nulli non repugnat*; 2.7.1 *non amat*; 2.10.5 *non diu durat*; 4.5.3 *non uult*), or provides *uariatio* in a list of different opinions (4.1.9 *non omittendus*). For general discussion and examples of litotes, see H–Sz Stilistik §32, Hoffmann (1989), Lausberg (1998) 268–9, Pinkster

686–8; and for a fuller study, Hoffmann (1987). For *leuis* in particular in litotes constructions, see Hoffmann (1987) 151. G. also commonly uses 'fossilized' or 'frozen' litotes expressions such as *nonnullus* (2.9.3, 2.11.1, 3.4.2, 4.1.8, 4.6.1): the process of fossilization has the result that '[such] expressions . . . are not used intentionally instead of a positive expression', as other cases of *negatio contrarii* are (Hoffmann (1987) 141).

affirmamus quod: For the *quod*-clause in place of accusative-and-infinitive syntax, see Introduction §v(c).

si . . . linantur . . . perseuerant . . . pollicentur: The subjunctive *linantur* in the protasis of this condition is best taken as iterative or frequentative (Nutting (1923) 151–3, (1925) 82–5, *NLS* §196, H–Sz §363), that is, as describing a condition for which the consequence in the main clause (apodosis) holds true iteratively: '*if* they are coated, then, etc.' = '*when(ever)* they are coated, then, etc.'. This kind of iterative subjunctive is well suited to a didactic context, for it allows the author to describe and vouch for a procedure that is taken to apply generally, not just in the experience of one person or another or at one time or another. Such mixed conditional statements are thus common in *Med.*: cf. 1.6, 6.7, 11.6 (*illinantur*), 17.3, 21.10, 29.15, 30.22, 31.3, 32.7, 35.5, 42.9, 52.12, 53.4, 54.2, 58.4. There are also a number of cases in *Pom.* where the subjunctive verb in the *si*-clause should be similarly taken as iterative, despite the fact that it might also be given a grammatical explanation on the basis of an indirect statement: cf. 1.1.10 *illiniri existimant . . . si . . . reponantur*, 1.1.11 *sufficere crediderunt, si . . . siccentur ac . . . contegatur*, 4.3.1 *Contendunt . . . procedere, si . . . seruetur*.[4]

melius perseuerant: Among Imperial authors *perseuero*, the most common meaning of which is to 'continue' or 'persist' in a particular action, state, condition, vel sim. (cf. 4.2.4n. *incolumitate perseuerandi*), is also used with a wider sense of 'continue to exist', 'last' (*TLL* x 1.1703.27 (Spoth) 'respicitur mera existentia continuata', *OLD* s.v. 4b). G. is alone among the agronomists in using it to refer to the preservation of fruit. The *iunctura* with *melius* here suggests a more precise meaning for the verb: we might say that the quinces 'keep better'.

ubi tempus erit . . . pollicentur: G.'s tendency to pair a future or future perfect verb in the temporal clause with a present verb in the main clause when the temporal clause specifies an action that must be fulfilled before that of the main verb is discussed in the Introduction §v(b). The examples there are explained on the basis of the imperatival force of the main verb. There are, however, other places where the same pattern is employed but without an imperatival main verb, as here and at 2.1.5 *tunc sentiunt, cum coeperint experimenta*

[4] Add also 2.5.6n. *affirment quod si pertranseant . . . tribuantur*, where the reason for the subjunctive *tribuantur* in the apodosis is obscure.

consulere (in spite of *tunc*); 3.7.5 *uidetur . . . contingere, cum . . . consederit.* While the main verbs in these clauses are not directive, they are future-looking, and the same '*x* only when *y*' logic applies in all cases. Thus, it is not difficult to understand how the verbs in such temporal clauses would also be put into the future and future perfect.

tempus: The 'proper or due time' (*OLD* s.v. 8), as also at 2.3.2 *tempore* and 3.6.4 *cumque iam tempus uidebitur* (cf. 3.6.4n. *cumque . . . praecipiunt*).

speciosa exire pollicentur: *polliceor* with an inanimate subject usually means 'lead one to expect, promise' (*OLD* s.v. 4; cf. *TLL* x 1.2553.29–43 (Reineke)), but there is an obstacle to the interpretation here: as Reineke points out in her *TLL* article (x 1.2553.10), *polliceor* in this sense is used of signs that indicate that something else will come to pass. The usage is well illustrated by e.g. Col. 9.14.17 *dum Arcturi ortus et hirundinis aduentus commodiores polliceantur futuras tempestates.* In the present case, the fruit would not be said to 'promise' anything *of themselves*; we would need rather a situation, fact, quality or other sign that would 'lead us' to 'expect' something *of them. polliceor* could be passive in sense ('are promised'; see *OLD* s.v. 5, *TLL* x 1.2553.52–63), but it is probably better to suppose that the fruit are personified and said to 'vouch for' or 'give assurance of' their own attractiveness (*OLD* s.v. 1). *polliceor* in this meaning may take a present infinitive (*TLL* x 1.2554.34–5, H–Sz §195.B.δ), and while the verb is restricted to human subjects in this sense, the required personification of the fruit is not difficult (cf. 2.3.5n. *subsidium teporis et umoris*): 'the quinces promise to emerge with a fine appearance' is an indirect way of saying that the quince is such a fruit as to become attractive when immersed in honey.

speciosa exire: The adjective *speciosus* indicates that the fruit will be pleasing to the eye but does not necessarily imply anything further, as shown later by the contrast 1.1.12 *aspera sunt sapore, quamuis et aspectui speciosa*; so cf. also Plin. *HN* 13.45 (sc. *adelphides*) *minus speciosae, sed sapore . . . sorores.* If we are to understand that the quinces have been wholly immersed in the honey, not merely coated, then *exire* may carry the literal meaning of 'come or go out' or 'emerge' (*OLD* s.v. 1–2). Otherwise, it will have its non-local predicative sense of 'turn out' or 'become', like *euenire* or *fieri* (*OLD* s.v. 11b, *TLL* v 2.1364.42–54 (Leumann)). For a parallel phrase with similar semantic ambiguity in the verb, cf. 1.1.12n. *speciosa procedant.*

exire pollicentur: The triple trochee (*exīrĕ pōllĭcēntur*), avoided in the Ciceronian system, is very rare in G.: for other examples cf. e.g. 3.8.2 īllĭcō rĕsōlues, *Med.* 6.1 *dicerēnt cŏāgŭlātam*, 18.4 *pōtĭōnĕ uīni.*

1.1.8 Quidam patina noua sicco gypso separata obruunt: G. condenses the recommendation of Columella for preserving the quince: 12.47.1

nonnulli haec eadem cydonea in patinas nouas sicco gypso ita obruunt ut altera alteram non contingat. When reporting the same advice, Palladius follows closely G.'s simplified version: 3.25.26 *alii in patina noua sicco gypso obruunt separata cydonea.* For the flat, open dish called the *patina*, see Hilgers (1969) 72–3, 245–7.

patina noua: The bare ablative of location (H–Sz §86, Pinkster 803–6) is generally restricted in earlier Latin prose to a handful of nouns (e.g. *mare, terra, locus, pars, regio*); it becomes more common outside of this restricted class of words from the time of Livy onwards. Although found regularly in technical authors from Vitruvius on (cf. Praun (1885) 92–3, Kottmann (1903) 21–2, Svennung (1935) 232, Önnerfors (1956) 116–17), it is hardly a distinctive feature of technical writing. Other examples of the bare ablative of location in *Pom.* include 1.1.13 *et scrobibus et in doleis*; 2.1.2 *calidioribus locis . . . posse disponi* (n.); 3.4.1 *altissimo scrobe recondant*; 4.6.4 *Iulius Atticus opacis et septentrionem spectantibus cliuis laetissimum fieri castaneae pollicetur* (n.).

sicco gypso: The word *gypsum* (Gk. γύψος) 'may mean either the mineral itself or the plaster obtained by heating gypsum at a suitable temperature (between 128° C. and 163° C.)' (Eichholz (1967) 108). It is the plaster (plaster of Paris) which is relevant here, and which was above all meant by the word *gypsum* when applications of the mineral were discussed by the ancients. When ground into a finely powdered form, gypsum could be used as a desiccating or insulating agent for fruit, as in the present instance. These properties were recognized in medical usage of the powder too, as shown for example by Oribasius' observation (*Eup.* 2.1.9) that *gypsum* is a desiccating (*desiccatiuus*) drug.[5] There were many other uses for gypsum, powdered or not, on the farm. Theophrastus already says that in Italy gypsum was added to wine (*Lap.* 67 περὶ δε Ἰταλίαν καὶ εἰς τὸν οἶνον), a claim that is perhaps borne out by the later appearance of such advice in the Roman agronomists (*TLL* VI 2.2384.42–6 (Brandt)).[6] Gypsum plaster itself could be employed as material for statues

[5] The mineral gypsum, more specifically anhydrous calcium sulfate ($CaSO_4$) obtained from heating gypsum to a high temperature, is still employed as a desiccant today (although no longer for preserving foods).

[6] Pliny (*HN* 14.120), for example, says that gypsum mitigates the asperity (*asperitas*) of wine (see André *ad loc.*). The *Geoponica* interestingly reports that, to the contrary, gypsum makes wine sharper at first, but once the wine 'breathes out' or 'disperses' the sharpness, it keeps for a long time: 7.12.5 γύψος ἐμβληθεῖσα κατὰ μὲν ἀρχὰς δριμύτερον τὸν οἶνον ποιεῖ, τῷ χρόνῳ δὲ τὸ μὲν δριμὺ διαπνέει, τὸ δὲ ἀπὸ τῆς γύψου χρήσιμον ἐπιπολὺ διαμένει. Cf. also Col. 12.20.8 on gypsum's preserving properties in wine. Eichholz (1965) 132 reported that gypsum 'is still added to crushed grapes in the production of "flor" sherry'.

(*TLL* VI 2.2384.20–8), mortar for construction (e.g. Theophr. *Lap.* 66) or sealant for preserving-jars (e.g. Cato, *Agr.* 39.2, Col. 8.8.7, 12.44.6, Pallad. 2.15.4), among other purposes.

obruunt: See 2.1.3n. *obruenda.*

1.1.9 Eadem alii milio siue palea separata demergunt: The advice is essentially identical to that for preserving the quinces in gypsum, but with the choice of two additional desiccating agents (millet or chaff). Pliny mentions that millet was widely recommended for other fruits too: *HN* 15.63 *aliqui omnia haec* (sc. *mala*) *in milio seruari malunt.* Palladius closely follows G.: 3.25.26 *alii milio obruunt uel paleis separata demergunt.* The *Geoponica* has similar advice but omits millet: 10.28.3 καλῶς δὲ καὶ ἐν ἀχύροις ταῦτα ἀποτίθεται καὶ φυλάττεται.

Eadem: It is possible that we should read *eādem* ('likewise, also'; see *OLD* s.v. *eadem* 2, *TLL* VII 1.208.63–5 (Hofmann)) rather than *eădem*, indicating that the methods are very close except for the desiccating agent employed. Cato has an instructive parallel for this usage in his discussion of preserving grapes: *Agr.* 7.2 *eae* (sc. *uuae*) *in olla in uinaceis conduntur: eadem in sapa, in musto, in lora recte conduntur* (cf. Varro, *Rust.* 1.58, Plin. *HN* 14.64). But how would a reader know that *eādem* and not *eădem* was intended? G.'s fondness for using *idem* as a means of adding precision to his statements (see the remarks on over-specification in the Introduction §v(b)), even where it might seem otiose, argues for *eădem* (roughly = *ea* (sc. *mala*)).

milio: '[M]illet, a standard grain-crop usually made into a rather inferior bread, grown both in N. Italy (Polybius 2.15.2) and in Campania (Pliny 18.100)' (Mynors on Verg. *G.* 1.216). See further *RE* (1) 8.1950.58–56.55 s.v. 'Hirse' (Orth), *LTBL* 209, *NPRA* 161–2.

palea: The husks left after winnowing or threshing grain – Varro claims that the name *palea* comes from the winnowing shovel *pala* (*Rust.* 1.50.3) – served many purposes (cf. White (1970) 450). Chaff had long been esteemed as a desiccant in the preservation of fruits and nuts (as here and at 3.8.2 *paleis obruuntur*), so much so that one of the Aristotelian *Problemata* asks, 'How is it that chaff ripens hard fruits, but does not allow them to rot once ripened?' (931a23–4 διὰ τί ποτε τὰ ἄχυρα τὰ μὲν σκληρὰ πέττει, τὰ δὲ πεπεμμένα οὐ σήπει;). The author speculates that it is because the chaff is warm and 'attractive' (ὁλκόν), both cooking the fruit to maturity and drawing off putrefying discharge (931a24–6 ἢ ὅτι τὰ ἄχυρα θερμόν τε καὶ ὁλκόν ἐστιν; τῇ μὲν οὖν θερμότητι πέττει, τῷ δὲ ὁλκὸν εἶναι τὸν ἰχῶρα τὸν σηπόμενον δέχεται, διὸ οὐ σήπει). For the ubiquity and importance of chaff in the process of preserving, one might well trace its various applications at Col. 12.44.1 (on the grape): *deinde labellum fictile nouum impleto paleis quam siccissimis cribratis, ut sine puluere sint, et ita uuas superponito; tum labello altero adoperito et circumlinito luto paleato atque ita* *in tabulato siccissimo composita labra paleis siccis obruito.* Among other uses, chaff could also serve as a manure, applied directly to the plants (e.g. Cato, *Agr.* 33.3 fertilizer for vines) or added to the compost heap (*stercilinum*; see Appendix II). When strewn on the floor of the *oporotheca,* chaff provided a moisture-absorbent base on which to rest fruit (e.g. Varro, *Rust.* 1.59.1, Plin. *HN* 15.59). A mixture of chaff and mud or clay also created an insulating layer that could be daubed on the exterior of containers used for preserving (e.g. Col. 12.44.7, 12.46.2, 12.46.4, 12.47.6); the same mixture also sealed grafts (e.g. Col. 5.6.13, 5.11.6 = *Arb.* 26.5, *Arb.* 8.2, Pallad. 4.1.2). Finally, in a blend with other plants it was used as feed for draught animals (e.g. Col. 11.2.99).

demergunt: In the immediate context *demergo* provides *uariatio* for 1.1.8 *obruunt,* but G. seems to have used this verb regularly in the context of preservation, at least judging from the evidence of Palladius. So cf. Pallad. 2.15.19, drawing on G.'s lost advice on the walnut: *Martialis expertum se ait uirides nuces tantum liberatas putaminibus suis melle demergi,* etc. Another example is Pallad. 12.7.16 *alii in aqua marina uel in muria feruente recens lecta pruna demergunt,* where, though G. is not mentioned, it is probable that Palladius' advice is based on *Pom.* G. uses the simplex *mergo* in the same way: see 2.2.2n. *mergi.* For *demergo* in application to planting, see 2.5.8n. *demergere.*

1.1.10 Sunt qui figularem cretam cum amurca subigunt soleque siccatis cydoneis illiniri existimant, si in loco sicco et frigido reponantur: The advice is based most closely on Col. 12.47.1 *nonnulli foliis ficulneis illigant, deinde cretam figularem cum amurca subigunt et ea linunt mala, quae cum siccata sunt, in tabulato frigido loco et sicco reponunt*; cf. also Plin. *HN* 15.60 *cetera mala et foliis ficulnis, praeterquam cadiuis, singula conuolui cistisque uitilibus condi uel creta figulinarum illini.* The reason for the absence of the fig leaves in G.'s account is unclear. Palladius recommends them for the quince even without the admixture of potter's clay (3.25.25 *alii* (sc. *cydonea*) *quae maiora sunt fici foliis inuoluta custodiunt*); the *Geoponica* has the fig leaves as well, but again with clay: 10.21.7 φυλάττεται τὰ μῆλα καὶ οὕτως· περικάλυπτε ἕκαστον μῆλον συκίνοις φύλλοις ξηροῖς, ἔπειτα περίπλασσε πηλῷ λευκαργίλλῳ, καὶ ψυγέντα ἐν ἡλίῳ ἀποτίθεσο, καὶ διαμένει οἷα ἐβλήθη. Despite the linguistic echoes with Columella, G. may have been using another source that omitted the *folia ficulnea,* or he may have dropped the detail of his own accord, believing that the mixture of potter's clay and *amurca* would alone suffice to protect the fruit (for the function of *amurca,* see 1.1.10n. *cum amurca*). The latter hypothesis is perhaps strengthened by G.'s continued focus on *creta* in the next sentence (1.1.11n.), which suggests that he took the core of this advice (1.1.10) and what follows (1.1.11) to be the use of clay as a preserving agent. It is at any rate true that both techniques rely on the same basic principle as other advice in the section, that is, of secluding the moisture and oxygen that would hasten the fruit's spoiling (1.1n.). If it is wondered how the quinces so coated with *creta* were to be prepared for consumption, Columella (12.46.5, quoted 1.1.11n.) and Pliny (*HN* 15.64) mention washing them, Pliny adding that wine may be used for pomes (fleshy fruits) in particular.

Sunt qui . . . subigunt . . . existimant: In such 'generic presentative clauses' (Cabrillana (2011) 34), the subjunctive is the preferred mood (consecutive or predicated; see Bennett (1910–14) I 298–9, *NLS* §157.2, H–Sz §300.a), even for authors as late as Firmicus (first half of the fourth century CE). The increasing appearance of the indicative in later authors (from Apuleius and Aulus Gellius on) is probably to be attributed to the more general reduction of the subjunctive in later Latin, especially in subordinate clauses: see Calboli (1995) 57–8 (with n. 1), Väänänen (1981) 133, Cabrillana (2011) 38. Yet as Cabrillana (2011) 38–9 emphasizes (cf. Väänänen (1987) 75–82), the subjunctive long continued to coexist with the indicative, and in many cases retained its vitality. In fact, G.'s use of the indicative in the present instance is anomalous in both *Pom.* and *Med.* Of seven further appearances of the construction (*Pom.* 6x, *Med.* 1x), only the subjunctive is found: 2.2.1 *Sunt qui . . . existiment*; 2.5.6 *Sunt qui . . . disponant* (*dispont* **N**) . . . *et affirment*; 2.7.5 *inuenti qui . . . existiment*; 3.8.2 *Sunt quibus . . . abscedat*; 4.4.4 *Inuenti tamen qui . . . audeant . . . credant*; 4.6.7 *Sunt qui putent*; *Med.* 47.5 *Sunt qui putent*. All such clauses are used as a means of introducing the opinions of authorities except for 3.8.2, which picks out a subset of almonds.

figularem cretam: For the agronomists, the word *creta* alone often designates a clayey sort of earth, but not necessarily a pure clay, and typically with reference to planting soil (*TLL* IV 1185.37–50 (Lambertz), *OLD* s.v. 1). When the adjective *figularis* is added, the point of the detail seems to be to pick out a clay of sufficient quality or purity for more specialized applications, not only for pottery (as the word immediately suggests) but also e.g. as a sealant for preserving fruit (here, and see further examples at *TLL* IV 1185.81–1186.7), making veterinary poultices (cf. Varro, *Rust.* 3.9.3, Col. 8.2.3) and many other purposes. A source of such good-quality clay on one's farm was thus potentially valuable: cf. Varro, *Rust.* 1.2.23, Wilson (2012) 137–8.

cum amurca: *amurca* (also *amurga*, Gk. ἀμόργη) is the dark, bitter liquid (lees) pressed out of the olive along with the oil: see Varro, *Rust.* 1.55.7, 1.64, *Geop.* 9.19.9, Niaounakis and Halvadakis (2006) 237 and *passim*, Janakat et al. (2015) 259. The substance was highly prized in Roman antiquity and had many important uses, from fertilizer (cf. 3.7.3n.) and pesticide to veterinary medicine and axle grease. Cato thus devotes 12 chapters to cataloguing these uses (*Agr.* 91–103), summarized by Pliny (*HN* 15.33–4), who accurately appends the statement *super omnia . . . celebrauit amurcam laudibus Cato.*

There come to mind two reasons for the recommendation to compound *amurca* with *creta figularis*. The first is that the addition of the liquid might strengthen the clay sealant itself, creating an impervious 'plaster-like finish' once dry that would protect the fruit (cf. Smith and Secoy (1975) 1052). The second is that *amurca* was thought by the Roman agronomists to be a powerful pesticide in its own right against disease, weeds and insects (see the overview of Smith and Secoy (1975)).[7] It was for this reason, that is, to ward off agents that could spoil agricultural produce, that the agronomists recommended that it be used in the construction of buildings for storage or processing: cf. Cato, *Agr.* 91–2, Varro, *Rust.* 1.57.1–2, Col. 1.6.12–15, Plin. *HN* 15.33, 18.295, Pallad. 1.19.2, *Geop.* 2.26.5. Varro (*ibid.*), for example, says that incorporating *amurca* into the plaster of a granary is beneficial *quod murem et uermem non patitur esse et grana facit solidiora ac firmiora.* Mixing *amurca* into the clay daubed onto the quinces might have been thought to offer similar protection.

subigunt: For the technical agricultural use of the verb *subigo*, see 2.5.8n. *subacto solo.*

soleque: The only instance in *Pom.* or *Med.* where G. attaches *-que* to a preceding short syllable (*solĕque*).

illiniri existimant: Already from the time of Vitruvius, *uerba sentiendi*, such as *censeo*, *sentio*, *existimo*, *credo*, *arbitror* and *puto*, could be used as *uerba uoluntatis* (i.e. with the sense 'think that . . . ought/should'): for a synopsis of the phenomenon, see H–Sz §191.1.D; and for examples and discussion, see Löfstedt (1907) 59–62, Pomoell (1931) 10–13, Wistrand (1933) 150–1, Linde (1936) 59–60, Tidner (1938) 142. Svennung (1935) 439 has an instructive parallel to G.'s usage of *existimo*: Ruf. *podagr.* 27.12 *et non oportet de subito remouere potiones aut de semel, neque diuretica existimo confestim resoluere quae consuetus erat bibere; sed paulatim subtrahendum est.* Note that *neque . . . existimo . . . resoluere* ('I am of the opinion that one *ought not* to remove, etc.') is flanked by the imperatival expressions *oportet remouere* and *subtrahendum est.* In G.'s case the construction is perhaps even more natural, for the action described by *illiniri* is virtually coordinated with that of the imperatival indicative *subigunt.* There is another example of *existimo* + infinitive with voluntative force in *Pom.* at 2.4.1 *Quolibet caelo et quamuis macro solo persicorum arborem poni Celsus existimat.* The nuance there is different, however, for when Celsus says that the peach tree is (or should be) planted (*poni*) in any conditions, he is not actually endorsing, e.g., poor soil (*macro solo*), only allowing that it is possible for the tree to survive (hence 2.4.1 *idem tamen*, etc.). Apart from the present instance and 2.4.1, G. prefers to use *existimo* with

[7] It is interesting to note that the efficacy of *amurca* in pesticidal applications has recently begun to be investigated again: see e.g. Janakat et al. (2015).

a gerundive rather than an infinitive: cf. 2.1.3 *poma ipsa . . . existimant obruenda*; 2.9.3 *Sparto ligandam* (*ligata* **N**) *aduersum plurima incommoda . . . existimant*; 4.6.6 *tenuem locum . . . existimant deligendum*. He also uses *existimo* in more obliquely directive expressions: cf. 2.2.1 *persicum serere melius existiment*; *Med.* 40.9 (sc. *pira*) *stomacho existimant conuenire*.

illiniri: The form of the infinitive is peculiar (as Reeve observes). Besides here, *illin(i)o* (25x in total in *Pom.* and *Med.*) is not a fourth conjugation verb for G. (though *illiniuntur* is a variant reading in some MSS at *Med.* 25.3). Granted that the infinitive *illinire* is often found in manuscripts for *illinere* (*TLL* VII 1.381.60–74 (Haffter), *OLD* s.v.), the passive *illiniri* is not at all common: from the *TLL* and a search of BREPOLiS, I count only two examples (Tert. *Pudic.* 20, Aug. *Serm.* 46) against some 47 occurrences of *illini* (the vast majority in Plin. *HN*). *illiniri* is the more remarkable when it is considered that the form *illini* would yield a better clausula (dicretic *īllĭni ēxīstĭmant*).

in loco sicco et frigido: Cf. Plin. *HN* 15.59 *in uniuersum autem de pomis seruandis praecipitur pomaria in loco frigido ac sicco contabulari*. For quince-specific instructions, cf. Plin. *HN* 15.60 *cotoneis in loco cluso spiramentum omne adimendum*; Pallad. 3.25.25 *alii tantum locis siccis reponunt, a quibus uentus excluditur*.

reponantur: One of the agronomists' standard terms for preserving (*OLD* s.v. 9a). Unlike other compounds of *pono* (see Introduction §v(d)), *repono* is not used of planting.

1.1.11 Multi uel solam cretam sufficere crediderunt, si cum ramulis suis mala detracta et oblita in sole siccentur ac, si quid extra hiauerit, luto contegatur: The basis of the advice is Col. 12.46.7 *multi cum ramulis suis arbori detrahunt et creta figulari cum diligenter mala obruerunt, in sole siccant, deinde, si qua rimam creta fecit, luto linunt et assiccata frigido loco suspendunt*. G. has changed the language, however, in order to emphasize what he understands to be the relationship between this recommendation and the previous one (1.1.10 *Sunt qui figularem*, etc.). In particular, G. unlike Columella juxtaposes the two recommendations in order to suggest that they are variations on a single premise, that of sealing the fruit in an insulating blanket of clay: that is the force of the detail *uel solam cretam* ('even clay alone'), which in its prominent position early in the sentence contrasts with *cretam cum amurca* in the previous. Probably for the same reason, G. does not repeat, as Columella does, the stipulation that the fruit be stored in a cold, dry environment once they have been sealed with mud: if it is only the substitution of pure *creta* for the *amurca*-compounded mixture that makes this advice different for G., then he would not need to repeat the rest of the instructions.

Presumably Palladius (3.25.25) is paraphrasing G. when he says (sc. *cydonea*) *seruantur . . . si luto ex omni parte claudantur*. A fuller version of the advice, with some additional details (cf. διὰ τριχῶν μεμιγμένῳ), is found in the *Geoponica*: 10.28.5 τινὲς δὲ τὰ κυδώνια φύλλοις συγκαλύψαντες, πηλῷ ἀργιλλώδει διὰ τριχῶν μεμιγμένῳ ἐπιμελῶς, ἢ γῇ κεραμεικῇ, ταῦτα περιπλάττουσιν, εἶτα ἐν ἡλίῳ ψύξαντες ἀποτίθενται. ὅταν δὲ ἡ χρῆσις ἀπαιτῇ, τὸν πηλὸν κατεάξαντες εὑρίσκουσι τὰ κυδώνια οἷα ἐνεβλήθη. Note finally that Columella reports another technique, very similar, for preserving the quince with clay alone: 12.46.5 *sed et idem* (sc. *Mago*) *auctor est creta figulari bene subacta recentia mala crasse illinere* [*illinire* codd.] *et, cum argilla exaruit, frigido loco suspendere, mox, cum exegerit usus, in aqua demittere et cretam resoluere. haec ratio tamquam recentissimum pomum custodit.*

cretam . . . luto: Columella (quoted in previous n.) specifies that the *creta* used for this purpose is *creta figularis* (see 1.1.10n. *figularem cretam*), which is how we should also understand G.'s *creta. lutum* often refers to moistened soil or mud used in building, patching, covering, etc. ('generatim adhibetur fere ad aedificia exstruenda, parietes oblinendos, fissuras obstruendas sim.', *TLL* VII 2.1901.47–1902.37 (v. Mess)) and is hence the appropriate term here.

cum ramulis suis: Columella says that as a rule it is best to gather fruit *cum ramulis* (see Plate 4), if it may be done without harm to the tree; otherwise, with the stalks (*pecioli*) alone: 12.46.7 *omne autem pomum quod in uetustatem reponitur, cum peciolis suis legendum est, sed si sine arboris noxa fieri possit, etiam cum ramulis.*

mala: For the usage of the word *malum* among the agronomists, see De Angelis (1997*a*) 82: 'Si tratta di frutti che richiamano per forma e dimensioni le mele o le pere e sono caratterizati dalla presenza di un grosso nocciolo centrale, il *semen*, oppure di un torsolo nel quale alloggiano semi di cospicue dimensioni.' It is the latter that is applicable to the quince, of course, but the former to the peach.

detracta: I.e. picked (Plates 5, 6).

oblita: Providing *uariatio* for *illiniri*, with the prefix *ob-* possibly influenced by Columella's *obruerunt.*

si quid extra hiauerit: Cf. 2.3.3 *cum extra hiare coeperint.* For remarks on the verb *hio*, see 3.1.1n. *hiascere coeperunt.*

1.1.12 Seruantur et gypso illita, sed aspera sunt sapore, quamuis et aspectui speciosa procedant: After the advice to coat (*illino/oblino*) the fruit with clay, G. mentions that gypsum plaster can be used for the same purpose. The advice therefore does not repeat 1.1.8 *Quidam patina noua sicco gypso separata obruunt*, for the point there is to sink the fruit in a desiccant (1.1.8n. *sicco gypso*), whereas here we are dealing with a wet gypsum plaster that will harden to form a shell around the fruit.

aspera sunt sapore . . . aspectui speciosa: For the stylistic qualities of this line, see Introduction §v(a).

speciosa procedant: As for 1.1.7n. *speciosa exire*, which *procedant* stylistically varies, it is not certain which idea is foremost: that of physically 'emerging' from the gypsum (*TLL* x 2.1495.58 (Terkelsen) 'in conspectum ueniendi, exeundi', *OLD* s.v. 3a), or predicatively of 'becoming' or 'turning out' such and such a way. For *procedo* in its common application to plant growth, see 2.5.5n. *planta procedit.*

1.1.13 Consentiunt aliqui posse et scrobibus et in doleis eodem modo quo punica supra uel alia m<ala> seruari: The detail *eodem modo quo punica supra* is important, for it shows that the quince was not the first of the fruits that G. discussed (1.1n.). Curiously, this sentence is patterned on Col. 12.47.1 *multi eadem ratione qua granata, in scrobibus uel dolîs seruant cydonea*; it is uncertain whether the shared order of the sections is coincidental or, if the evidence may be pressed so far, whether G. followed Columella (up to a point) in his order of discussion of fruits. In any case, Columella preserves the rather elaborate advice for the pomegranate to which G. should be referring: 12.46.3–4 (sc. one should not follow the previous methods,) *praesertim cum liceat etiam detracta arboribus eadem innoxia custodire; nam et sub tecto fossulae tripedaneae siccissimo loco fiunt, eoque cum aliquantulum terrae minutae repositum est, infiguntur sabuci ramuli; deinde sereno caelo granata leguntur cum suis pediculis* <*et ramulis*> *inseruntur, quoniam sabucus tam apertam et laxam medullam habet ut facile malorum pediculos recipiat. sed cauere oportebit ne minus quattuor digitis a terra absint et ne inter se poma contingant; tum factae scrobi operculum imponitur et paleato luto circumlinitur, eaque humus quae fuerat egesta superaggeratur. hoc idem etiam in dolio fieri potest, siue quis uolet resolutam terram usque ad dimidium uas adicere seu, quod quidam malunt, fluuiatilem harenam, ceteraque eadem ratione peragere.*

et scrobibus et in doleis: For the doubled *et*, disjunctive here, see 2.1.3n. *et . . . praebeat et . . . acquirat.* The variation in local expressions (bare ablative vs. *in* + ablative) could be for stylistic effect, reworking and making G.'s own the Columellan source *in scrobibus uel dolîs* (quoted in previous n.). It may however simply follow conventional usage with these words: *dolium* (and the plural *dolia*) require *in*, but *scrobis* (and *scrobes*) are often used with the bare ablative: in *Pom.*, cf. 3.4.1 *altissimo scrobe recondant*; perhaps also 3.3.2 *modicis scrobibus excitatis . . . nuces mergunt.* Pliny prefers *in* + ablative with *scrobis*, Columella and Palladius the bare ablative. G. uses *in* + ablative at 3.1.5 (*Quidquid in summo scrobe induruit leui sarculo relaxetur*), where the preposition *in* is necessary for attaching *summo scrobe* to *quidquid.* For remarks on the bare ablative of location, see 1.1.8n. *patina noua.*

scrobibus: Simply referring to a hole here, with no further specific reference, but for *scrobis* of a 'planting hole' in particular, the most common application of the word among the agronomists, see 2.7.1n. *Scrobes.*

doleis: *doleum* (so also 2.4.2 *doleo*) is an orthographic variant of *dolium* and is printed also at Cato, *Agr.* 23.2, Varro, *Rust.* 1.61.1, 3.15.2: see *TLL* v 1.1832.26–33 (Lackenbacher). For use of the *dolium* in preserving, see 1.1.6n. *dolium*.

punica . . . uel alia m<ala>: Scotti's suggestion *alia m<ala>* is the best way of handling the impossible MS reading *aliam*. It is to be commended on four grounds, none of which is decisive in itself but which together make a case for the emendation: (1) it has the benefit of making the passage clearer by specifying that *alia* refers to other fleshy fruits; (2) it explains the presence of *m* in the manuscript, and (3) does so with a good *ratio corruptelae*, for *ala* could plausibly have fallen out after *alia*; (4) it yields a slight improvement in rhythm, giving G.'s favoured cretic-trochee (*m<ālă> sēruāri*).

supra: Used also by G. at *Med.* 42.15, 44.3, 50.4 when adverting to earlier parts of his writings.

1.1.14 Quintilii cydonea in scobe seruata diuturna promittunt: Yet another method for desiccating, now employing sawdust rather than chaff, millet or gypsum. Columella gives a more detailed version of the technique, for the pomegranate, that is attributed to Mago: 12.46.6 *idem iubet Mago in urceo nouo fictili substernere scobem populneam uel iligneam et ita disponere, ut scobis inter se calcari possit, deinde facto primo tabulato rursus scobem substernere et similiter mala disponere, idque sic facere, donec urceus impleatur; qui cum fuerit repletus, operculum imponere et crasso luto diligenter oblinere.* The *Geoponica* is the only other text to mention the sawdust method specifically in connection with the quince: 10.28.3 φυλάττεται ὁμοίως τὰ κυδώνια ἐπὶ πλεῖστον χρόνον ἐν πρίσμασι κατορυττόμενα. ὑπὸ γὰρ τῶν πρισμάτων ἀναξηραινόμενα ὠφελεῖται. From Columella onwards, sawdust is indeed widely recommended for preserving a variety of fruits: grapes (Col. 12.44.4, Plin. *HN* 15.67, *Geop.* 4.15.9), pomegranates (*Geop.* 10.38.6), apples and other pomes (Col. 12.47.6, Pallad. 3.25.18, *Geop.* 3.13.8), and pears (*Geop.* 10.25.1). Palladius twice elsewhere recommends it (4.10.18 citrons, 13.4.2 *hypomelides*), both times in passages where he is following G.

scobe: Sawdust, lit. 'scrapings' or 'filings', from the verb *scabo*, 'to scratch' (E–M 1054, de Vaan 541).

1.1.15 Diligentioribus rectissime uisum est praemonere ne cum ceteris pomis cydoneum misceatur: sibimet censent reponendum, austeritatem eius timentes, quae corrumpere solet, ut necesse est, quidquid de proximo adfuerit: The caution for the quince is found also at *Geop.* 10.28.4 φυλακτέον δὲ ταῦτα μὴ ἐν τῷ οἴκῳ, ἔνθα αἱ ἄλλαι ὀπῶραι ἀπόκεινται. τῇ γὰρ αὐτῶν δριμύτητι καὶ ὀσμῇ λυμαίνεται τὰ παρακείμενα, καὶ μάλιστα τὰς σταφυλάς. The medium for the corruption in the *Geoponica* seems to be the odour that the quince exudes (δριμύτητι καὶ ὀσμῇ implies δριμείᾳ ὀσμῇ), which attaches to and spoils the flavour of nearby fruits.

A similar principle is expressed in connection with grapes by Columella: 12.44.8 *illud in totum maxime praecipimus, ne in eodem loco mala et uuae componantur neue in uicino unde odor malorum possit ad eas peruenire; nam huiusmodi halitus celeriter acina corrumpunt.* For a more general version of the connection between odour and exhalation such as is suggested by Columella (and is implicit in the *Geoponica*), cf. Theophr. *Od.* 1.3 τὸ γὰρ τῆς ὀσμῆς ἐν ἀναπνοῇ (= ἀποπνοῇ). The worry about the quince's odour is legitimate: '[q]uinces should always be stored alone and not with other fruit, which their strong aroma can taint' (*EFN* 640). For the causal asyndeton between the two sentences (marked by the colon), see 2.4.6n. For the brachyology of *austeritas*, 'pouvoir astringent', see Maire on *Med.* 4.1 (n. 7).

Diligentioribus rectissime uisum est praemonere ne: The structure of the sentence with the impersonal passive use of *uideo* is of the didactic type exemplified by Col. 3.12.6 *nobis in uniuersum praecipere optimum uisum est ut*, etc. It is better to take *rectissime* as a sentence adverb (i.e. an adverb that qualifies the entire sentence), expressing G.'s approval of the caution issued by the *diligentiores*, than as modifying *uisum est*: though it is in general common to attach an adverb in the positive, comparative or superlative degree to *uideor* in this construction, *recte* is not frequently so used, and *uideor* alone has the proper sense (*OLD* s.v. 23–4).

Diligentioribus: The adjective *diligens* (also 3.5.2, 4.3.5) and its adverb *diligenter* (4.1.4, 4.5.2) can have relatively strong directive force (as here), denoting as they do the application of careful, even painstaking, attention to a labour. Especially Columella but also Pliny use both the adjective and the adverb commonly in providing instructions (cf. *TLL* v 1.1181.48–50 (Gudeman)).

The comparative deserves some comment: why *diligentioribus* rather than *diligentibus*? Here, as elsewhere in *Pom.* (a version of the following remarks applies also to *Med.*), it can be difficult to determine the exact significance of the comparative form where there is no explicit contrast at hand: cf. 2.1.2 *calidioribus*; 2.5.2 *Crebrior*; 2.5.3 *pars acutior*; 2.6.2 *propius*; 2.7.3 *maiora*; 2.8.1 *molliore*; 3.7.2 *latior*; 3.7.4 *crebrius*; 3.7.4 *dulcior*; 3.7.6 *tenuiora*; 4.1.4 *diligentius*; 4.3.5 *diligentioribus*. Three explanations for such comparatives can be distinguished. (1) Where there is no hint of comparison with what precedes and no standard of comparison assumed, the comparative may be taken as equivalent in meaning to the positive form, as is evidently the case with 2.5.3 *pars acutior* = 3.1.4 *Acuta pars*. This is a well-attested phenomenon in later Latin and Latin of lower registers: see Svennung (1935) 279–80, Stefenelli (1962) 24, H–Sz §100.b, Cuzzolin (2011) 597, Adams (2016) 184–5, 286–7. (2) In other cases, an implicit standard for the comparative can be retrieved without great difficulty. At 4.1.4, for example, the claim that Columella writes *diligentius* very likely takes its point from the list of authors that immediately precedes. Slightly more ambiguous is the present *diligentioribus* (also at 4.3.5), which has no such standard in its vicinity; one could argue however that in both instances there is implied an unspecified group who would provide a standard ('the more diligent [of those who have written on the matter]' or 'the more diligent [cultivators]'; cf. Tarriño (2011) 397). Cf. also 3.7.4n. *dulcior sucus hauriatur.* (3) In still other cases, the comparative may have a purely intensive force ('elative' or 'absolute' comparative, H–Sz §96, Cuzzolin (2011) 557–8), translated by a variety of expressions such as 'rather *x*', 'quite *x*', 'too *x*', etc. Regarding categories (1) and (3) in particular, there is often no *prima facie* means of deciding whether a comparative has the intensifying function or merely stands for the positive; in such cases I translate as elative if the sense permits it, although the positive is possible.

ceteris pomis: I.e. the other kinds of pome fruits that are to be stored (so τὰ παρακείμενα of *Geop.* 10.28.4, quoted 1.1.15n.).

sibimet censent reponendum: *sibimet* has the most point if it can be interpreted as equivalent to *per se* (*sibi*), 'by itself' (*OLD* s.v. 6b). It is not impossible, however, that it should be construed rather with *censent*, for there are a number of examples in late Latin of the reflexive pronoun used with *uerba sentiendi et dicendi*; according to H–Sz §164 it is an extension of the ethical dative: see also von Morawski (1881) 4, Löfstedt (1911) 142, *Syntactica* II 392–3.

corrumpere solet, ut necesse est: We might expect either *solet corrumpere* ('tends to spoil') or *corrumpit, ut necesse est*, vel sim. ('spoils, as is inevitable'), but not both. The problem is that a claim of inevitability (*necesse est*) is logically stronger than one of habituation or frequency (*solet*), and it is not easy to see at first how the two qualifications can be reconciled. (Perhaps for this reason I find no comparable instance of *necesse* in the vicinity of *soleo*.) But probably what G. means is that the *austeritas* of the quince *inevitably* produces a situation in which nearby fruit are *often* corrupted. That is, the necessity applies to a corrupting property of *austeritas*, not to any particular instance of corruption.

de proximo: The adverbial phrase occurs twice in Plautus and then not again until Apuleius, from whose time it becomes more common (*TLL* x 2.2035.29–45 (Ramminger)).

1.1.16 Cydoneorum pondus adeo iumentis praeter cetera infestum est ut remedia quoque contra haec nullas uires habere compertum sit: The statement may seem like a *non sequitur*, but it is in fact a logical extension of the discussion of preserving the quince: the next concern for the professional cultivator will be the transportation of produce to the market (sc. on pack animals, *iumenta*), something which prompts G. to bring up this caution about the special problem presented by the quince. For the idea that beasts of burden balk at transporting fruit, cf. Plin. *HN* 23.116 *mala piraque* *iumentis portatu mire grauia sunt uel pauca. remedio aiunt esse, si prius edenda dentur aut utique ostendantur*; 24.2 *pomorum onera a iumentis statim sentiri, ac, nisi prius ostendantur iis, quamuis pauca portent, sudare ilico.* That G. assumes some belief such as that expressed by Pliny is shown by his denial – as a special case (*praeter cetera*) – that the ordinary *remedia* (presumably those mentioned by Pliny) are of any avail when it comes to quinces.

adeo . . . praeter cetera infestum: With *cetera* supply *poma* or *mala* from *cydoneorum. praeter* joined with *adeo* makes a compendious comparison (*OLD* s.v. *praeter* 2); so cf. also *Med.* 40.12 *annotatum in piris quod praeter alia poma corpus nutrire credantur.*

remedia quoque contra haec nullas uires habere compertum sit: For *uires* of the efficacy of a drug, see *OLD* s.v. 23b. For *comperio* denoting experiential discovery, cf. *Med.* 1.1 *raphano calidam inesse uirtutem omnium medicorum opinione compertum est*; 27.19 *alopecias tunsis cepis infricare inter efficacia remedia compertum est.* For *remedium* + *contra*, cf. *Med.* 1.6 *contra uenena unicum remedium est*, etc.

<II DE PERSICIS>: Fr. 6 Condorelli. Ch. II Mai. Ch. IV Scotti. Section numbers (1–12) follow Mazzini, not those of Scotti, who counts 11 sections, or Mai, who counts 13 (Introduction §IX).

2.1–7 G. begins with a thorough and expertly informed discussion of questions regarding the planting of the peach and cultivation of the young tree (Plate 7). He treats from several angles and with considerable nuance questions such as planting time (2.1–3), seed storage and preparation (2.2–3), climatic conditions and soil quality (2.3–4), establishment of the plants in the nursery (2.5), alternative methods of propagation (2.1, 2.5), nourishment and protection of newly sprouted plants (2.4–6), and transplantation and establishment of the trees in the orchard (2.5, 2.7). Although most of the information in 2.1–7 is concerned with growing the peach from seed, G. also briefly considers the possibility of planting with a shoot taken from a mature tree (2.5.7n. *Planta . . . quae de codice auellitur*); in 2.10–11 (n.) he will return to alternative methods of propagation via a discussion of grafting. Although he does not directly compare the efficacy of the different methods of propagating the peach, he at least entertains a greater variety of techniques than in the case of the almond, where only cultivation from seed is mentioned. For more detailed consideration of the status of different methods of propagation in antiquity and in recent history, see 2.10–11n., 3.1–4n.

2.1 For an overview of the contents of 2.1, see Introduction §III(a). The detail *et mense Ianuario* ('even/also in January') may suggest that additional planting times were offered in the preceding lines; it could however do no more than anticipate a contrast with the preference of the Quintilii. We might have hoped that Palladius' remarks on the peach could help us to estimate the contents of the lacuna, but unfortunately they cannot be mapped onto the present passage without difficulty. Palladius suggests that November is the ideal planting time in warm locations, with January serving for other climes: 12.7.1 *hoc mense* (sc. *Nouembri*) *locis calidis, ceteris uero Ianuario persici ossa in pastinatis areis sunt ponenda*, etc. This statement not only *prima facie* contradicts the recommendation of Curtius Iustus that G. transmits (i.e. that a January planting is for warmer climes), but also presents a greatly simplified picture in comparison with the range of planting times that G. considers (cf. 2.1.3 *autumnalis satio*; 2.2.1 *ab exortu Fauoni* (spring); 2.3.2 *In calido merito sub ipsum tempus autumni*). To judge only from G.'s thoroughness, Palladius' November planting time may have been mentioned somewhere in the lacuna, though it is not certain how he would have presented it. For a synopsis of the planting times for the peach mentioned by earlier agronomists, see 2.1.3n.

2.1.1 ossa: Peaches belong to the class of fruits known as drupes (stone fruit), which are distinguished from other kinds of fruit (e.g. pomes) by a hard endocarp that encloses the seed, i.e. by its possession of a 'pip', 'stone' or 'pit' (*ENF* ix; see Plate 7). Other drupes include olives, dates and pistachios, as well as all fruits from plants of the genus *Prunus* (to which the peach itself belongs), such as almond, cherry and apricot, among others. Although in *Pom.* the regular word for the stone is *os* (as here), the Roman agronomists in fact used a variety of terms for the stone of the drupe, and not always consistently. The diachronic and synchronic variation is hardly surprising, for the Romans obviously did not have the category of 'drupe' that would collect seemingly quite different fruits under one heading and impose on them a unifying description of their morphologies. This is not to say that they did not recognize structural affinities among fruits, only that the terminology was not made scientifically precise; the chief criterion of a word was its descriptive efficacy, and it did its job if it identified with a minimum of ambiguity the part it was meant to. Nothing in fact prevented a single term from referring to different parts of fruit in different contexts, provided that it engendered no confusion.

A case in point is the term *nuc(u)leus*, lit. 'little nut', which before Pliny's time is the preferred word for the stone of a drupe. In Cato, Varro and Columella, the only fruit whose stone or pit is regularly described is the olive. The word for it is *nucleus*: cf. Cato, *Agr.* 37.1, 37.2 *bis*, 66.1, Varro, *Rust.* 1.41.6, Col. 12.51.2, 12.52.6. That is not because the word *nucleus* itself means something like 'stone', but rather – as Varro (*ibid.*) shows when he stresses that the *nucleus* of an olive is just its *semen* (*oleae semen cum sit nucleus*) – because the putative olive seed is nut-like in bulk and shape. In other words, it is the physical appearance and not the function that makes the word *nucleus* suitable for the olive stone. For the same reason, the word was also used for other nut-shaped seeds, such as the edible ones of the stone pine (from Varro on, *OLD* s.v. 1b). By the time of Pliny and Scribonius Largus, *nucleus* was extended further to cover the stones of other drupes, including the cherry (23.141) and peach (Scrib. Larg. 184 *prosunt et nucleorum persicorum interiora ex uino trita*), as well as the pips and seeds of non-drupe fruits (e.g. Plin. *HN* 16.154 berries, 23.106 pomegranate). As in the earlier agronomists, the word continues to be first and foremost physically, not functionally, descriptive, used to name any part of a fruit that resembles a 'little nut'. Three examples, all from Pliny, will suffice to show the breadth of applications and versatility of the word. The first is *HN* 15.89, where Pliny describes the structure of an almond. He names three parts of the fruit: the outermost covering (*summum operimentum*), the shell (*putamen*) and finally the *nucleus*, which is the edible kernel of the nut itself. The key point is that whereas in other contexts *nucleus* may describe the hard pit or stone, here it is used rather to distinguish the seed from its hard shell (cf. Pallad. 2.15.13). A second example is *HN* 16.28, where the addition of the genitive *oliuae* suggests that *nucleus* itself was not sufficiently precise to refer to the stone of an olive, or any other fruit, without further context: (sc. *roboris*) *mora . . . quibus fructus inest nucleis oliuae similis*. The third is *HN* 17.59, where *nucleus* is used not to contrast a larger kind of seed (i.e. one like a 'little nut') with a smaller one, as is often the case, but rather to achieve the opposite. The *nuclei* here are the small seeds of grapes, apples and pears, and are distinguished from *fructus*, meaning the whole walnuts and chestnuts themselves: 17.59 *ac pleraque ex his natura ipsa docuit et in primis semen serere, cum decidens exceptumque terra uiuesceret. sed quaedam non aliter proueniunt, ut castaneae, iuglandes, caeduis dumtaxat exceptis; et semine autem, quamquam dissimili, ea quoque quae aliis modis seruntur, ut uites et mala atque pira. namque his pro semine nucleus non, ut supra dictis, fructus ipse.* The description of the seeds as *nuclei* may seem less apt, but the usage presents no difficulty, for Pliny's *ad hoc* distinction between *fructus* and *nucleus* fulfils his purpose of distinguishing different means of propagating from seed.

The word *os* (cf. Gk. ὀστέον), which in *Pom.* and Palladius entirely supplants *nucleus* in application to the stone of a drupe, is found first in Pliny (*HN* 23.98 palm), and then sporadically but widely thereafter (*TLL* IX 2.1098.80–1099.20 (Baer), *LAT* 89; see *REW* no. 6114 for Romance survivals).[8] In *Pom.* we find *os* only in connection with the peach, for this is the single stone fruit that G. discusses, but in Palladius, besides the peach (3.17.8, 12.7.1, 12.7.3, 12.7.8), it is applied also to the plum (3.25.32, 12.7.13), jujube (5.4.1), palm (11.12.1), cherry (11.12.7) and olive (12.17.1). The word *nucleus* has not lost its versatility

[8] If Vollmer (1918) 25 is right to defend the reading of **F** at line 95 of the ps.-Ovidian *Nux*, it will be the earliest use of *os* in the approximate sense: *lamina mollis adhuc tenet os in lacte quod intro est* ('die noch weiche Schale hält den Kern in dem Milchsafte, der drinnen ist'). But that reading is doubtful, and the true one is probably *lamina mollis adhuc, tenero est in lacte quod intra est* (adopting the punctuation of Courtney (1988) 276).

by G.'s time, however, and where it appears in *Pom.* it refers to the seed of the peach in contrast to the stone: 2.3.2 *cum ossibus exterius tempore resolutis totam germinis sani uitam nuculeus experitur.* The application is already familiar from Pliny (see the first example cited above), and is found also in Palladius: 12.7.3 *affirmantibus Graecis persicus scripta nascetur si ossa eius obruas et post septem dies, ubi patefieri coeperint, apertis his nucleos tollas et his cinnabari quod libebit inscribas.* The late Christian author Salonius offers a striking example of the same contrast in application to the almond (for which an *os* is not elsewhere distinguished): *in eccles.* p. 1009[D] *sicut nux amgydali constat ex corio, osse et nucleo, ita Christus ex carne, anima et diuinitate; quoniam per os amygdali diuinitas designatur.* Finally, for a loose application of *semen* to the stone of the peach (perhaps motivated by *uariatio*), cf. 2.4.2 *semina.*

2.1.2 Curtius Iustus et mense Ianuario calidioribus locis putat posse disponi; Quintilii armenia sola consentiunt Ianuario mense ponenda: The asyndeton is adversative: see 2.4.6n.

mense Ianuario . . . Ianuario mense: For the repetition, see the remarks on over-specification in the Introduction §v(b). Before the first century CE, the bare ablative is preferred for such temporal expressions; *in* + ablative is increasingly found in poetry and in later Latin more generally (Pinkster 837–40). In *Pom.*, however, there is a clear preference for the bare ablative: cf. also 3.2.2 *eodem mense*; 3.8.1 *omnibus annis*; 4.3.5 *tertio mense.*

calidioribus: *calidus* is eventually overtaken in spoken language by the syncopated *caldus* (Stefenelli (1962) 66–7), but the syncopated form is nowhere found in G. The lacuna makes it difficult to assess the exact force of this comparative: for the possibilities, see 1.1.15n. *Diligentioribus.*

disponi . . . ponenda: For *pono* ('to plant') and its compounds, see Introduction §v(d). Here, *ponenda* picks up *disponi* with no change in meaning: the sequence is an example of the iteration of a compound verb by a simplex, an old Indo-European phenonemon that long survived in both Greek and Latin. For the construction in prose, see Shipp (1944). Adams (1992*a*) develops Shipp's thought and adds additional examples; see further Adams (1995*a*) 635–7 for discussion of the phenomenon in the technical context of the *Mulomedicina.* On such iteration more generally, see Watkins (1966); and for iteration in poetry, Clausen (1955) 49–51, (1965) 97–8, Renehan (1977), Coleman on Mart. *Spect.* 26.9. Further examples in *Pom.* include 3.3.5 *describit . . . scribit*; 4.1.4 *conscribere . . . scribit.*

armenia: Although most often directing his remarks in *Pom.* to the *persica* (peach) without qualification, G. also mentions three specific varieties: the *armenium* (also 2.9.2, 2.10.7), the *praecoquum* (2.9.2, 2.10.4, 2.10.7) and the *duracinum* (2.10.7). Of these, only the *duracinum* is in truth a sort of peach; *armenium* (or *armeniacum*; on the different forms see *LTBL* 41, *NPRA* 25) and *praecoquum* (or *praecox*) refer to the apricot (*Prunus armeniaca* L.). The apricot is grouped variously with the peach (as here and at *Med.* 44.2, Pallad. 12.7.4 (possibly interpolated), Isid. *Etym.* 17.7.7), or with plums (Plin. *HN* 15.41): see *LTBL* 41, André on Plin. *HN* 15.40. Originally the *praecoquum* is distinguished from the *armeniacum* as a precocious variety of apricot (thus *Pom.* 2.9–10, and cf. *Med.* 44.2–4), but later it comes to name the species of fruit more generally (*LTBL* 260, André on Plin. *HN* 15.40); from here are the Romance survivals, including (via Arabic) It. *albicocco*, Fr. *abricot*: see *REW* no. 6712. On the apricot in general, see *RE* (1) 2.270.42–271.55 s.v. 'Aprikose' (Orth), de Angelis (1995) 119. For remarks on the *duracinum* variety, see 2.10.7n. *duracina*.

2.1.3 Graecis omnibus ferme autumnalis satio ratione naturali uidetur electa: Of the planting times for the peach that G. reports, this autumn planting (cf. also 2.3.2 *In calido merito sub ipsum tempus autumni ossa merguntur*) most closely agrees with the opinions of the extant earlier agronomists. Pliny and Columella thus both state that the peach should be planted 'during autumn before the commencement of winter': Col. 5.10.20 = *Arb.* 25.1 *siliquam Graecam, quam quidam ceration uocant, et* [*Arb.*: *item*] *Persicum ante brumam per autumnum serito*; Plin. *HN* 17.136 *communis quidem Italiae ratio tempora ad hunc modum distribuit . . . siliquae Graecae et Persicis ante brumam per autumnum.* These specific recommendations in fact agree with Columella's general advice that in warm climates trees and vines should be planted in autumn: 5.6.20 *melius autem locis apricis, ubi caeli status neque praegelidus neque nimium pluuius est, autumni tempore et arbores et uites post aequinoctium deponuntur.* One notes in passing the similarity in language and content of Pliny's and Columella's statements (peach joined with carob, the common phrase *ante brumam per autumnum*), which makes it likely that they are drawing on a common source; if it is not Mago, it could be one of the *Graeci* whom G. mentions (though this group might itself depend on Mago). Finally, it is interesting to observe that the autumn planting time recommended by the Greeks is attributed to them *mutatis mutandis* for the chestnut, where indeed a very similar explanation based on the observation of natural growth is adduced: 4.2.1–2nn. *Sed et quibusdam Graecis placet ab Arcturi* [*et*] *exortu serere castaneam. Nec immerito naturam secuti quibuscumque tempus hoc placuit: tunc enim seminis, id est nucis, iusta maturitas creditur adeo ut hiante tegumento, si legent, tum manus condat in terra, cum caduca labuntur.*

Graecis . . . satio ratione naturali uidetur electa, siquidem: For *uideo* with a participle, see *OLD* s.v. 20b. The construction deserves further comment: what does *uidetur* contribute? It may be suspected that it is an example of the phenomenon relatively common in later Latin, and often alleged to be Vulgar, of the use of certain verbs (e.g. *debeo*, *uolo*, *possum*, *uideor*, *inuenio*, etc.) as more or less colourless auxiliaries, whether for stylistic effect or to produce a periphrastic verb form for another reason: see the discussion and many examples at Löfstedt (1911) 207–11 (*uideor* on 210), Svennung (1935) 451–6 (*uideor* on 456), *Syntactica* II 181–2, H–Sz Stilistik §39.III. If we believe that *uideo* is weakened here, *uidetur electa* should be semantically equivalent to *electa est*. A better explanation, however, is to understand *uidetur* as motivated by the causal ablative phrase *ratione naturali*. The basic thought will therefore be something like *a Graecis autumnalis satio electa est*, but the addition of *uidetur* with the predicate *ratione naturali*, which anticipates the *siquidem*-clause, allows G. to speculate on the basis for the recommendation of an autumn planting time. The nuance is the difference between statement (1) 'an autumn planting time has been chosen by the Greeks' and statement (2) 'an autumn planting time seems to have been chosen by the Greeks for a natural reason'. The addition in Latin of *uidetur* of course influences the syntax of the sentence, demanding a dative *Graecis* (rather than the ablative *a Graecis*) and the past participle *electa* (rather than finite construction *electa est*).

An illuminating parallel to the usage of *uideri* and *ratio* here can be found at *Med.* 13.1 *natura nasturcio caustica est et tamen usum uenerium uidetur inhibere eadem fortasse ratione qua*, etc. As the addition of *fortasse* in particular makes clear, G. uses *uideri* to speculate on causes about which he is not certain. There are a handful of other instances of *uideri* in *Pom.* and *Med.* where weakening could be alleged (cf. *Pom.* 3.7.5 *uidetur . . . contingere*; *Med.* 28.8 (sc. *anethum*) *ideo uentrem uidetur inhibere*; 34.4 (sc. *semen napi*) *urinam uidetur inhibere nisi*, etc.), but it is unnecessary to insist in any of these cases that the verb is colourless, and I prefer to see them as similarly expressing G.'s tentative position on the relevant issues.

In general, some care must be exercised in determining whether a verb is truly weakened: cf. 2.1.5n. *coeperint . . . consulere*, 2.6.4n. *debeant germinare*. A more credible case of weakening is offered by *mereo* at 4.3.4 (*placere meruerunt*), where it seems to mean no more than 'happen' and may be used only *metri causa*: see Introduction §v(a).

omnibus ferme: The postposition of *ferme* is regular (*OLD* s.v. 2d), but the reverse order *fere omnes* is found at 4.6.3.

autumnalis satio: Among the agronomists, the adjective *autumnalis* is found most commonly attached to *aequinoctium* (as one might expect), but the present *iunctura* occurs at Col. 2.9.6, 2.12.7, 11.3.63; cf. similarly, e.g., Col. 4.14.2, 4.17.2 *autumnalis ablaqueatio*.

satio: The word (<*sero*) has two major applications among the agronomists in reference to sowing or planting (*OLD* s.v., Westerath (1938) 35–6). The first and most common is the act or process of planting itself (here, 2.3.4, 2.5.1, 4.1.5, 4.1.7, 4.1.9, 4.3.1). In some cases, a more precise reference to a 'method' of planting can be understood, as at 2.3.4 *Huic sationi potest prodesse quod generaliter Celsus agi scribit*, etc. The second application is a concrete extension to the crop itself that has been sown, well illustrated by Col. 11.3.63 *locis frigidis ubi timor est ne autumnalis satio hiemis gelicidiis peruratur*, etc. ('in cold places where there is fear that *what has been sown in autumn* will be consumed by winter frost, etc.'). At 4.1.5 (*quamquam ipse sationem caeduam assentiatur*) it is possible that *satio* is used in this concrete sense ('timber crop'), but the abstract meaning is better suited to the context ('a sowing [of trees] designated for felling'). At 4.7.4 *singulis* (sc. *mensuris*) *satio consistat* there is likewise some ambiguity (n.): *satio* could refer conceivably either to the method of planting (with a view to the depth of trenching, in the context) or to the growing chestnut itself.

siquidem . . . uideamus: The subjunctive is that of attributed reason (K–S II §182.3, Handford (1947) 152–4, *NLS* §240, §242.b, §285, Pinkster, 619–21, 646, 669), for G. is giving the explanation to the Greeks – which is not to say that he is casting doubt on the reality of the maturation and germination of the fruit described. For another such subjunctive of attributed reason, cf. 2.7.2 *causam propter quam . . . consenescat*.

cum pomis caducis ossa delapsa: The pairing of *pomis* and *ossa* with the near-synonymous *caducis* and *delapsa* may be for stylistic effect, but the adjectives are not strongly pleonastic: the adjective *caducus* brings an association, appropriate here, with the natural ripening of fruit (see following n.).

pomis caducis: *caducus* may mean either 'dropped' or 'fallen' (*OLD* s.v. 3), as here, or 'prone to drop' (*OLD* s.v. 6a), as at 4.2.2 (*caduca labuntur*). In both cases, when applied to fruit the adjective refers to a state of ripeness, often excessive, that either makes the fruit prone to detach or has caused it already to detach (i.e. by natural abscission consequent to maturation of the fruit; Plate 5): cf. Gell. *NA* 10.11.3 = Macrob. *Sat.* 6.8.8 *nam et in frugibus et in pomis matura dicuntur quae neque cruda et immitia sunt neque caduca et decocta, sed tempore suo adulta maturataque*. Cato thus uses the word in application to dropped olives, which must be harvested swiftly before they rot (*Agr.* 23.1, 58.1, 64.1; cf. Col. 12.52.22, Plin. *HN* 17.230); Varro (*Rust.* 2.2.12) recommends that the dropped ears of wheat (*caduca spica*) in the field be used as fodder for sheep, and Columella (7.9.4) adds dropped beans (*leguminum caducis frugibus*) to hog feed.

Palladius more often uses *caducus* to refer to the undesirable quality of a tree that loses its fruit prematurely (3.25.13, 3.25.15, 12.7.5, 12.7.14), for which see *OLD* s.v. 6b.

in nouas plantas . . . fruticare: *fruticare*, a normal word for the germination of plants, otherwise appears with *in* + accusative only twice, in both cases metaphorically (*TLL* VI 1.1447.32–6 (Bacherler)). The expression is not difficult, however, for *in* often indicates that into which something passes (*TLL* VII 1.760.46–761.29 (Bulhart-Hofmann), *OLD* s.v. 19a).

nouas plantas: The word *planta* is used by the agronomists in several specific senses. The earliest attested meaning is a shoot, slip or cutting for propagation (*TLL* X 1.2325.5–58 (Ottink), *OLD* s.v. 1a). That is the meaning at 2.5.7 *Planta . . . quae de codice auellitur* (n.); and, referring specifically to suckers (*uiuiradices*; see 4.4.2–3n.), at 4.4.2 *in plantis iuxta arborem sponte nascentibus*; 4.4.3 *eiusmodi planta*; 4.4.4 *hoc genere plantarum*; 4.4.4 *plantas . . . nucibus procreandis esse utiliores*. From the time of Columella the word is also used to refer to the entire plant conceived of as a growing entity (*TLL* X 1.2325.61 'significatur potius tota arbor, herba sim., quae progignitur'). Ottink subdivides this meaning into (1) growing plants conceived of as young or new, i.e. recently grown from a slip or seed (X 1.2325.60–2326.6); and (2) the plant specifically as a subject of transplantation (X 1.2326.6–2326.23). Besides the present instance (cf. *nouas*), another clear example in *Pom.* of (1) is 2.5.5 *planta procedit ex osse*; examples of (2) are 2.5.6 *plantas adultas . . . transferant*; 4.7.4 *plantas post biennium recessuras*.

saepe: The appearance of *saepe* here and at 2.10.4, 2.12.4, deserves note, for, as Adams (2016) 312 remarks, '*saepe* faded from use in non-literary imperial Latin', and was already in decline by the time of Petronius: see further Löfstedt (1911) 276–8, Lundström (1948) 119, Stefenelli (1962) 23–6, Adams (1976) 120, (2016) 372. The comparative *saepius*, often used merely for *saepe*, survived longer (Stefenelli (1962) 24, Adams (2016) 312); it is used once by G. (*Med.* 43.14).

culturae: Among the agronomists the word has both a narrow application in reference to the tillage of soil (*OLD* s.v. 1a) and a broad application in reference to the general care of plants, encompassing pruning, manuring, watering and other forms of maintenance (*OLD* s.v. 1b). All of the appearances of the word in *Pom.* (also 3.3.3, 3.5.1, 4.1.2, 4.1.8) are used in the broader sense.

adminiculis: The word *adminiculum* properly refers to a physical prop or support for plants (so used at 2.6.2, 3.2.1), but here it is applied metaphorically to a support or aid of any kind – perhaps punningly, since we are prepared for the concrete meaning. The metaphorical usage is well attested from the time of Plautus (*Mostell.* 129–30 *adminic[u]lum eis danunt / . . . aliquem cognatum suom*); the reference to a physical prop as such is not found in the agronomists before Columella and Pliny, in both of whom the usage is frequent (*TLL* I 727.68–80 (Vollmer)). Cicero is in fact the first to use the word in its strict sense: *Nat. D.* 2.120 *iam uero uites sic clauiculis adminicula tamquam manibus apprehendunt* (cf. *Sen.* 52–53). Columella emulates the line: 4.6.2 *dum clauiculis suis quasi quibusdam* *manibus adminicula comprehendat.*[9] For discussion of the actual props recommended by the ancient agronomists, see 2.6.2n.

ideoque: It is possible that *ideoque* = *ideo*, a common fossilization resulting from a weakening of *-que*: see Löfstedt (1920) 96–7, (1936) 36–46 (*ideoque* on 36–40), Svennung (1935) 489 n. 3, *TLL* VII 1.220.24–7 (O. Prinz), H–Sz §254.c (α *ideoque*), Adams (1976) 77–8. Löfstedt (1920) 97 remarks on the frequency of sentence-initial *ideoque* = *ideo* in particular, noting that it is a function of 'die ausserordentlich verbreitete Gewohnheit, einen neuen Satz durch *ideoque* anzuknüpfen'. Yet it is preferable to allow *-que* its full force here when we incorporate the evidence of *Med.*: of 17 occurrences in total there, *ideo* appears only twice with *-que*; bare *ideo* is used sentence-initially six times, *ideoque* only once; and there are no pleonastic expressions with *ideoque* that would suggest that the meaning of *-que* has been entirely bleached. For a more likely case of semantic weakening in *-que*, see 4.5.2n. *itemque*. Cf. also 2.1.4n. *hodieque*.

non ossa sed poma ipsa matura integra existimant obruenda: Palladius has a method for planting the sorb apple that is very similar and he also makes an appeal to first-hand experience in support of it: 2.15.1 *sorba seruntur egregie, ita ut matura in seminario ipsa poma pangantur. ego expertus sum multas arbores ex pomis sponte progenitas et in crescendo et in ferendo extitisse felices.*

poma ipsa matura integra: The adjectives *matura integra* are not in asyndeton bimembre (for which see 4.6.2n. *calidas . . . credit*), for *integra* is predicated: that is, the point is not that the 'ripe [and] whole' fruit are buried (*obruenda*), but that the 'ripe fruit' (*poma ipsa matura*) are 'buried whole' (*integra . . . obruenda*). (The cretic-trochee *īpsă mātūra* suggests a pause between *matura* and *integra*, helping to construe the adjectives appropriately.) *ipsa* is contrastive, as usual (cf. Meader and Wölfflin (1900) 388–9), and the entire phrase thus forcibly complements the preceding *non ossa*, emphasizing the importance to the Greeks' method of using the entire fruit rather than the stone.

obruenda: *obruo* has a common technical application among the agronomists with the meaning of 'sow' or 'plant' (from Varro on, esp. common in Columella; see *TLL* IX 2.152.76–153.2 (Lossau), *OLD* s.v. 2b): this is the sense here and at 2.5.2 (*ossa . . . obruta*). This usage derives from the basic meaning of the word, 'bury', as at 3.3.3 (*Stercora in terra obruunt*). A common application of *obruo* among the agronomists in its general meaning is to describe the process

[9] The similes *tamquam manibus* (Cic.) / *quasi quibusdam manibus* (Col.) may hint at an ancient etymology *ad* + *manus* (which is adopted by L&S s.v.), but if this was so, it was never explicitly remarked: the *TLL* reports no ancient etymologies, and *adminiculum* is absent from Maltby. Other derivations are now preferred: see de Vaan 24.

of sinking a fruit into a desiccant or other preserving agent: cf. Cato, *Agr.* 143.3 *uuas in uinaceis et in urceis in terra obrutas et nuces Praenestinas recentes in urceo in terra obrutas habeat*; Col. 12.44.4 *sicco flore gypsi obruunt*; 12.46.7 *creta figulari . . . mala obruerunt*; Pallad. 2.15.19 *nuces seruantur uel paleis obrutae uel harena uel foliis suis aridis*. The verb is so used by G. at 1.1.8 (*sicco gypso . . . obruunt*), 3.8.2 (*paleis obruuntur*).

et . . . praebeat et . . . acquirat: The emphatic repetition of *et* (K–S II §155, H–Sz §283.e) is something of a stylistic feature in *Pom.* In the present case, it emphasizes the benefits that the fruit confers upon the seed; the same force is evident at 3.3.4 (*et robore et fructu pulcriore*), where the doubled *et* stresses the advantages claimed by Mago and Diophanes for their planting method. G. employs repeated *et* for other specific didactic purposes: sometimes in order to emphasize that both elements of a conjunction of qualities are significant (3.4.1 *Et natura et mediocritas*; 4.3.2 *harena fluuialis, quae et pinguis et dulcis*) or must obtain (4.6.6 *montana . . . et umida et frigida*), and sometimes like *uel* to mark rather a disjunctive recommendation, both of whose members are equally satisfactory choices (1.1.13 *et scrobibus et in doleis*; 2.10.6 *Persicus et in semet ipsam inseritur et in amygdalum et prunum*).

ossibus uicem stercoris praebeat: For *praebeo* + *uicem* + genitive + dative, cf. Ov. *Ars am.* 3.666 *saepe uicem dominae praebuit illa* (sc. *ancilla*) *mihi*; Sil. *Pun.* 6.48 *ferrique uicem dens praebuit irae*; also Plin. *HN* 8.31, 16.161. There may be an echo of G.'s idea at *Geop.* 10.13.3 (also on the peach): εὐαυξῆ δὲ γίνεται τὰ φυτὰ ἐὰν ἅμα τῷ φαγεῖν σπείρωμεν τὸ ὀστοῦν, ἐγκαταλιπόντες τι τῆς σαρκὸς περὶ τὰ ὀστέα. The idea that the flesh will serve as manure is of course different from the old observation that a portion of the seed nourishes the sprout and its root: cf. Arist. *Gen. an.* 731a7–9 ἐκ τοῦ σπέρματος ἐκ μέρους γίγνεται τὸ φυόμενον, τὸ δὲ λοιπὸν τροφὴ γίγνεται τῷ βλαστῷ καὶ τῇ ῥίζῃ τῇ πρώτῃ.

stercoris: For manure in the agronomists, see Appendix II.

pomis futuris illibatum saporem . . . acquirat: For the participle *futuris*, see the remarks on over-specification in the Introduction §v(b). The adjective *illibatus* picks up and provides *uariatio* for the synonymous *integer*. In *Med.* G. is fond of the construction *acquiro* + accusative + dative to indicate the positive effects of a medicine: cf. 21.4 (sc. *porrum*) *fecunditati plurimum acquirere*; 27.6 (sc. *cepae*) *gratiam nobis boni coloris acquirant*; 57.2 (sc. *nucem*) *stomacho aliquid uigoris acquirere*.

2.1.4 Tenet quosdam . . . persuasio ut putent: The elaboration of *persuasio* with an explanatory *ut*-clause (i.e. consecutive noun clause) can be paralleled by examples from Quintilian (*Inst.* 8.2.21 *peruasit . . . multos ista persuasio, ut id . . . putent*, etc.), Pliny the Younger (*Paneg.* 63.6 *haec persuasio superbissimis* *dominis erat, ut sibi uiderentur principes esse desinere, si*, etc.) and Tacitus (*Dial.* 26.9); see further *TLL* x 1.1771.38–45 (Keulen). The construction is also found with other abstract nouns, such as *opinio*, for which several examples from the prose of Cicero and Caesar can be adduced: see *TLL* ix 2.718.55–68 (Teßmer), Adams (2016) 185. *puto* is strictly redundant in the *ut*-clause after *persuasio* (hence omitted in the English), but the *abundantia* is not irregular in Latin; as Adams (2016) 185 remarks on *opinio*, 'in explanatory *ut*-clauses . . . there is often a redundant verb of the semantic field "think", as is noted at *TLL* 718.55f. "epexegeticum q. d. (saepe per abundantiam cum verbis sentiendi . . .)"'. This kind of *abundantia* is well attested for *persuasio* too (and observed by *TLL* x 1.1771.39 (Teßmer)); see in particular the examples from Quintilian and Pliny quoted above.

The next difficulty raised by the present construction is the transmitted *putant* in the explanatory *ut*-clause, where the subjunctive *putent* might be expected. Mazzini 49 (and *app. crit. ad loc.*) defends the indicative as a Vulgarism, but Condorelli emends, as would Winterbottom. A similar instance of the indicative in a consecutive *ut*-clause is transmitted at 4.4.3 *ut demum uel illam sucus recreet* (*recreat* **N**). There can be no doubt that the indicative is at least possible in consecutive *ut*-clauses of the period; an example – but a very early one – is found already in a surveyor's inscription dated 153 CE from Algeria (*CIL* VIII 2728 = 18122 = *ILS* 5795): *apparuit fossuras a rigorem errasse, adeo ut . . . petit.*[10] The phenomenon is discussed generally at Löfstedt (1911) 254–5, *Syntactica* II 431–2, H–Sz §346.1.b, Kiss (1982) 72–3, Adams (2016) 298. As Adams notes, the usage, which anticipates Romance practice (Salvi (2011) 377), is encountered most often in late, low-register texts (an exception is the jurist Gaius). We note that G.'s normal practice is to use the subjunctive in both consecutive and explanatory *ut*-clauses:[11] 18x in *Pom.* (1.1.16, 2.2.2, 2.3.4, 2.4.2, 2.4.5, 3.1.2, 3.2.2, 3.3.4, 3.3.5, 3.4.3, 3.6.4, 4.1.4 *bis*, 4.2.2, 4.3.2, 4.4.1, 4.4.4, 4.7.1); 30x in *Med.* (2.11, 3.14, 3.16, 6.1, 6.4, 6.13, 13.5, 14.10, 15.8, 17.8, 17.12, 23.7, 27.6, 28.12, 29.4 *bis*, 30.30, 30.36, 40.5, 41.9, 41.14, 41.15, 42.1, 42.15, 42.19, 46.6, 52.4, 52.8, 53.18, 56.4).[12] Since there is no meaningful difference between the clauses presented in this register and those of 2.1.4 (the present instance) and 4.4.3, if the indicatives are to be defended they must be attributed to a lapse from a standard that G. elsewhere follows. Granted that, as Adams (2016)

[10] On the inscription generally, see Cuomo (2011), with text and translation 144–5; a more critical translation, and the most recent into English, can now be found at Adams (2016) 293–5, with linguistic commentary 295–306.

[11] Final *ut*-clauses are excluded from these counts, but anyway, G. uses only the subjunctive in them.

[12] At *Pom.* 4.1.4 (*nunciat . . . ne quid intersit*) and *Med.* 17.12, 30.36, 42.15, *ne* = *ut non*, for which see 4.1.4n. *nunciat . . . intersit.*

298 (cf. 306) remarks of the surveyor, 'technical writers, however correct (on the whole) and complex their Latin, were more prone to departures from standard syntax than were littérateurs and pedants', G.'s *usus scribendi* does not suggest that he was responsible for the slips (cf. Introduction §v(a), §v(d)), and the exchange of neither *putant* for *putent* nor *recreat* for *recreet* is at all challenging to accept among so many scribal misspellings in **N** (Mazzini 25–7; cf. Appendix III).

hodieque: *hodieque* appears again at 2.9.3 and *Med.* 23.3, but *hodie* appears nowhere in G.'s writings. Better than supposing a weakening in *-que* in these instances (i.e. *hodieque* = *hodie*; cf. 2.1.3n. *ideoque*, 4.5.2n. *itemque*) is to understand them simply as an example of the common substitution of *hodieque* for *hodie quoque* (i.e. *etiam nunc*): see Löfstedt (1936) 39, Blomgren (1944) 130, Lundström (1948) 132, H–Sz §254.b, *OLD* s.v. 2c, *TLL* VI 3.2853.4–29 (Ehlers).[13] This interpretation would give good sense here and at 2.9.3 – in both cases G. will be mentioning an 'old-fashioned' magical or superstitious belief – as well as at *Med.* 23.3 (*quod hodieque diacalaminthes a medicis praedicatur*), where Maire understands *hodieque* in the same way ('qu'encore aujourd'hui'). In any case, with Winterbottom 142 n. 3 we should reject the claim of Mazzini 53 that *et hodieque* (2.9.3) is a Vulgar pleonasm.

ossa quae dente contacta sunt aut prorsus extingui aut infeliciter nasci: G. warns that the touch of animals' teeth (3.5.2 *pecorum . . . dente*; 3.7.5 *ex morsu pecorum*) will cause an almond tree to produce bitter fruit: see further 3.5.2n. It may therefore be that the *dens* mentioned here is human, not animal, for in view of the situation with the almond it is difficult to see why else G. would dismiss the dangers for the peachstone so casually (so *frustra id credidisse*, etc.). For the stylistic qualities of the words *aut prorsus extingui aut infeliciter nasci*, see Introduction §v(a).

infeliciter: 'Unfruitfully' is the oldest meaning of the word: see *TLL* VII 1.1361.11–21 (Fleischer-Ehlers), L&S s.v. I, *OLD* s.v. 1; cf. also Verg. *Ecl.* 5.37 *infelix lolium et steriles nascuntur auenae.*

2.1.5 tunc . . . cum: An increasing preference for *tunc* against *tum*, the former originally a more emphatic word, is documented from the first century CE; along with this change were lost some distinctions in the usage and meaning of the words. For the historical development, see especially Svennung (1935) 407–18; also H–Sz §285.A.a Zusatz, Gaertner (2007) 215–16, Adams (2016) 178–9. In *Pom.* and *Med.* *tunc* is found five times (*Pom.* 2.3.3, 4.2.2, 4.3.2,

[13] For *-que* = *quoque* when attached to other words, cf. Prop. 2.33.35, where *tuque* = *tu quoque* (with Fedeli for additional examples).

Med. 43.15) and *tum* twice (*Pom.* 3.8.3, 4.2.2).[14] The significance of G.'s continued usage of both words is hard to assess. There is not enough information to conclude what distinction G. made between them, but some basic facts about their distribution in his writings are recorded here: *tum* does not appear before a vowel, *tunc* not before a guttural; here and at 2.3.3 *tunc* may be emphatic when paired with *cum* (thus note the addition of *denique* at 2.3.3[15]); both *tum* and *tunc* are used sentence-initially (3.8.3, 4.2.2, *Med.* 43.15); *tunc* is used progressively at least twice (4.3.2, *Med.* 43.15), on which see Svennung (1935) 413–16, and *tum* is probably so used at 3.8.3; the presence of *tunc* and *tum* in the same sentence at 4.2.2 is noteworthy, but ambiguous between explanations of *uariatio* and intentional word choice before vowel or consonant: *tunc enim seminis, id est nucis, iusta maturitas creditur adeo ut hiante tegumento, si legent, tum manus condat in terra, cum caduca labuntur.*

sentiunt: The verb draws attention to the appeal to sensory experience that G. offers in the *cum*-clause.

coeperint . . . consulere: *coepi* has been implicated in various claims of semantic weakening to the effect that *coepi* + infinitive is sometimes 'equivalent to the simple perfect' (Galdi (2016) 247; cf. Adams (2016) 127).[16] The traditional view is that weakened *coepi* + infinitive is a phenomenon of low-register texts which shows continuity from Plautus all the way through to a flowering in late Latin (*Syntactica* II 450–2, H–Sz §178.c.α). These claims have been subjected to revision: see concisely Adams (2016) 126–7; more fully Galdi (2016), also Rosén (2012*a*). It is not that one does not find instances of semantic bleaching with *coepi*, especially in later texts, only that neither the continuity of the phenomenon nor its association with low-register Latin stands up to closer inspection (summary remarks at Galdi (2016) 264). Moreover, as Adams (2016) 217 fundamentally cautions, 'automatic use of the term "periphrastic" [sc. in reference to *coepi* + infinitive] is to be avoided until objective criteria are established for identifying weakened uses'. Certainly, none of the instances of *coepi* in *Pom.* (cf. 2.3.3 *hiare coeperint*; 3.1.1 *hiascere coeperunt*) shows any signs of semantic bleaching: they all mark the inception of a continued or repeated action, as *coepi* often does (*FRH* III 302 (Briscoe), Adams (2016) 126; Rosén (2012*b*) 128, however, remarks on the 'indifference' of *coepi* 'to the contrast durative :

[14] *tum* must be deleted at 2.4.6 *contrarium* [*tum*] *crediderunt* (n.).

[15] The order 2.3.3 *denique tunc* is unusual, and may stress the correlation of *tunc* with *cum*. Moreover, one might have expected *tum* for *tunc* because of the formulaic nature of *tum demum/denique* constructions; is *tunc* used because *inseruire* follows?

[16] For general remarks on semantic weakening into auxiliary verbs, see 2.1.3n. *Graecis . . . siquidem.*

punctual'). For some nuances of *coepi*, see also 3.1.1n. *hiascere coeperunt*. For the tense of *coeperint* here and at 2.3.3, see 1.1.7n. *ubi tempus erit . . . pollicentur*.

experimenta: *experimentum* can refer to 'experiment' in the sense of a test or trial (*OLD* s.v. 1, *TLL* v 2.1654.45 (Meyer)), not necessarily scientific, or to 'experience', in the sense of knowledge or disposition gained through observation (*OLD* s.v. 2, *TLL* v 2.1657.37). G. uses *experimentum* in the latter sense at *Med.* 53.16 *uehemens hoc* (sc. *medicamentum ex amaris amygdalis*) *esse etiam domesticis in uxore seruata experimentis probaui*; probably the present instance should be understood in the same way (so also *TLL* v 2.1658.40–1 (Meyer)), though that is not to exclude the notion of testing. For the extent to which G.'s appeal to *experimenta* touches on the broader question of theory versus practice in *Pom.*, see Introduction §IV; and for biographical gleanings, §I(a).

2.2 This section shows the same digressive structure as 2.1 (cf. Introduction §III(a)). Although the guiding theme is still planting time, G. again turns to the discussion of an issue that seems off topic but is in fact prompted by the main subject. That is, the mention of a spring planting time (2.2.1 *ab exortu Fauoni*) implicitly raises the problem of seed preservation: how are the peach stones to be preserved unspoiled from the time of the fruit harvest (in autumn) until the commencement of spring? G. reports a solution of dehydrating and burying them (2.2.2), later echoed by Palladius: 12.7.1 *ossa uero quae ponenda sunt aliqui siccata prius paucis diebus cineris mixtione terra soluta in qualis reseruant. ego uero usque ad serendi tempus sine ulla cura saepe seruaui.* This method relies on the same basic principles that underlie the preservation of fruit for consumption, namely, the provision of a cool and stable environment: see 1.1n., and compare in particular the advice to bury the quinces 'in both holes and vats' (1.1.13 *et scrobibus et in doleis*).

2.2.1 Sunt qui ab exortu Fauoni persicum serere melius existiment: No other extant source on the peach mentions a spring planting time, though it is conditionally recommended for other fruits such as pomegranate (Pallad. 4.10.1), fig (Pallad. 4.10.23) and grape (Col. 3.14, *Arb.* 3.5, Pallad. 3.9).

ab exortu Fauoni: *Fauonius* is the gentle west wind (*Zephyrus*/Ζέφυρος) which commences blowing in the spring and promotes the growth of plants: see *TLL* VI 1.382.42–66 (Hofmann) for references to its life-bringing aspect (whence also a fanciful derivation from *foueo*). *exortus* is not commonly used to describe the initiation of winds, but there are clear precedents for G.'s expression: cf. Col. 2.17.7 *secundum Fauoni exortum*; Plin. *HN* 7.10 *ab ipso Aquilonis exortu*. The whole phrase *ab exortu Fauoni* is in fact simply a periphrasis for the beginning of springtime: cf. Col. 8.11.7 *cum Fauonii spirare coeperunt, id est tempus ab Idibus Februariis ante Martium mensem*; Plin. *HN* 18.222 *inter brumam et aequinoctium die XLV flatus Fauoni incohat uernum tempus* (with André).

persicum: For the name of the fruit, cf. Plin. *HN* 15.44 *in totum quidem persica peregrina etiam Asiae Graeciaeque esse ex nomine ipso apparet atque e Perside aduecta*; see also *LTBL* 244, *NPRA* 193. The peach was brought under cultivation in China over 4,000 years ago, and indeed spread to the West along trade routes through Persia (*EFN* 717).

tamen: *tamen* underwent considerable weakening in later Latin and came to acquire several functions apart from its original adversative use, which it clearly possesses at 4.1.5 *Gaius Plinius, quamquam ipse sationem caeduam assentiatur, hiemali tamen primum tribuit.* The present instance is a good example of its 'connecting' or 'progressive' function (≈ *autem*, δέ), whereby *tamen* annexes another idea to the sentence but does not offer a real contrast. Another important usage is 'restrictive' (≈ *et quidem, id est*; in conditionals with the force of *si quidem, si modo*), as at 4.6.6 *si tamen in campo necesse est ponere, tenuem locum, et hunc harenosum, existimant deligendum, aut iuxta uiciniam fluminis, si tamen limus ex eo stare non soleat.* See on both usages Löfstedt (1911) 27–33, Wahlén (1930) 65, H–Sz §267.a; for aspects of *si tamen* in particular, Schmalz (1909), H–Sz §371.e, Bertocchi and Maraldi (2003) 459–65. Maire (2002*a*) xl also notes the appearance of weakened *tamen* in *Med.*, but in general, one may remark that the word is not yet so attenuated in G.'s writings as it would be in later literature (consider thus the examples in works cited above).

anxie atque sollicite: The adverbs are obliquely directive, indicating the application of painstaking and meticulous care (as also of preserving at 4.3.1). If the paradosis *anxius* were to stand, it would be the sole instance in Latin of the comparative adverb formed from the adjective *anxius*. I emend to *anxie*, although the case is not clear-cut. There is nothing impossible or even especially unusual about G. using a comparative in place of a positive form (see 1.1.15n. *Diligentioribus*). Moreover, while the parallel 4.3.1 *anxie atque sollicite* carries no little weight, one should note that G. does not elsewhere scruple to vary a construction by substituting a comparative for a positive without any difference in meaning: cf. 2.5.3 *pars acutior* = 3.1.4 *Acuta pars.*[17] But despite whatever *uariatio* might be achieved through the use of *anxius* here, the juxtaposition with *sollicite* would be too harsh: *atque* is a strong connector in *Pom.* as well as *Med.*, and G. elsewhere maintains parallelism in elements joined by it (see Introduction §v(c)).

2.2.2 Et quidem: The instructions that follow obviously explain what it means to guard the stones 'with painstaking care' (*anxie atque sollicite*), but the transmitted *quidam* obfuscates this connection by making two groups of recommenders out of one. The easy emendation to *quidem* (*et quidem* also at *Pom.*

[17] For this substitution, see the remarks on *uariatio* in the Introduction §v(a).

2.11.3, *Med.* 50.2) provides a happy transition, attaching this method as an elaboration of the general statement *anxie atque sollicite ossa custodiunt.*

recentia . . . mergi patiuntur: For the meaning of *patior* here and at 3.6.3 *patiuntur inolescere* (cf. also *Med.* 28.2), see *TLL* x 1.728.37–55 (Kruse): 'dicuntur pati, qui auctores sunt alicuius rei, aliquid iubent, efficiunt, sim.'. *recens* means 'fresh' or 'green'; for the adjective in reference to plant parts, see L&S I, *OLD* 3c.

paucis diebus: The use of the ablative instead of the accusative in expressions of duration (cf. also 2.2.2 *biduo*; 2.3.4 *diebus quinque*; probably 2.3.5 *hiemalibus brumis*) is common from the Imperial period on, and indeed the ablative becomes the preferred case in later Latin: see generally Löfstedt (1911) 51–6, H–Sz §46.b (cf. §87.d), Adams (2016) 304. The bare accusative is found only once in *Pom.*, at 4.1.4 *nunciat Kal. Martias in Idus easdem ne quid intersit* (n.). Note that G. also avoids the bare accusative elsewhere by using the preposition *per*: 2.9.1 *per triduum*; 3.7.1 *per aliquos dies*. Adams (1995*a*) 446 notes Pelagonius' similar avoidance of the bare accusative by using *per* or the ablative. For the ablative of duration in *Med.*, see Maire (1999) 432.

mergi: The simplex *mergo* shows the same range of meanings in *Pom.* as *demergo* (see 1.1.9n. *demergunt*; 2.5.8n. *demergere*). Thus, it refers here to the submersion of peach stones in a desiccant (ash), but at 2.3.2 and 3.3.2 it refers to planting, for which see 2.3.2n. *ossa merguntur.*

quidquid . . . continent: G. uses the indicative with *quidquid* (also 1.1.15, 2.10.3, 3.1.5, 4.1.3), although the subjunctive would be possible through modal attraction, relatively common in classical Latin and increasingly so later: see H–Sz §302, and cf. §299.

uentilata: Here *uentilo* simply means to expose to the wind (or air), as often (*OLD* s.v. 1a, de Vaan 662).[18] The breeze (a cooling one, which is the force of *in umbra*) will continue to desiccate the stones that have been removed from the ash.

cum terra soluta: The soil should be 'loose' (both *solutus* and *resolutus* may be used for this; cf. Col. 2.10.4, 3.14.3, 4.33.1, 5.6.10, 10.140, 12.46.4, White (1970) 100–1) because it will serve better thus as an insulating blanket than it would if tightly packed.

[18] *uentilo* is also found with a technical meaning among the agronomists in reference to the wind-assisted winnowing of grain, whence the word *uentilabrum* (the winnowing shovel, also called *uallus*): clear descriptions can be found at Varro, *Rust.* 1.52.2, Col. 2.10.14; see also Plin. *HN* 18.322, Pallad. 1.18.3, White (1967) 32–5, *LAT* 176, White (1970) 185–7, Spurr (1986*b*) 78, Halstead (2014) 128–31.

in qualo uel in fictili uase: For similar *iuncturae*, cf. Plin. *HN* 17.97 *in qualis statim aut uasis fictilibus*; 17.98 *per uasa fictilia uel qualos* (both instances in connection with vessels for layering plants).

qualo: A rigid basket used for countless purposes on the farm, including picking fruit (see Plates 5, 6), straining honey from a comb, layering plants and so forth: see concisely White (1975) 59–61. Both *qualus* and *qualum* are found for the nominative, but there is no telling from the evidence in G. (2.5.6 *qualis . . . repletis*) which he favoured.

fictili uase: On the *uas*, see White (1975) 203: 'a generic term for a container of earthenware or metal, without differentiation of size or shape . . . [i]n more restricted applications the term is used by the technical writers to denote either the material of which the container is made (e.g. *uas plumbeum*) . . . or the class of equipment to which it belongs (e.g. *uas oleare*)'. As White (1975) 203–4 notes, 'by a useful and natural "shorthand" process some of the adjectives came to be used quite regularly without the substantive', and this is in fact true of *fictile* as well, which can stand alone for the earthenware jar itself: see *TLL* VI 1.647.80–648.12 (Vollmer), Hilgers (1969) 181. *fictile uas* (or *uasum*) appears three further times in *Med.* (54.2, 56.2, 57.6).

idque tectum atque oblitum: Winterbottom's compelling conjecture of *idque* for *adque* better establishes the relationship between this and the previous clause. Although the instructions to 'cover and seal' (*tectum atque oblitum*) the container before burying it should apply in principle both to *qualus*(*-m*) and *uas*, the language shows that G. has in mind primarily the *uas*, for *oblino* is the *uox propria* among the agronomists for stopping up or sealing a jar: see *TLL* IX 2.97.39–55 (Lumpe).

infodiunt: The verb is technical among the agronomists for setting or planting vegetation (*TLL* VII 1.1472.55–65 (Wuhrmann-Ehlers)), but can of course be used of anything that is buried in the earth (*TLL* VII 1.472.65–80): for *dolia*, cf. Plin. *HN* 14.33 *mitiores plagae doliis condunt* (sc. *uinum*) *infodiuntque terrae*; Paul. *Dig.* 19.1.54.1 *dolia . . . quae infossa essent.* Wuhrmann and Ehlers include the present instance under lemma 1.A.1.a.α (VII 1.1472.55, 'herbas sim.'), but that is somewhat misleading: it is the container that is being sunk, not the fruit, and only as a temporary expedient to preserve the desiccated nuts until the true planting time in the spring.

accessum neque sol neque uentus inueniat: For a reasonably close parallel, cf. Apic. 1.17 on storing grapes: *uas picari et gypsari facies, et in locum frigidum, ubi sol accessum non habet, reponi facies.* For *accessus* of winds, cf. Verg. *Aen.* 3.570 *portus ab accessu uentorum immotus*; and for the phrase *accessum inuenio* in particular, Sen. *Dial.* 11.8.1 *nullum illa* (sc. *tristitia*) *ad te inueniet accessum.*

A similar idea is expressed at *Med.* 49.16 *hanc* (sc. *aquam*) *diligenter liquatam et in uasa picata diffusam eiusmodi loco seruare solemne est ad quem . . . soli negetur accessus*.

neque sol neque uentus: *neque* appears only here (twice) in *Pom.* and four more times in *Med.* (40.7 *neque frigidi neque calidi*; 57.7 *neque umido neque nimium sicco*), against 10 occurrences of *nec* in *Pom.* and 14 in *Med.* The decline of *neque* in favour of *nec* in the history of Latin is well documented: see Svennung (1935) 489, *Syntactica* I 331–8, Löfstedt (1959) 178–9, H–Sz §241.B.ε, Väänänen (1981) 151, Pinkster 694. As is true for other connectors that see a decline in later Latin, such as *-que*, *atque/ac* and *-ue* (see Introduction §v(c), 2.9.3n. *truncis ramisue*), *neque* is slower to retreat in certain contexts, among them texts of higher register (see references above; on Tacitus add Eriksson (1934) 78–80, *Syntactica* II 286–8, Löfstedt (1948) 2), or those that show the influence of Greek (for Irenaeus see Lundström (1948) 125–7). It also persists in some formulaic patterns such as *neque . . . neque*, *neque enim* and *neque tamen* (H–Sz §241.B.ε, *Syntactica* I 333–6; for the jurists add Nelson (1981) 408–9): in G.'s writing the word appears thus only in its formulaic doublet.

2.3 G. extends and further qualifies the considerations of 2.1–2. (None of the information in this section is found in the extant earlier agronomists or reported by Palladius or the *Geoponica*.) The wordy proleptic statement *Interest autem*, etc., marks an impressive return to the question of the planting time for the peach after the digression on preserving the stones in 2.2. In warm climates – we might think of Italy here – the beginning of autumn is recommended, a timeframe earlier attributed to the Greeks (2.1.3n.). G. provides a reassurance that the young peach plants will not be damaged by the winter weather if planted in autumn; to this he adds a visual indication of the stones' readiness for planting (2.3.3) and a means of soaking to prepare them (2.3.4).

2.3.1 Interest autem cuiusmodi statum caeli condicio regionis, calidum frigidumne, sortita sit: The sentence is reminiscent of, though less elaborate than, the introductory statement of Col. 3.1.6 *sed et post haec non ignorabit dominus loci plus prodesse qualitatem caeli frigidam uel calidam, siccam uel roscidam, grandinosam uentosamque uel placidam, serenam uel nebulosam.*

cuiusmodi statum caeli: The phrase *status caeli* in reference to the climatic situation of a place is a favourite of Columella (32x): see further Svennung (1935) 608, *OLD* s.v. *caelum* 7, *TLL* III 93.25–38 and *passim* (Bannier). *caelum* is thus regularly adjoined to other words too, such as *condicio*, *positio*, *qualitas* and *situs*.

condicio regionis: *condicio* perhaps does not add much here, but the phrase *condicio regionis*, like *status caeli*, has a Columellan pedigree (8x).

sortita sit: For this use of *sortior*, see *OLD* s.v. 4b; cf. also Ov. *Met.* 2.241 *sortita loco distantes flumina ripas*; Apul. *De dog. Plat.* 1.7 *tesserarum figura, quae cum sit immobilis, terrae constantiam non absurde sortita est*; and esp. Pallad. 12.7.2 *locis quidem qualibusque proueniunt, sed et pomis et frondibus et durabilitate praecipua sunt, si caelum calidum, solum harenosum et umidum sortiantur.* Cf. also 2.7.3n. *eiusdem folia sortita.*

2.3.2 ossa merguntur: See *TLL* VIII 833.61–4 (Lausberg-Bulhart) on this technical usage of *mergo*: 'in serm. agricolarum surculos propagandos in terram i. q. deponere'. The same usage is found at 3.3.2 (*nuces mergunt*). (The appearances in *Pom.* update the definition of the *TLL*, however, for it is *ossa* and *nuces*, not *surculi*, which are planted.) *mergo* is not found in this sense before Columella, who seems in fact to offer a muted apology for the metaphorical extension of the word at 4.26.3 *palmae . . . uelut mersae cacuminibus in terram despiciant.* See also 2.2.2n. *mergi.*

metus enim non est gelum: immo si qua corrupta depereant, frigida relapsa hieme, cessabunt cum ossibus exterius tempore resolutis totam germinis sani uitam nuculeus experitur: These statements are difficult to construe and can sustain multiple interpretations (cf. Antonelli, col. 1678). The two most plausible are: (1) there is no danger of frost because, if any of the seeds begin to perish during the autumn, they will cease to do so (with *cessabunt* supply *deperire*) at the return of chill winter (*frigida relapsa hieme*), when they germinate [because of it]; (2) there is no danger of frost because, if any of the seeds begin to perish at the return of chill winter (*frigida relapsa hieme*), they will cease to do so when they germinate [in spite of it]. The first construal attaches *frigida relapsa hieme* most closely to what follows, the second to what precedes; both provide the corroborating statement that we would expect after *immo*, but in contradictory ways, for according to (1) the cold winter is a source of aid, according to (2) a threat.

A clue to deciding the question can be offered by establishing the context of the remarks in this section. G. has just said that 'it makes a difference' (2.3.1 *Interest*) whether the climate is warm or cool (2.3.1 *calidum frigidumne*). The following sentence (2.3.2), as shown by the words *in calido merito*, gives an example of that principle: when it is warm (*calido*), an autumn planting is chosen for the peach with good reason (cf. *merito*). If the principle stated in 2.3.1 is to have any significance, we must suppose that in a cool climate (i.e. *in frigido*) a different planting time would be appropriate. Thus, the focus of these statements is specifically the climatic conditions as they affect the planting of the peach, and the following *metus enim non est gelum* must explain (cf. *enim*) the prudence of the autumn recommendation precisely in terms of the 'warm climate' (*in calido*) that is the subject of discussion. Such an explanation is confirmed if we suppose that when G. gives the reason that there is 'no fear of frost' (*metus . . .*

non est gelum) he is adverting to the fact that *in calido* there is simply less of the frost that would argue against the planting time in a colder climate. In other words, G. would surely acknowledge that *gelum* is a danger that the grower must take into consideration when deciding when to plant the peach, but he would allow that where frost is not severe in the autumn and winter, i.e. *in calido*, the grower may safely plant at the beginning of autumn.

If these points are granted, our presumption should be that *immo si*, etc., will offer additional remarks confirming that, in a warm climate, the peach need not fear cold weather. But interpretation (1), according to which the cold winter weather aids the seeds, shifts the subject of discussion from the warm climate to a surprising property of the peach that would allow it to benefit from the chilly temperature.[19] Why would G. dilate on the rejuvenating effects of 'cold winter' (*frigida hieme*) when he has just said that the peach is well planted in autumn in a warm climate precisely because 'there is no fear of frost', implying, as we remarked, that *gelum* is a danger? Add that 2.3.5 also suggests harmful consequences from wintry weather (note thus *Vt asseruentur hiemalibus brumis*, etc.). Interpretation (2) avoids these difficulties. The words *si qua corrupta depereant* will refer not vaguely to rot sustained during the autumn season and repaired by the winter, as on interpretation (1), but to the damaging effects of cold and frosty weather.[20] When G. says that the seed that was planted before autumn will shrug off the injury and return to life when it finally germinates (*cum*, etc.), he may have in mind the remarks at 2.3.5, where he states (2.3.5n.) that the plant's ability to withstand winter is a result of the mild weather and moisture it enjoys in the (warm) autumn months after it is set out.

metus . . . non est gelum: For *metus* indicating the source of fear, i.e. danger ('i. q. periculum'), see *TLL* VIII 911.23–65 (Bulhart). G.'s whole expression poses a novelty of one kind or another: either *metus* is a nominative predicate after a nominative *gelum* (*OLD* s.v. *sum* 16a),[21] although the

[19] One might think at first of Pliny's remarks (*HN* 17.10–11) that exposure to cold benefits trees when budding, but he is speaking there of the grown plant and not the *ossa* or buds.

[20] For *corrumpo* of environmental damage, cf. Col. 3.2.8 (evaluating two types of vine): (sc. *uitis minor*) *longe . . . praecedit maiorem* (sc. *uitem*), *quia et imbres et uentos fortius patitur. nam maior celeriter in flore corrumpitur et magis in iugis quam in arboribus*, etc.

[21] *gelum* is one of three nominative forms of the word: there are also found *gelus* (masculine, used e.g. by Cato) and *gelu* (neuter, used e.g. by the author of the ps.-Ovidian *Nux*; the least common). For the ancient grammarians' doctrines and a synopsis of forms of the word, see *TLL* VI 2.1732.9–42 (Geissler).

word *metus* is not elsewhere so used (cf. *TLL* VIII 912.63–72 (Geissler)); or *gelum* is the genitive plural (considered by Geissler, *TLL* VI 2.1733.24), as though *gel(u)um* from *gelu* – but a plural form of the noun (in any declension) is otherwise not attested.

corrupta depereant: The point of the present tense *depereant* is that the plants 'are perishing' or 'begin to perish' as a result of being 'damaged' (*corrupta*); they are not killed outright. That is why they may still 'cease' to perish (*cessabunt* (sc. *deperire*)), that is, by germinating. For *depereo* of plant death, cf. Col. 3.5.3 *malleolorum pars maior deperit* (also 3.16.3); Plin. *HN* 17.11 *in totum poma depereunt*; Gell. *NA* 12.1.16 *saepe uideas arborem laetam et nitentem in locum alium transpositam deterioris terrae suco deperisse.*

ossibus exterius . . . resolutis totam germinis sani uitam nuculeus experitur: With *exterius*, *ossibus* stands by an easy synecdoche for the hard outer covering surrounding the seed (rather than the entire stone). For more detailed remarks on the flexibility of the words *os* and *nuc(u)leus*, see 2.1.1n. *ossa*.

germinis: Absent from Cato and Varro but an increasingly common word for 'sprout' or 'bud' in Columella (8x), Pliny (21x) and Palladius (25x): see *TLL* VI 2.1920.78–1921.30 (Schuster). It is a favourite of G. in this sense (8x), besides here found also at 2.4.2, 2.5.2, 2.6.1, 2.10.4, 2.10.8, 3.2.1, 3.2.2. Cf. also 3.6.4n. *germinare*.

nuculeus: The form is much less common than *nucleus* (*OLD* s.v.), appearing among the agronomists only elsewhere at Cato, *Agr.* 119, Varro, *Rust.* 1.41.6. In six appearances in *Med.* only forms in *nucl-* are transmitted in the MSS.

2.3.2–3 experitur. Mala: Condorelli xxxii observes what may be a space for a rubric here (see 3.6n.); of six such spaces in *Pom.* (three in *persic.*), only this one does not agree with the section division.

2.3.3 Mala denique tunc inseruire arboribus debent: For *inseruio*, see *TLL* VII 1.1882.62–72 (Hugenschmidt): 'laxius de statu apto fere rerum (fere i. q. adhiberi, usui esse)'. *debent* is not weakened (as may with somewhat more justification be suspected in other cases of *debeo*; cf. 3.6.4n. *debeant germinare*), but rather adds directive force: when G. says that the *mala* (fruit/stones; see following n.) are to be 'put to the use of trees', that is an oblique way of saying that they are ready, indeed ought now, to be planted.

Mala: The transmitted text (*Mali*) is difficult, because *mali* (feminine plural, from *malus*) should refer properly to the peach trees. Mazzini 50–1 takes *mali* to be G.'s own slip for *mala*, perhaps a colloquial one; Winterbottom 143 n. 2 points the reader to *TLL* VIII 211.33–43 (Richter) for *malus* = *malum*, but the examples there do not seem to enforce the equivalence.[22] It is better to emend with Scotti to *mala*, although even this choice does not wholly resolve the difficulties, for the subject of *inseruire . . . debent* and *hiare coeperint* must logically be the peach stones (*ossa*), not the fruit (*mala*). This is shown by several facts: (1) from 2.2.1 on, G. has maintained the assumption that it is the stones which are being handled, not the fruit, and it would be extremely confusing to alter the method (e.g. in order to follow the *Graeci* of 2.1.3) without a statement to that effect; (2) the stones are what in the previous sentence were said to be shed (*ossibus exterius . . . resolutis*) at the germination of the seed, and this is apparently the same process referred to by *extra hiare coeperint* in the present sentence; (3) the next sentence, which picks up (cf. *Huic sationi*) and advances this one, indicates that the *ossa* are the planting material under discussion. If *malum* is granted, it will be an unusual synecdoche for the stone; I can find no good parallels for the usage. Reeve's suggestion that *mali* originated as a gloss on and subsequently ousted a word such as *ossa* is an attractive one; perhaps even *Mali <ossa>* would be preferable to *Mala* alone.

denique tunc . . . cum: See 2.1.5n. *tunc . . . cum*.

cum extra hiare coeperint: Cf. 1.1.11 *si quid extra hiauerit*; this phrase is later varied by 3.1.1 *cum hiascere coeperunt* (n.). For the tense of *coeperint*, see Introduction §v(b).

2.3.4 Huic sationi: That is, an autumn planting. The meaning of *satio* here verges on a 'method of planting' (see 1.1.3n. *satio*).

quod generaliter Celsus agi scribit, ut . . . committantur: G. is fond of such noun clauses (on which see *NLS* §168, H–Sz §349, Pinkster 623–5), which allow him to introduce a measure of syntactic variety into his writing: in *Pom.*, cf. 2.4.2 *sic enim fieri ut . . . ament*; 3.6.4 *cum . . . tempus uidebitur ut debeant germinare*; 3.7.3 *utile est si effundatur*; 4.4.4 *fas est ut admittere in animum debeamus*; 4.6.3 *adiunxit quod . . . conseri posset*. The passive infinitive *agi* is probably obliquely imperatival, as though an imperatival indicative were put into accusative-and-infinitive syntax: '[ought] to be done'.

generaliter: Probably meaning not that Celsus recommends the advice for other fruits too, but that he does so for the peach without reference to any particular method or timeline, hence 'in general'. G.'s adaptation of the advice and his remarks on its presentation in Celsus are suggestive of his independent judgement.

[22] I do not know whether Winterbottom has in mind e.g. Verg. *G.* 2.70 *steriles platani malos gessere ualentes*, but *malus* there refers to the grafted apple branches (not the apples themselves).

priusquam . . . committantur: The subjunctive in forward-looking *antequam* and *priusquam* clauses is always regular where the time is either 'intended' or 'designated' on the one hand, and thus also where the action of the main clause must of necessity be completed before that of the temporal clause (*NLS* §227), and 'expected' or 'likely' on the other (Pinkster 638); cf. 3.6.4 = 3.71 *antequam floreant*, and also *Med.* 7.5, 41.17. We might expect the subjunctive anyway, however, for it is widely generalized in such clauses in the first century CE and later, even where the reference is to factual time: see Handford (1947) 161–4, Heberlein (2011) 326–7, Pinkster 640–1.

terrae committantur: This somewhat metaphorical use of *committo* in reference to planting (*OLD* s.v. 12b) is found most often in connection with *sulcus* (Verg. *G.* 1.223, 2.289, Col. 4.33.2, *Arb.* 4.2, 7.1), but also with *solum* (Col. 3.15.5) and *terra*/*harena* (Pallad. 1.6.16). Columella is fond of using the word in other contexts to refer to the environmental conditions to which one should or should not 'entrust' a plant: cf. e.g. 3.12.6 (sc. *uineae*) *melius Aquiloni uel Fauonio committentur.*

teneantur . . . innantia: For the quasi-auxiliary *teneo* with a participle, 'to cause to remain, keep, maintain (in a given condition)', see *OLD* s.v. 20.

aquae innantia: Most words dependent on *inno* are ambiguous between the dative and the ablative (*TLL* VII 1.1700.22–3 (Schmeck)), but it is found securely with the dative at Liv. 21.26.9, where it is also joined with *aqua* (*alueos . . . innare aquae*).

2.3.5 Vt asseruentur hiemalibus brumis, subsidium teporis et umoris habemus in ceterum autumni: G. corroborates the recommendation of an autumn planting by arguing that the mild temperature (*tepor*) and moisture (*umor*) of the season – we should still be speaking of a specifically warm climate here (2.3.2 *In calido merito*; cf. 2.3.2n. *metus . . . experitur*) – will aid the stones that have been planted; strengthened, they will be able to endure the freezing temperatures and storms of winter. The subject of *asseruentur* is *ossa*, but with reference to the sprouts growing from them. An illuminating comparison with G.'s argument is Columella's recommendation regarding the planting time for lupines: 2.10.2–3 *teporem tamen autumni desiderat, ut celeriter confirmetur, nam si non ante hiemem conualuit, frigoribus affligitur.* (Columella's *teporem* enforces Reeve's easy emendation of *temporis* (**N**) to *teporis*.) For G.'s *umor*, cf. also *Geop.* 10.2.2–3 stressing the superiority of an autumn planting time because of the moisture.

hiemalibus brumis: The ablative probably expresses duration, as it does elsewhere in *Pom.*: cf. 2.2.2n. *paucis diebus. bruma* refers properly to the winter solstice and appears commonly with this meaning in the agronomists (thus

2.10.4 *inter brumam et exortum Fauoni*). The extension to 'winter', 'wintry weather' or even simply 'cold' is largely poetic (especially in the uncommon plural): see *OLD* s.v. 2, *TLL* II 2208.23–2209.5 (Bannier). *hiemalibus* is not simply redundant, but stresses the seasonality of the dangers: the *brumae* which occur 'during the winter season', i.e. the *brumae* which come soon after the autumn planting.

subsidium teporis et umoris: The phrase belongs to a class of agricultural metaphors originating in a medical or military context – but it is not certain how 'metaphorical' these would really have felt, especially to someone writing as late as G. The possibility of such metaphors, which are very common in all of the agronomists, ultimately rests on the unmarked and nearly ubiquitous personification of plants by those authors: see esp. Svennung (1935) 73–87; cf. also Hardy and Totelin (2016) 108. Other examples of these metaphors in *Pom.* include 2.5.7 *largus umor et frequens pluuia ei subuenerit* (*OLD* s.v. *subuenio* 1a, 2b);[23] 2.6.1 *medicina continuae runcationis adhibenda est* (cf. Col. 2.13.3); 2.9.3 *solias in itinere derelictas pro remedio* (*prae medio* **N**) *truncis ramisue suspendunt*; 3.1.5 *umoris alimenta sumministrentur*. The frequent appearance of *sub*-prefixed words in these phrases is noteworthy. While in some cases *sub-* retains a strong spatial sense of '(from) underneath', and is thus attached to words that denote supporting or standing (cf. e.g. *subsidium* at Col. 4.6.1, *Arb.* 5.1), in others it seems only to mark the abstract idea of aid (as here).[24]

in ceterum autumni: For *ceterum* in its neuter singular substantive use, see *OLD* s.v. 3. It should be taken closely with the partitive *autumni*, with temporal *in* + accusative contributing the sense *per* or *intra*: see *TLL* VII 1.754.1–24 (Bulhart-Hofmann).

idem facere superuacuum est, cum iam [iam] gelidae hiemis aduentu frigus imberque ueniat: Probably a narrowly parenthetical comment on the words *subsidium . . . umoris habemus*, for *superuacuum* suggests that *idem facere* is something good for the plant. The idea would thus be that it is 'unnecessary' or 'superfluous' (*superuacuum*) to continue irrigating once winter comes with its freezing rain (*frigus imberque*). *idem facere*, quite vague even if one supposes *idem* is weakened (i.e. equivalent to *id*; but see the remarks on over-specification in the Introduction §v(b)), will thus refer to the supply or application of water (≈ *subsidium . . . umoris habere*). G. should not be implying that the cold of winter *per se* is a good (as shown by 2.3.5 *Vt asseruentur hiemalibus brumis*; cf. 2.3.2n. *metus . . . experitur*), only that, incidentally, there is water enough and more to make further irrigation pointless.

[23] *subuenio* used thus is frequent in *Med.* (11x): cf. *CG* s.v.

[24] Note the (unsurprising) frequency of words such as *succurro*, *subuenio* and *sumministro* in *Med.*, in reference to therapies: see *CG* s.vv.

cum iam [iam]: The transmitted *iam iam* (dittography) is precluded equally by meaning and by the preceding *cum*. Mazzini (who prints *iam*) reports in his apparatus that *iam* is the reading preferred by Scotti and Mai; in fact, they print *iam iam* (as does Condorelli), and it is only in Mai² that the emendation *iam* is encountered. Whether fortuitous or not – it seems to belong to the category of errors introduced into the text of the German editions of Mai (see Introduction §VIII)[25] – , it is an improvement on the transmitted text. For other instances of syllabic dittography in *Pom.*, see 2.4.6n. *contrarium* [*tum*] *crediderunt.*

gelidae hiemis: Varies 2.3.2 *frigida . . . hieme.*

hiemis aduentu . . . ueniat: For *aduentus* thus of the beginning of a season, see *OLD* s.v. 6b. *aduentus* may influence the choice of the verb *uenio* by a kind of virtual iteration of compound by simplex (ordinarily between two verbs): see the remarks at 2.1.2n. *disponi . . . ponenda.*

2.4 G. supplies still further details on suitable conditions for growing the peach; a special focus of the information is on soil quality and access to water, whether naturally or through irrigation.[26] There are no precedents in the extant earlier agronomists for the recommendations contained in the passage, though they are echoed by Palladius (see below). The suggested environmental conditions vary greatly: Celsus believes that the peach can flourish in any climate and even in poor soil; Curtius Iustus prefers a cold and rainy environment, albeit taking special precautions to avoid damage to the plants (2.4.2 *doleo aliquo,* etc.); the Quintilii like warm weather and moist soil (2.4.3); the Greeks prescribe a controlled irrigation system with coarse soil that is not rich (2.4.6). As often in *Pom.*, there are some details whose significance is difficult to assess. Consider the statement that a moderate climate is necessary to avoid the excesses of drought or overwatering (2.4.4 *Demissa temperies necessaria,* etc.). It is not clear whether this should be attached to the Quintilii, to G. himself, or to the Greeks; as when this difficulty presents itself elsewhere in *Pom.*, however, it is probably best to construe the lack of qualification or attribution as a mark of the fact that G. simply regards it as uncontroversially true (cf. Introduction §III(b)). The statement is consonant at least with the advice of the *Graecorum celeberrimi auctores* (2.4.5), who wish that the water be moderated by hand so that it can be led into or away from the orchard as required. It

[25] This spot may have escaped Condorelli's notice, for he neither includes it in the list of errors common to the German editions (xxxviii–xxxix) nor in his apparatus.

[26] The concern about a water source for the peach is later reflected at *Geop.* 10.13.1 τὰ δωρακινὰ χαίρει ὑδρηλοῖς χωρίοις. εἰ δὲ μή, συνεχῶς ποτιζομένοις. οὕτω γὰρ μείζων γίνεται ὁ καρπός.

may be that G.'s introduction of their advice after the remark is intended to corroborate it (by 'intensive' didactic layering; see Introduction §III(b)), as well as provide a solution to the problem.

As noted above, the remarks at Pallad. 12.7.2 seem to rely on this section: *locis quidem qualibuscumque proueniunt, sed et pomis et frondibus et durabilitate praecipua sunt, si caelum calidum, solum harenosum et umidum sortiantur; frigidis uero et maxime uentosis, nisi obiectu aliquo defendantur, intereunt.* We cannot know how far Palladius is drawing on other sources too, but it is at least possible to construe his statements as an original synthesis and simplification of the varied recommendations reported by G.: Celsus' *Quolibet caelo . . . poni* becomes *locis . . . qualibuscumque proueniunt*, to which is joined the Quintilii's *calidum aerem et umidum solum* (= Pallad. *caelum calidum, solum . . . umidum*) along with a detail from the Greeks (*calculosum et harenae et sabulo proximum* = Pallad. *harenosum*). Finally, the recommendation of Curtius Iustus to plant in cold and watery places with a *dolium* for protection is transformed into a vague cautionary statement: *frigidis uero et maxime uentosis, nisi obiectu aliquo defendantur* [cf. G.'s *doleo aliquo supra posito quod ab Aquilone defendant*], *intereunt.*

2.4.1 macro: Like some other words that refer to soil quality, *macer* is ambiguous and may indicate '*either* physical quality *or* degree of natural fertility' (White (1970) 105). Synonyms for it among the agronomists include *ieiunus*, *gracilis* and *exilis*: see White (1970) 99–100. *macer* here is obviously a negative quality (cf. Col. 2.2.7, Pallad. 1.5.6), hence *quamuis*, but its particular valence cannot be teased out: perhaps only the poverty or infertility of the soil is intended, but if physical thinness is meant, this could be undesirable too. Cf. also 4.3.2n. *pinguis . . . dulcis*.

For the Roman agronomists' knowledge of and terminology for different soil types, see the comprehensive study at White (1970) 86–109, and especially the tables at 97–102, 108–9. Although White's discussion is oriented towards cereals and legumes, his careful attention to the descriptive words for soils from Cato to Columella, where there is much terminological variation, is invaluable. White makes the salient point that while Roman understanding of soil types remained firmly 'pre-scientific' – there is nowhere an attempt to systematically investigate the texture or structure of soils – it was nevertheless sound in many of its general precepts. De Angelis (1995) 25–30 collects the agronomists' recommendations regarding the qualities of soil thought to be especially advantageous for fruit trees.

persicorum arborem . . . laetissimas: When the name of the fruit tree is joined to *arbor* as a specifying genitive, the singular is regular (thus 2.7.1 *arbor persici*); where the plural is found, it almost always accompanies the plural *arbores* (e.g. Col. 11.2.59 *arbores ficorum*, Hyg. *Poet. astr.* 2.40.1 *arbores complures ficorum immaturas*). Examples of the singular *arbor* with the genitive plural are extremely rare, and I find only two others: Serv. *ad* Verg. *Georg.* 4.145 *prunorum arbor spinus uocatur genere masculino*; Hier. *in. Zach.* 3.12. l. 374 *dicamus, et iuxta lxx* ῥοῶν *appellatur, non una arbor malorum granatorum, id est mali punici, sed locus his arboribus consitus*. A reading *arbores* (Reeve) is attractive, perhaps especially so in view of the following *laetissimas*, but the number of the latter word is not by itself decisive, for variation across sentences between the (collective) singular and plural is not uncommon in *Pom.*: cf. 2.4.3–4 *persico . . . eis* (unless *est* should be read); 3.4.3 *amygdalas . . . eius*; 4.3.1–2 *nux . . . translatae eriguntur*. For the common use of the collective singular with plant names, see *Syntactica* I 14.

poni . . . existimat: See 1.1.10n. *illiniri existimant.*

idem: For *idem* here and at 2.4.6, see the remarks on over-specification in the Introduction §v(b).

laetissimas: The word *laetus* is the *uox propria* for describing fertility, whether of soil or plants: see *TLL* VII 2.883.79–68 (v. Kamptz). With *laetissimas* supply *arbores* or *persicos* from *persicorum arborem*; for the variation in number, see 2.4.1n. *persicorum arborem . . . laetissimas*.

si remotione loci blandimentum ceperunt: The phrase is obscure because *remotio loci* is not well understood; commentators have attempted to ascribe various meanings to it in order to make sense of the statement. Mazzini translates 'quando hanno ricevuto la cura del trapianto', but does not explain how *remotio loci* comes to mean 'transplantation' (a sense found neither for *remotio* nor for *remoueo*). Antonelli translates 'qualora abbiano avuto l'allettamento d' un ben divelto terreno'. In support he quotes two examples of *moueo* where it refers to ploughing or trenching: Lucan. 7.861 *nec terram quisquam mouisset arator*, Col. 11.2.95 *his diebus qui religiosius rem rusticam colunt, nisi si uinearum causa pastines, negant debere terram ferro commoueri*. (To these add the references in de Angelis (1997*b*) 229 n. 58.) He argues that the *re-* prefix adds the idea of repetition (indeed a possible function of *re-*; see Schrickx (2015) 271–3), giving the meaning 'trench repeatedly', and ultimately concludes that *remotio loci* = *repastinatio loci*. The argument is ingenious, but does not fully convince, for *re-* does not generally add the sense of mere repetition to compounds from *moueo*; what is more, *loci* would be very vague in the statement, and *ager*, *terra*, *solum*, vel sim., would be desirable to make *remotio* more concrete.

Until it can be shown that *remotio* may bear a different meaning, we should hesitate to assign an extravagant meaning to the phrase. There is a way to give a sense to the text and allow *remotio* its ordinary meaning (a 'removing' or 'removal') if *remotio loci* may be construed as roughly equivalent to *remoto loco*; the latter phrase would be easier to understand (L&S s.v. *remoueo* II.A, *OLD* s.v. *remotus* 1a), referring to a 'remote location' or 'a location that is removed'. By *remotio loci* G. would have in mind not, however, the physical distance from the farmstead, but rather the fact that the spot is 'removed' from harmful environmental conditions, such as excessive wind, rain or sun: cf. Pallad. 1.37.1 *apibus stationem non longe a domnicalibus uel in horti parte secreta et aprica et a uentis remota . . . locare debemus*; 6.9.2 *sit autem* (sc. *caseus*) *loco clauso et a uentis remoto.* The word *blandimentum*, 'pleasantness', 'mild' or 'pleasant relief' (*OLD* s.v. 2b, *TLL* II 2030.35–40 (Münscher)), helps to guide the interpretation. Finally, if we adopt this reading of *remotio loci*, we may give an adversative (not progressive) force to the preceding *idem tamen*: although Celsus says that they may be planted anywhere, yet he believes (*idem tamen*) that they are most fruitful when the particular spot of their planting is sheltered.

2.4.2 irriguis et frigidis locis mallet †aquam† et, adiungit, doleo aliquo supra posito quod ab Aquilone defendat: The words *irriguis et frigidis locis mallet aquam* give the strange meaning 'he would prefer water in well-watered and cold places'. Scotti would emend *aquam* to *aptari* or *apponi* (sc. *persicos*); other editors (following Antonelli, col. 1680) understand a form of *ponere* with *mallet* and take *aquam* with *et adiungit*: 'he would prefer to plant them in well-watered and cold places; he also adds water, with a vat placed over them for protection against the north wind'. *aquam* seems no less otiose on this interpretation, however. I am not confident in emending *aquam* with Scotti; I obelize the word and as a stopgap opt to understand *poni* with *mallet.* When not construed with *aquam*, the verb *adiungit* will mean 'he adds' in the sense of 'stipulates' or 'mentions in addition' (*OLD* s.v. 9); *et* thus coordinates not *mallet* and *adiungit*, but *irriguis et frigidis locis* and *doleo aliquo*, etc. (This construal and punctuation are due to Reeve.)

irriguis: *irriguus* refers to irrigated land (*TLL* VII 2.420.75–421.23 (Kortekaas), *OLD* s.v. 1a), whether watered by a natural source such as a river or stream, or by artificial means (so the 2.4.5 *irrigua . . . loca* of the Greeks). The insistence on irrigation of the peach by Curtius Iustus and the *Graeci* was not the norm in Greek and Roman arboriculture: for the general question of the irrigation of fruit trees, and some methods used for that purpose, see 3.3.2n. *quo . . . immoretur.*

quod ab Aquilone defendat: *Aquilo* is the storm-bringing northwind (*NP* s.v. 'Boreas' (Hünemörder)) and hence poses a threat to the young plants. When the word appears in 3.1.4 (*bis*) it simply picks out the compass point north (*OLD* s.v. 3).

sic enim fieri ut semina pecorum stercus ita ament ut alia germina: The statement follows from *irriguis et frigidis locis mallet*, not the parenthetical phrase *doleo aliquo supra posito quod ab Aquilone defendat.* Its relevance to what precedes can be established by considering the Roman agronomists' theory of manure: (1) Manures warm the earth. (2) In order to achieve the same warming effects, cold climates will require more manuring than warm ones, in which there is to the contrary an acute danger of making the soil too 'hot' and thus burning the plants (see the remarks at 2.5.5n. *Multi et fimo ossa circumdant*; 3.5.3n. *cum . . . concrematae*). (3) Watery climates can handle more manuring than arid ones, for in an arid climate the manure becomes excessively concentrated and again poses a risk of burning the plants (White (1970) 131). Here are the principal passages supporting these statements:

- Col. 2.15.2–3 *nec ignorare colonos oportere reor, sicuti refrigescere agrum qui non stercoretur, ita peruri si nimium stercoretur, magisque conducere agricolae frequenter id potius quam immodice facere. nec dubium quin aquosus ager maiorem eius copiam, siccus minorem desideret, alter quod assiduis umoribus rigens hoc adhibito regelatur, alter quod per se tepens siccitatibus hoc assumpto largioribus torretur; propter quod nec deesse ei talem materiem nec superesse oportet.*
- Plin. *HN* 18.194 *ager si non fuerit stercoratus, alget; si nimium stercoratus est, aduritur, satiusque est id saepe quam supra modum facere. quo calidius solum est, eo minus addi stercoris ratio est.*
- Pallad. 10.1.3 *nec prodest nimium stercorare uno tempore sed frequenter et modice. ager aquosus plus stercoris, siccus uero minus requirit.*

It seems clear therefore that the reason why the peach seeds become 'as fond of the manure of livestock as are other buds' in cold (*frigidus*) and watery (*irriguus*) conditions is that the temperature and moisture make it possible to apply manure without fear of burning the buds, as might happen in dry and arid conditions.[27] If *enim* of the manuscript is correct (there is no reason to believe it is not; cf. 2.9.2 *sic enim*) and it has true explanatory value for the preceding statement, then Curtius Iustus is recommending a cold and watery climate for no other reason than that he believes (the indirect discourse of *fieri* imputes the explanation to him; compare the same construction at *Med.* 47.3 and, more extended, *Med.* 40.5–8) that in this way (*sic*) the peach sprouts will be able to benefit from manuring.

pecorum stercus: See Appendix II.

2.4.3 aerem: Besides meaning 'air', *aer* can refer more broadly to the climate of a spot (*OLD* s.v. 5), like *caelum*, for which it is probably *uariatio* here.

2.4.4 Demissa temperies: *temperies* can refer generally to the climate (*OLD* s.v. 2) and thus provides stylistic variation for *caelum* and *aer*, but it is especially appropriate in the present instance because it can bear the more specific meaning of a 'moderate' temperature or climate (L&S s.v. I, *OLD* s.v. 2b), a sense reinforced by *demissa* ('i. q. modicus', *TLL* v 1.494.15–16 (Kieckers)).

[27] Scotti adduces in explanation Pallad. 12.7.4 *nunc etiam contra pruinas stercus ingeratur persico uel faeces uini cum aqua permixtae uel, quod magis prodest, aqua in qua faba decocta est.*

eis: A referent in the plural (*persicis*, *arboribus persicorum*, vel sim.) may be supplied from *persico*: so cf. 2.4.1 *persicorum arborem . . . laetissimas* (n.).

ne siccitas anni ruat et admodum exigua poma constringat ac frequens umor enormi uastitate et inani sapore distendat: For the stylistic qualities of this line, see Introduction §v(a). *admodum exigua poma* is proleptic, stating a result of the stunting (cf. *constringat*) of the fruit, just as *enormi uastitate et inani sapore* presents a consequence of their distension. (The ablatives *enormi uastitate* and *inani sapore* must be construed *ad sensum*, for it is not the size or flavour with which the fruit are filled, but rather the water.)

siccitas anni ruat: *annus* is the year 'as subject to seasonal changes, esp. in the weather' (*OLD* s.v. 7); cf. e.g. Sen. *QNat.* 4b.13.3 *contra anni feruorem*. With the verb *ruo*, which in application to natural phenomena often denotes a forcible or violent onset (*OLD* s.v. 3b; cf. 6b), *siccitas* will refer to a drought rather than a general dearth of moisture.

enormi uastitate: With *uastitas*, *enormis* refers to excessive size, whereas at 3.5.1 (*enormiter natos*) the idea is irregularity of growth: see 3.5.1n. *ramos . . . praefractos*. For the degree of size conveyed by *uastitas*, see 3.2.1n. *uastior palus*.

2.4.5 Graecorum celeberrimi auctores: Apparently a more selective group than 2.1.3 *Graecis omnibus ferme*, but it is impossible to know their identity.

hortensia huic arbori loca destinanda: The MS reading is *hortensiadhuic*, etc. Although it is palaeographically easier to restore an *a* than to delete *d*, the resulting *ad huic arbori* (a slip for *ad hanc arborem*) is unlikely. In the first place, it should be noted that when G. uses *destino* elsewhere (3.3.1) he does so with the bare dative (a fact stressed by Önnerfors (1993) 264 n. 68; cf. also 3.3.1n. *omnem . . . destinatum*). But even setting aside the evidentiary value of that occurrence – might G. simply have varied the construction? – , there is little to warrant printing *ad* + dative. It would be far the earliest attested instance of a rare construction that is usually explained as a conflation of newer *ad* + accusative constructions and older dative constructions following verbs of giving; but the explanatory basis for those occurrences (e.g. Chiron. 462, Diosc. 3 PA p. 421.21–2, *Vita sanctae Euphrosynae* 9.16) is lacking in G.'s case (the regularity of *destino* + *ad* + accusative notwithstanding): see Svennung (1935) 337, 469–70, 576–7, Adams and de Melo (2016), esp. 126–8. Önnerfors (1993) 264 n. 68 argues forcibly to the same conclusion.

hortensia: Either from *hortensis* or *hortensius* (the latter found only in Pliny). *hortensis* is used by Columella to refer to a mound located in a garden plot (9.4.4 *at in hortensi lira consita nitent candida lilia*); otherwise, *hortensis* and *hortensius* refer almost exclusively to plants which are grown in a garden (*TLL* VI 3.3006.20–30, 3006.34–46 (Ehlers), *OLD* s.v. *hortensius*). G.'s example is not registered by L&S, *OLD* or *TLL*, but it clearly means 'of' or 'for a garden'. What is meant is that the same set-up that is desirable for a garden plot is desirable for the peach; the context (collocation with *irrigua* as well as the *ubi*-clause) shows that the set-up is conceived of primarily in terms of irrigation. Pleasure and vegetable gardens were indeed associated by Roman authors with intensive watering (note the frequent collocation of *hortus* and (*ir*)*riguus*, vel sim.: *TLL* VI 3.3015.74–6, 3016.6, 3016.13–15, *passim* (Ehlers)), and were indeed more likely to be irrigated than other parts of the farm: cf. Plin. *HN* 19.60 *hortos uillae iungendos non est dubium riguosque maxime habendos, si contingat, praefluo amne, si minus, e puteo rota organisue pneumaticis uel tollenonum haustu rigatos*; also Oleson (2000*a*) *passim* (partly quoted 3.3.2n. *quo . . . immoretur*).

destinanda senserunt: *consentio* is one of G.'s favoured ways to report the consensus of authorities (1.1.13, 2.1.2, 4.6.5), but it is not necessary to emend here. *sentio* means 'hold or express a given belief or opinion' (*OLD* s.v. 6; cf. L&S s.v. III) and is found in the same meaning (again paired with a gerundive) at 4.6.3 *anteponenda senserunt*; cf. also *Med.* 6.1, 42.1, 51.4.

ubi umor manu temperatus nocens ita auerti potest ut et possit induci: Two small and plausible emendations are required to make sense of the transmitted *ubi umor manu temperatus nocens ita uerti potest et ut possit induci*. The first is Scotti's restoration of *a-* before *uerti*, which could easily have been lost by haplography after *ita*. This provides the required sense of 'turning away' or 'diverting' water, sc. that which poses a threat to the trees (*nocens*); it is moreover an appropriate contrast to *induci*, which refers to water's being channelled in to the plants. The second is Winterbottom's transposition of *et* and *ut*. Perhaps one could accept *et* as grammatically superfluous in the phrase *et ut* (i.e. 'apodotic' *et*; see 3.1.4n. *spectet . . . ipse*), but the more significant difficulty would remain that *et ut* does not correctly relate *possit induci* to the preceding (*a*)*uerti potest*. Suppose, for example, that the text were to be read without emendation: *ita uerti potest et ut possit induci*. We might translate this statement, 'l'acqua, regolata dalla mano dell'uomo, se dannosa, può essere giudata così da poter essere condotta sulla pianta' (Mazzini). But what would be the sense in providing harmful (*nocens*) water to the plants? *induci* requires rather a positive sense that will oppose and balance *nocens*. Adopting *auerti* and transposing *et* and *ut* opens the possibility for a more felicitous contrast: 'the water, controlled by hand, can be *turned away* (*auerti*) when harmful in such a way that it could *also* (*et*) be channelled in [when it is necessary, beneficial, vel sim.]'.[28] This reciprocal action would

[28] For the meaning of *ita . . . ut*, 'in such a way . . . as not to preclude (some other situation, course of action, etc.)', see *OLD* s.v. 34a.

be appropriate to describing the workings of an artificial irrigation system, for example, with channels that may be opened or closed.

The artificial irrigation described here may be through furrows (see 3.3.2n. *quo . . . immoretur*), hence *induci* (cf. 3.3.2 *canalibus . . . infusus*), but there were many technologies for irrigation in antiquity (Plate 8): see Tölle-Kastenbein (1990), esp. 21–42, 106–29, Thomas and Hodge (2000), Oleson (2000*a*), Bruun (2015); and regarding the related question of water-moving technology, see White (1970) 156–7, Oleson (2000*b*), Wilson (2000), Margaritis and Jones (2008) 161–2, Wilson (2008), Malouta and Wilson (2013), Jansen (2018).

2.4.6 Idem persico solum pingue contrarium [tum] crediderunt: calculosum et harenae et sabulo proximum conueniet, sicuti ceteris omnibus quae gummis glutina excludant: There is an explanatory or causal parataxis between the two clauses (marked by the colon): the ground for the Greeks' rejection of rich soil (*Idem . . . crediderunt*) is given by the statement (*calculosum . . . excludant*) that the peach is best suited to a coarse sand, insofar as it belongs to the class of gummy trees. On paratactic asyndeton in general, see K–S II §177, H–Sz Stilistik §55.II, Halla-aho (2009) 69–75, Adams (2016) 244–5. For the present 'asyndeton explicativum', in which the second clause offers a ground for the first, see the examples and discussion at K–S II §177.9, Tidner (1922) 77–8, Wahlén (1930) 82, Blomgren (1937) 37–8, Hallén (1941) 109–10, Halla-aho (2009) 74. Another good example of explanatory asyndeton is 2.5.2 *ossa deponi . . . non altius debent quam quod amplectitur trium mensura palmorum: obruta ulterius germine infirmo ad superna non scandent.* The sentence *obruta . . . scandent* explains why it is that the stones should not be planted deeper than three palm-lengths: '[because] those buried any further will not ascend to the surface owing to the weakness of the bud'.

Other varieties of asyndeton are also found in *Pom.* Clear examples of the adversative type ('asyndeton adversativum', K–S II §177.7–8, Tidner (1922) 75–7, Wahlén (1930) 79–80, 83, Blomgren (1937) 43) are 2.1.2 *Curtius Iustus et mense Ianuario calidioribus locis putat posse disponi; Quintilii armenia sola consentiunt Ianuario mense ponenda*; and 3.3.5 *Sic et Columella describit ferularem; alio modo scribit.* A probable example of 'asyndeton conclusivum' (Tidner (1922) 77, Andersson (1938) 141–2, Hallén (1941) 105–6, Halla-aho (2009) 73), where the first clause supplies the ground for the second, is 3.4.2 *Illa interualla etiam simili ratione moderanda sunt: uiginti et quinque nonnulli, quidam et uiginti pedum iusta pronuntiant.* Far and away the most common asyndetic parataxis, however, is found when G. (1) enumerates or introduces the opinions of other agronomists or (2) reports in steps the instructions for an agricultural procedure. These cases of unmarked asyndeton are so common in *Pom.* (and in all varieties of technical writing) that it is almost unnecessary to call attention to any particular passage, but for very clear cases of (1), follow the proper names in 2.1, 3.1 or 4.6;

and for (2), cf. 3.2, 3.5. This parataxis is as well suited to providing instructions as it is to quoting the opinions of other agronomists; it is naturally at home in technical writing, where high value is often placed on swift and unencumbered communication.

On the difficulties of determining the stylistic level of paratactic asyndeton in general, see Halla-aho (2009) 71–2, 75; for earlier treatments cf. e.g. Jeanneret (1918) 151, Svennung (1935) 486, Hofmann (1978) 110–13.

solum pingue: See 4.3.2n. *pinguis . . . dulcis.*

contrarium [tum] crediderunt: *tum* is difficult to construe. At best it might be given a connecting or progressive function ('moreover'), but the position seems to make that impossible. As Mazzini 140 notes, *tum* could easily emerge by dittography following *-ium*; for other instances of syllabic dittography (some reported only in Appendix III), cf. 1.1.13 *punica* (*punicaca* **N**); 2.3.5 *iam* (*iam iam* **N**); 3.1.4 *spectet,* [*et*] (see 3.1.4n. *spectet . . . ipse*); 4.1.8 *sererentur* (*serere rentur* **N**); 4.3.3 *segregare* (*segregarere* **N**).

calculosum et harenae et sabulo proximum: The solution to the transmitted *calculosum et harena et sabulo proximata* is uncertain, and the current text is an intermediate measure. Two alternative approaches adopted by editors are (1) to emend *proximata* to *proximum* but retain *harena* (Scotti, Mai in his text); (2) to emend *proximata* to *proxima terra,* also retaining *harena* (Condorelli xviii–xix). Both of these readings produce more or less harsh variations in subject, which can be alleviated by the assumption of a relatively easy haplography from an original *harenae et* (unless *harena* should be more radically altered, e.g. to *harenosum,* which would give a not un-Gargilian rhyme and sibilance: *calculosum et harenosum et sabulo proximum*). For confusion of forms of the verb *proximo* (whence, presumably, *proximata*) and *proximus* in MSS, see *TLL* x 2.2368.34–7 (Grossardt).

The adjective *calculosus* means 'pebbly', i.e. replete with small stones (*calculi*); it is not a term used precisely by the agronomists to refer to fractions of coarseness in the soil. (*h*)*arena* and *sabulum/o*, on the other hand, refer more definitely to degrees of coarseness, (*h*)*arena* being a finer and *sabulum/o* a coarser sand (White (1970) 97, 108–9). There are at least three ways to put the terms together: (1) *calculosum et harenae et sabulo proximum* is an inclusive, disjunctive series (i.e. one may use pebbly soil *or* fine sand *or* coarse sand, all being equally suitable); (2) *harenae et sabulo proximum* 'correct' *calculosum*, indicating that what is meant by pebbly soil is either fine or coarse sand (punctuate with a comma after *calculosum*); (3) *harenae et sabulo proximum* are simply additive with *calculosum* (i.e. sandy soil that also contains larger pebbles). It is perhaps relevant to observe that modern agricultural authorities hold the peach to do best in sandy, well-drained soils (Wickson (1891) 295, *EFN* 723).

harenae: Though *harena* need not denote only sand (it can also refer to a light soil: see Mynors on Verg. *G.* 1.69–70, Coleman (2014*b*) 9, *TLL* vi 3.2526.53–2527.25 (Brandt)), the stricter sense is appropriate here.

sicuti ceteris omnibus quae gummis glutina excludant: The verb *excludant* yields appropriate sense in the context, 'emit' or 'send out' (*TLL* v 2.1271.39–71 (Rehm)), but the easy emendation to *ex(s)udant* (Reeve) also merits consideration: cf. thus *Med.* 53.17 *gummi, quod amygdali arbor exsudat.* For *gummis glutina*, see the following n. The class of trees described here (cf. Pallad. 11.12.6 *omnibus gumminatis*) includes, besides the peach (*Med.* 44.11), also plum (*Med.* 46.10, Pallad. 2.15.20, 12.7.15), cherry (*Med.* 52.12, Pallad. 11.12.6) and almond (*Med.* 53.17; see also 3.7.2n.). The word *arboribus* may be supplied with *ceteris omnibus* from *persico* earlier in the sentence, which has established the tree as the subject of discussion.

gummis glutina: *gummis* is my emendation for *gummi* of **N**. Both *gummi* and *gluten* are regularly used of the sticky secretions of trees: see respectively *TLL* iv 1379.70–1380.6 (Lommatzsch), vi 2.2111.62–75 (Blatt). *gummi* in particular is a favourite of G. in this application: cf. *Med.* 44.11 *gummis gutta, quam persici truncus illacrimat*; 52.12 *gummi, quod truncus* (sc. *cerasi*) *illacrimat*; 53.17 *gummi, quod amygdali arbor exsudat.* The noun *gluten* is not found elsewhere in G., although the adjective *glutinosus* is used at *Med.* 32.4 in reference to sticky phlegm (*pituitas glutinosas*) in the body.

There is no difficulty *per se* in the collocation of *gummi* and *gluten* (cf. Isid. *Etym.* 17.7.10 *haec* (sc. *prunus*) *sola arbor gummi glutinosum et compactiuum distillat, quod et medici et scriptores utuntur*), but the transmitted text *gummi glutina* is difficult to accept. The first problem is the lack of a connector. We might take the nouns in asyndeton bimembre (see 4.6.2n. *calidas . . . credit*), perhaps one of them adjectivized (cf. 2.11.2n. *in glarea argilla*), but the plural *glutina* seems to argue against that interpretation. The second problem is indeed this word *glutina*, which, if correct, would be one of perhaps only two instances of the plural form of the noun in Latin, the other occurring at Tert. *Pudic.* 5 *concorporauit nos scriptura diuina, litterae ipsae glutina nostra sunt.* Tertullian's use of the plural *glutina* seems well motivated (cf. *litterae ipsae*), but what would be the point of G.'s? There is a way to obviate both difficulties if we print *gummis* for the paradosis *gummi*. Although there is a great deal of variation in the spelling, gender and forms of *gummi* (*TLL* iv 1379.40–69 (Lommatzsch)), the evidence of *Med.* (quoted above) shows that for G. the nominative is *gummi* (neuter) and that the genitive singular is *gummis*.[29] The genitive *gummis* attaches to *glutina*, which may then be taken distributively: 'trees which put out [their several] sticky secretions (*glutina*) of gum/sap (*gummis*)'.

[29] For many authors the word *gummi* is indeclinable; the form *gummis* is also found not uncommonly for the nominative.

2.5–7 These three sections concern the propagation of the peach from seed and slip. The first part of 2.5 (up to 2.5.5) deals with the basics of creating planting holes and planting the peach stones. Although it is not made explicit until 2.6.3 (cf. *translata*), these instructions apply to a *seminarium*, i.e. a nursery for raising young plants, from which they will be transplanted once grown; that much is already indicated by the creation of planting trenches (2.5.2), however, and especially by the small distances between the planted seeds, which would not be feasible for the grown trees. Compare thus Palladius' treatment (12.7.1), which, though following G., makes the fact of their eventual transplantation explicit from the beginning: *persici ossa in pastinatis areis sunt ponenda binis a se pedibus separata, ut, cum ibi plantae excreuerint, transferantur.* One should therefore note that the detail mentioned in G.'s alternative method (2.5.6) that the plants will be transferred *in seminarium* is not contrastive (i.e. implying that the other peach seedlings will not be grown in the nursery), but simply indicates that the basket-grown plants will soon join the plants started out in the nursery plot. The remainder of 2.5 contains remarks on the proper way to start a peach tree from a cutting (2.5.7n. *Planta . . . quae de codice auellitur*). 2.6 focuses on the sustenance of the young trees while in the nursery; 2.7 returns to questions of spacing, but now in relation to the trees' establishment in the orchard.

The progression and structure of 2.5–7 are close to those of 3.1–4 in fundamental respects (cf. also Introduction §III(a)), though each tree receives at least one method of propagation not mentioned for the other: for the peach, planting from a cutting (2.5.7–9),[30] and for the almond, the fennel method given at 3.3.4–5 (n.). Moreover, G. exhibits closer attention to the mechanics of transplanting for the almond than for the peach; the reader is encouraged to consider 3.1–3 and nn. *ad loc.* as well as the references given at 2.5.6n. *seminarium* for the opinions of the Roman agronomists on this subject, which are of equal relevance for the peach. 2.5–7 and 3.1–4 are also in contact on many points of language and technique, as observed *passim* below.

2.5 See previous n. for the contents of this section. Palladius' initial remarks on the peach are based partly on sentences 2.5.2–3: cf. Pallad. 12.7.1 *hoc mense locis calidis, ceteris uero Ianuario persici ossa in pastinatis areis sunt ponenda binis a se pedibus separata, ut, cum ibi plantae excreuerint, transferantur. sed ossa ponantur acumine deorsum uerso et non amplius quam duobus aut tribus palmis obruantur.* As is also the case for 2.4 (n.), Palladius freely reworks his source. He drops certain details that G. includes (not to mention all the advice after 2.5.3), adds some of his own (trenching, a notice about transplantation), and otherwise recombines

[30] An attempt is made at 3.1–4n. to speculate on the reasons for the absence of this method of propagation, as well as of grafting, in *amygd.*

or modifies G.'s instructions. An example of recombination is *persici ossa . . . ponenda binis a se pedibus separata*. This apparently conflates G.'s advice on the distance between trenches (*duobus intermissis pedibus*) with his statements on the separation of the stones (*quattuor palmis inuicem separata ponuntur*). An example of modification is *non amplius quam duobus aut tribus palmis*, which alters G.'s *non altius debent quam quod amplectitur trium mensura palmorum.*

2.5.2 sulcus . . . aperitur: For *sulcus*, see 2.7.1n. *Scrobes*. The verb *aperio* here means to 'open' a hole or trench in the ground, a common usage: see *OLD* s.v. 3a, and cf. Pallad. 4.10.11 *scrobem . . . aperies*. Different is *aperio* of soil loosened by trenching or ploughing (Col. 2.21.3, Pallad. 2.1, 6.3.1) or through seasonal weather changes (*OLD* s.v. 3b).

duobus intermissis pedibus: Not referring to the width of the trench (which is left vague) but to the distance between the *sulci* dug in the nursery (*OLD* s.v. *intermitto* 4a). The expression is stylistically varied by the addition of *spatium* at 3.3.2 (*intermisso pedis spatio*) in the instructions for the almond's *scrobes.*

pedibus: The Roman foot (*pes*, Gk. πούς) was slightly shorter than ours: a current estimate is 11.65 in. or 29.6 cm. It was divided into 12 inches (*unciae*) or 16 digits (*digiti*, Gk. δάκτυλοι). See Richardson (2004) 29–32, *NP* s.v. 'measures' IIa (Schulzki).

ossa deponi . . . non altius debent quam quod amplectitur trium mensura palmorum: obruta ulterius germine infirmo ad superna non scandent: For the causal asyndeton, see 2.4.6n.

deponi: For the significance of *depono* here ('planting down' into the earth), see Introduction §v(d).

siccata erecta: The participles are not in asyndeton bimembre, for they differ in rank (see 4.6.2n. *calidas . . . credit*): that is, the meaning is 'the dried-out stones that have been set upright' rather than 'the dried-out, upright stones'. The detail *siccata* is a good example of G.'s practice of 'extensive' didactic layering (Introduction §III(b)): it implies that the stones have been preserved through desiccation, thus recalling and reinforcing the earlier advice to that effect (cf. 2.2.2 *quidquid umidi continent illo siccante deponant*). It is less certain whether it also implicitly deemphasizes Celsus' recommendation to soak the stones in water; probably not, since these could in some sense still be called *siccata* in reference to the long period of desiccation before their 'reactivation' in water. The detail *erecta*, on the other hand, anticipates the later statement 2.5.3 *Infigenda autem terrae pars acutior, unde radicis exordium est.*

non altius . . . quam quod amplectitur trium mensura palmorum: The *quod*-clause in the comparison is a nice amplification, later simplified by Palladius with an ablative: 12.7.1 *non amplius quam duobus aut tribus palmis*. For *amplector* of the amount that something embraces, see *OLD* s.v. 11a; cf. also *complector* at 2.6.3 *Quae sollicitudo non amplius quam unius anni spatio complectitur.*

trium mensura palmorum: The *palmus* (Gk. παλαιστή) comprised 4 *digiti*, i.e. one-quarter of a *pes* (see 2.5.2n. *pedibus*); 3 *palmi* are thus equivalent to a *dodrans*, i.e. 9 *unciae* (about 22.2 cm).

germine infirmo: Functioning like a noun phrase *infirmitate germinis.*

ad superna non scandent: The expression *ad superna*, lit. 'to what is above', refers to the surface; *scandent* thus describes simply the upward growth of the plant. (The verb is used more vividly among the agronomists of a plant climbing an object, for which see 3.2.1n. *scandere . . . consuescere*). For the neuter plural *superna* used substantively, cf. Gell. *NA* 9.1.2 *cum procliuior faciliorque iactus sit ex supernis in infima quam ex infimis in superna.*

2.5.3 Infigenda . . . terrae pars acutior, unde radicis exordium est: For the *uariatio* between this sentence and 3.1.4 *Acuta pars nucis . . . prima defigitur: radices exitum moliuntur*, see Introduction §v(a).

pars acutior: I.e. the tapering end of the stone (see Plate 7).

2.5.4 Commodissime quattuor palmis inuicem separata ponuntur: Placing plants too close to one another incurs risks: cf. Col. 5.10.7 = *Arb.* 20.2 *si* [*Arb.*: *in*] *eodem scrobe duas aut tres arbusculas uoles constituere* [*Arb.*: *ponere uoles*], *curato ne inter se constringant* [*Arb.*: *contingant*]*; nam ita uermibus interibunt*; Plin. *HN* 17.70 (sc. *in seminario plantas*) *interuallo sesquipedum seri – nam si inter se contingant, praeter alia uitia etiam uerminosa fiunt*; Pallad. 3.19.3 *ubi duas in una scrobe plantas deponis, cauendum est ne se contingant, nam uermibus interibunt.*

2.5.5 Multi et fimo ossa circumdant: Cf. 3.1.2 *terram congerunt . . . cum stercoris admixtione*; 3.3.3 *Stercora in terra obruunt.* The advice here differs possibly in two small respects from those places: first, *fimus* very likely refers to cow dung in particular (see 3.1.1n. *fimum bubulum*); second, the dung here is apparently placed directly around the stones (*circumdant*) rather than mixed into the soil.

The agronomists agree on the value of manuring young plants, but they are also anxious about the dangers posed by potent manures or excessive application, which, as they observe, run the risk of 'burning' or 'drying out' the tender shoots (cf. generally Theophr. *Caus. pl.* 3.6.2 ἡ γὰρ δριμεῖα (sc. κόπρος) καὶ ἰσχυρὰ διαθερμαίνει μᾶλλον ἢ καὶ ξηραίνει):[31] thus Theophrastus states that 'the weakling requires the lightest manure' (*Caus. pl.* 3.9.5 τὸ δ' ἀσθενὲς (sc. κόπρου)

[31] For the Roman agronomists' views on the heating effects of manure, see also 2.4.2n. *sic . . . germina.*

κουφοτάτης δεῖται). See also Theophr. *Caus. pl.* 3.6.1–2, 3.9.1–2, 3.9.4–5, Col. 2.14.2, *Arb.* 5.4, Plin. *HN* 17.258, Alcock et al. (1994) 156.

aqua leuiter adrorant: *adroro* is found only twice elsewhere, in both cases similarly joined with an ablative of what is sprinkled (*TLL* II 655.50–2 (Vollmer)). *ad-* is not superfluous here: it clearly adds the force of '(on)to', sc. the stones in the hole. The root *roro* (<*ros*) is joined effectively with *leuiter* to indicate only a slight or moderate application of water. For the force of *leuiter* in such applications to watering, compare the similar use of *leniter.* Cato, *Agr.* 151.4 (of the cypress) *si quando non pluet, uti terra sitiat, aquam irrigato leniter in areas; si non habebis unde irriges, gerito inditoque leniter*; Plin. *HN* 17.74 *leniter* [*leuiter* al.] *rigatur a solis occasu in trinis diebus, ut aequaliter bibat, donec erumpant.* The adverb *leuiter* is also used by Palladius of lightly sprinkling a seedling with vinegar: 11.11.5 (sc. *ocimum*) *citius nasci fertur hoc tempore, si aceti imbre leuiter spargatur infusum.*

robustior: Referring to the vigour or stoutness of the plant: note the instructive contrast at Pallad. 3.25.29 *plantam* (sc. *morum*), *si robustam, transferes mense Octobri et Nouembri, si teneram, Februario et Martio.* When used of grain in particular, *robustus* indicates hardiness, i.e. capacity to tolerate difficult conditions: see *OLD* s.v. 5b, and cf. Verg. *G.* 1.219 *robusta . . . farra* with Mynors, Col. 2.9.8.

etsi tardior: I.e. though the unaided seedling will turn out stronger, it will grow up more slowly. For *etsi* with an adjective or adverb, see *TLL* V 2.977.61–71 (Hofmann); cf. *OLD* s.v. 1b.

planta procedit: *praecepit* gives no sort of meaning in the context (rejected also by Baumgartner, *TLL* X 2.441.62). Scotti's *procedit* is a relatively easy fix, and the word is common in application to the growth of plants: see *TLL* X 2.1498.72–1499.12 (Terkelsen), *OLD* s.v. 5a; cf. 4.3.1 *castaneam melius ex uerna satione procedere.* Compare also the compositionally and semantically similar *prouenio* in this application (see 1.1.2n. *melius proueniunt*).

simpliciter exponatur: I.e. with no extra care such as manure or water.

2.5.6 Sunt qui in gastris uel qualis cribrata terra repletis ossa disponant, unde mox plantas adultas in seminarium cum cespite suo transferant: Compare the recommendation of Palladius to start pistachios in perforated cups (the language is strikingly close to G.'s in some places): 11.12.3 *qui diligentius facere* (i.e. *ponere*) *uoluerit, pertusos caliculos et stercorata terra repletos parabit et in his pistacia terna constituet, ut ex omnibus germen quodcumque procedat. quod, ubi conualuerit planta, hinc facilius transferatur mense Februario.* Also similar is Palladius' recommendation for the pine (12.7.11). Gleason and Palmer (2018) 398 note the increasing archaeological evidence for the pot- and basket-transplant method of starting plants in Roman gardens, especially elite ones; see more fully Jashemski (2018*b*) 441–5.

Sunt qui . . . disponant: *dispont* is transmitted by **N**. It is in theory possible to restore either the indicative or subjunctive, since both moods are found in G.'s generic *qui*-clauses, but the subjunctive is preferable both as being the more usual (see 1.1.10n. *Sunt qui . . . subigunt . . . existimant*) and as agreeing with the mood of the coordinated verb *affirment* (see 2.5.6n. *transferant, et affirment quod*).

gastris: A 'large-bellied' container (Gk. γάστρα), for which see Hilgers (1969) 188. The word is rare in Latin, besides here attested only twice in Petronius (70.6, 79.3), once in Marcellus Empiricus (8.23), and once in a gloss: see Stefenelli (1962) 137.

qualis: See 2.2.2n. *qualo*.

cribrata terra: Cf. Pallad. 10.14.2 (on azarole) *prope cribrata terra*; *Geop.* 10.22.1 (on pear) γῇ σεσησμένῃ. The point is equally to obtain finely textured soil (cf. 3.1.2n. *terram . . . minutissimam*) and to remove stones and other obstructions.

plantas adultas: *adolesco* refers especially to the growth in size that necessitates the plants' being moved to new plots: cf. *OLD* s.v. 4a.

seminarium: The standard technical term for a nursery among all the agronomists (*OLD* s.v. a). For the creation and maintenance of the *seminarium*, cf. 3.1–3 with nn. *passim*. Important passages discussing the *seminarium* include: Cato, *Agr.* 46–8, Col. 5.6.6–9 (elm), 5.9.1–6 (olive), *Arb.* 1.3–6, Plin. 17.69–70, Pallad. 3.10.1, *Geop.* 5.3, 9.5. Pliny's jocular remarks on the origins of *plantaria* (nurseries for suckers) are also worth quotation here: *HN* 17.65 *natura et plantaria demonstrauit multarum radicibus pullulante subole densa et pariente matre, quas necet: eius quippe umbra turba indigesta premitur* . . . 17.66 *omnia ea* (sc. trees that cannot grow under their mother's shade) *non statim moris est in sua locari, sed prius nutrici dari atque in seminariis adolescere iterumque migrare, qui transitus mirum in modum mitigat etiam siluestres, siue arborum quoque, ut hominum, natura nouitatis ac peregrinationis auida est, siue discedentes uirus relinquunt mansuescuntque tractatu ceu ferae, dum radici auellitur planta* . . . 17.69 *namque ad id praecipuum eligi solum refert, quoniam nutricem indulgentiorem esse quam matrem saepe conuenit.*

The word *seminarium* gained some currency outside of the agronomists in evocative figures of speech. It was a favourite, for example, of Cicero (5x), who puts it to vivid effect in metaphors inspired by agriculture: cf. *Pis.* 97 (of Piso's foreign commands) *ex illo fonte et seminario triumphorum cum arida folia laureae rettulisses*; Cat. 2.23 (of the corrupted *res publica*) *hoc . . . seminarium Catilinarum futurum* (with Dyck). Outside of his *De re rustica*, Varro also found a charming use for it in a few lines on wine (*Sat. Men.* 111; on the theme, see Deschamps (2000)): *uino nihil iucundius quisquam bibit / hoc aegritudinem ad medendam inuenerunt, / hoc hilaritatis dulce seminarium, / hoc continet coagulum conuiuia.*

cum cespite suo: *c(a)espes* refers to the composite of roots and caked soil: see *TLL* III 111.78–85 (Maurenbrecher). Mynors on Verg. *G.* 4.273 quotes Martyn on this meaning: 'the root consists of a great bunch of fibers . . . for I take *cespes* in this place not to signify the earth, but *radix cespitosa*, a root whose fibres are thick matted together, so as to form a kind of turf'. Preserving and transferring the clump of earth along with the roots of the *planta* was considered the best practice in transplantation: cf. Cato, *Agr.* 28 *cum radicibus eximito cum terra sua*; Col. 5.9.8 (on the olive) *deinde ut arbusculae spatium pedale in circuitu relinquatur atque ita cum suo caespite planta eruatur*. Columella goes on to provide a method for avoiding the break-up of the root-soil mass, which he also calls a *pila* ('[root] ball'): 5.9.8–9 *qui caespes in eximendo ne soluatur, modicos surculos uirgarum inter se conexos facere oportet eosque pilae quae eximitur applicare et uiminibus ita innectere ut constricta terra ueluti clausa teneatur, tum surruta parte ima leuiter pilam commouere et suppositis uirgis alligare atque plantam transferre.*

transferant, et affirment quod: *affirment* cannot be easily understood in coordination with *transferant*; it should rather continue the *Sunt qui*-construction: 'There are some who plant (*disponant*) . . . and assert that (*et affirment quod*), etc.'[32] The same fact will explain the mood of *affirment*, for although both the indicative and subjunctive are found in G.'s *sunt qui*-clauses, the subjunctive is more common (see 1.1.10n. *Sunt qui . . . subigunt . . . existimant*).

transferant: The standard technical term for transplanting (as at 2.6.3, 3.2.2, 3.3.3; non-technical at 3.1.1, 4.3.2): see White (1970) 265, *OLD* s.v. 2b. The proper techniques and precautions for transplanting are much discussed by the agronomists: for brief remarks see De Angelis (1995) 30–6; further, cf. Cato, *Agr.* 28, Varro, *Rust.* 1.40.3 with Flach, Col. 3.5.1–4, 4.16.1, 5.9.8–10 (olives), Plin. *HN* 17.79–87, 17.198, Pallad. 1.6.4 and index of Fitch s.v. 'transplanting', *Geop.* 10.85 with Lelli. The essentials can be summarized in brief compass. It is of the highest importance that the transplant be made into a soil of similar or superior quality (Theophr. *Caus. pl.* 3.5.2, *Hist. pl.* 2.5.1, Verg. *G.* 2.265–7, Col. 3.9.9, Plin. *HN* 17.79, Pallad. 1.6.4, 3.9.5): Columella (3.5.2) goes so far as to sententiously declare that *prudentis . . . coloni est ex deteriore terra potius in meliorem quam ex meliore in deteriorem transferre*. The trees should face the same direction in the orchard as they did in the nursery and are often marked with a reddish pigment (ruddle) before transplantation to aid in determining the proper orientation (Col. 5.6.20, 5.9.8 = *Arb.* 17.4, 5.10.7 = *Arb.* 20.2, Plin. *HN* 17.83–4). The timing of transplantation (usually spring) and the climatic conditions must be carefully coordinated, and the soil and planting holes that will receive the trees must be prepared well in advance by trenching and

[32] This in spite of the rhythmical sequence *transferānt ĕt ādfirment*, a cretic-trochee.

manuring (see 2.7.1n. *Scrobes*). During transplantation the soil and roots should be carefully preserved (cf. 2.5.6n. *cum cespite suo*).

affirment quod si pertranseant . . . tribuantur: For the *quod*-clause introducing reported speech, see Introduction §v(c). The subjunctive *pertranseant* is iterative (cf. 1.1.7n. *si . . . linantur . . . perseuerant . . . pollicentur*). Harder to explain is the mood of *tribuantur*, for it does not seem to be required by the condition, and G. uses the indicative for main verbs in his other *quod*-clauses: see further Introduction §v(c).

pertranseant: For *pertranseo* here ('ad finem attingendum'), see *TLL* x 1.1821.19–55 (Breimeier). The verb takes its point from the preceding *in seminarium . . . transferant*, as Breimeier notes. (Mazzini translates 'se sopravvivono', less literal but capturing the gist.)

gratioris saporis et suci amplioris poma tribuantur: Lit. 'if the *plantae* come through, fruit of a more pleasant taste and more abundant juice are granted [to them]'. With *tribuo* is understood a dative *plantis*, vel sim., and with *poma* the genitives are description or quality (*NLS* §84, H–Sz §56, Pinkster 1002–3). For the phrase *gratioris saporis*, cf. *Med.* 48.2 (sc. *zizyphi*) *saporis non nulla gratia.*

suci: *sucus* here is 'the juice of a thing (as determining its flavour)' (*OLD* s.v. 1c), as also at 2.12.3, 3.3.4. At 2.10.3, 2.11.1, 3.7.4, 4.4.3, however, it refers to the nutritive element of the plant, i.e. its sap: see further 2.10.3n. *quidquid . . . damnatur.*

2.5.7 Planta . . . quae de codice auellitur: This is the method of propagating ἀπὸ παρασπάδος (Theophr. *Hist. pl.* 2.1.1), i.e. from a shoot that has been 'torn off' the tree along with a portion of the stock and (ideally) roots (*Hist. pl.* 2.5.3).[33] With G.'s language and description should be compared Verg. *G.* 2.23–4 *hic plantas tenero abscindens de corpore matrum / deposuit sulcis*; Plin. *HN* 17.67 *et aliud genus simile monstrauit* (sc. *natura*), *auolsique arboribus stolones uixere, quo in genere et cum perna sua auelluntur partemque aliquam e matris quoque corpore auferunt secum fimbriato corpore.* (G.'s *codex* is the lower trunk or stock of the tree, corresponding to *corpus matris* of Virgil and Pliny: see *LAT* 63–4.) In his account Pliny identifies the slip with the *stolo*, i.e. a sucker growing from the roots of a tree: for the name and meaning, see Varro, *Rust.* 1.2.9 *qui* (sc. *Licinius Stolo*) *propter diligentiam culturae Stolonum confirmauit cognomen, quod nullus in eius fundo reperiri poterat stolo, quod effodiebat circum arbores e radicibus, quae nascerentur e solo, quos stolones appellabant.*

[33] Cf. also *Geop.* 10.3.4 παρασπάδες δὲ ἤτοι μοσχεύματα καλεῖται ὅσα προσήνωται τοῖς δένδροις, καὶ ἐξ αὐτῶν ἀποσπᾶται μετὰ φανεροῦ μέρους τῆς ῥίζης.

For the ancient agronomists' appreciation of techniques for propagating trees from a live portion of the plant (rather than seed), see 3.1–4n. There are several other such methods not mentioned by G. for any of the fruit trees he discusses, including planting from branches, layering and so forth. On these, see the references in 3.1–4n., plus e.g. Theophr. *Hist. pl.* 2.1.1 with Amigues (n. 1), Cato, *Agr.* 45, 133, Varro, *Rust.* 1.39.3, 1.40.4, Verg. *G.* 2.17–34 with Mynors, *Geop.* 10.3 with Lelli, de Angelis (1995) 44–7, Hardy and Totelin (2016) 134–5. For further details, cf. Pallad. 3.25.21 (quince), 3.25.27 (carob), 3.25.28 (mulberry), 4.10.1–2 (pomegranate), 4.10.12–13 (citron), 4.10.19 (medlar), 4.10.23–4 (fig), 10.14.1 (azarole), 11.12.1 (date palm), 11.12.5 (cherry), 12.7.13 (plum); also *Geop.* 10.8 (citron), 10.12.3 (pistachio), 10.22.3–5 (pear), 10.63.2–3 (chestnut).

de codice auellitur: In the earliest period *auello* is constructed with *ab*, but already from the time of Cicero also with *ex* and *de*, and (especially in the poets) with the dative and bare ablative: see *TLL* II 1305.44–8 (Ihm). There is nothing *per se* substandard or late about G.'s usage of *auello* + *de* (one might compare Verg. *G.* 2.23 *abscindens de corpore*, quoted in previous n.), and it is very doubtful whether we can attribute the construction here to the increasing prevalence of *de* in later Latin at the expense of *ab* and *ex* (Löfstedt (1911) 103, H–Sz §146, Adams (1995*a*) 435, 442, (2003) 567, (2011) 259, (2016) 240, 311). In general *ab* and especially *ex* are still vigorous in G.'s writings: against 40 instances of *de* (to exclude the 60x where *de* appears in the title of a chapter of *Med.*, e.g. *de raphano*) are found 151 of *e*(*x*) and 31 of *a*(*b*).

autumnale tempus expostulat: huic namque arescendi periculum incumbit nisi largus umor et frequens pluuia ei subuenerit: Columella advocates an autumn planting time for olive slips in dry places, probably also thinking of the wetness of the season, as G. is here: 5.9.6 *plantae autem in olіueto disponuntur optime siccis et argillosis agris per autumnum, laetis et umidis uerno tempore, paulo antequam germinent* (cf. *Arb.* 17.1).

periculum incumbit: The usage of *incumbo* is late but well attested, 'fere i. q. imminere' (*TLL* VII 1.1075.55–72 (Rehm)). It is found first in G.'s writings, here and at *Med.* 40.16 *aiunt et ipsa poma remouenda cum feminae partus incubuit*.

largus umor: For the *iunctura*, cf. *Arb.* 1.5 *proderit . . . uitem largo consuescere umore*.

2.5.8 fimo oblitam: The agronomists widely recommend coating the cut or torn portion of a slip, transplant, etc., with manure or a mix of manure and another substance such as ash or mud. *fimus* here is probably *fimus bubulus* (for which see 3.1.1n. *fimum bubulum*), the variety of manure typically recommended by the agronomists in this application: cf. Cato, *Agr.* 40.2, 46.2, Col.

5.6.8, *Arb.* 3.4, Pallad. 2.15.16, 4.10.13, 10.14.1, *Geop.* 4.12.16.[34] (Exceptional is Pallad. 4.10.1 recommending pig dung for pomegranate cuttings, but note there *porcino stercore*, not *fimo*.) The collocation of *fimus* and *oblino* is very common: cf. Cato, *Agr.* 28.2, 46.2, Varro, *Rust.* 3.16.16, Col. 5.6.8, 5.9.3, 11.2.42, Plin. *HN* 17.87, 19.133, Pallad. 3.24.10, 3.24.12, 3.25.28, 12.7.13.

inferioribus foliis deputatam: The verb *deputo* is used of pruning from Cato on, especially of pruning 'back' or cutting off top growth (White (1970) 264, *OLD* s.v. 1a) and of cutting away a shoot (*OLD* s.v. 1b). In the present instance, however, *de-* simply adds separative force.

subacto solo: In agricultural contexts *subigo* is technical for the careful tilling of soil (*OLD* s.v. 7b, Summers on Sen. *Ep.* 79.6), as also at 4.5.1 *solum . . . molle et subactum*. The word is stronger than simply 'plough', focusing on the action of turning the soil over and rendering it friable (so 4.5.1 *molle*): compare in this regard Cic. *De or.* 2.131 *subacto mihi ingenio opus est, ut agro non semel arato, sed nouato et*[35] *iterato, quo meliores fetus possit et grandiores edere*. The recommendation to work over the soil into which a slip, branch or cutting is to be transplanted is standard. For the olive, cf. Cato, *Agr.* 45.1 *locus bipalio subactus siet beneque terra tenera siet beneque glittus siet*; Col. 5.9.1 *solo modice ualido et sucoso neque denso neque lutoso solo* [trad. *modo*], *potius tamen resoluto . . . quam* (sc. *terram*) *cum in tres pedes pastinaueris*, etc. For fruit trees, cf. Col. 5.10.21 = *Arb.* 25.2 *omnium autem generum ramos . . . in hortis, <ubi et> subacta et stercorata terra est, per puluinos arearum disponere conuenit*. Palladius says of an azarole cutting (10.14.1), *statuatur pingui terra et subacta*.

praecedente punctorio: The verb *praecedo* is used also at 2.10.1 (*Inseri persicus utroque modo potest, uel in cortice uel in ligno praecedente fissura*), again in an ablative absolute, in order to establish that one action is to be completed before another. There is obviously a temporal priority intended in the sequence, but the didactic context establishes a logical one, too: for this aspect of *praecedo*, 'ordini temporis simul inest ratio quaedam logica, condicionalis', see *TLL* x 2.403.49–50 (Somazzi).

punctorio: *punctorium* is a ἅπαξ: for the deverbal derivation from *pungo* (*punct-*), cf. *deuersorium* (<*deuerto*) or *seclusorium* (<*secludo*). Leumann §278 observes that nouns in *-torium*/*-sorium* generally refer to 'Aufentshaltsräumen, Behältern, Gefäßen' (add Rodgers (1975) 86 on *calcatorium*, a 'pressing area'), but *punctorium*

[34] We should probably also understand *fimus* to stand for *fimus bubulus* in such application at Col. 5.9.3, Pallad. 3.25.28, 12.7.13.

[35] Wilkins secludes *nouato* in the OCT and in his commentary argues that the meaning is inappropriate, claiming that it is probably a gloss on the words *semel arato*. Be that as it may, the key detail is the repetitive ploughing implied by *subigo*, which could be denoted by either *nouato* or *iterato* alone.

refers rather to an instrument for pricking or puncturing the soil. (There is a Sardinian survival, *puntordzu*, 'spike': see *REW* no. 6846.) Mazzini 54 says of the word: 'possiamo tutt'al più dire che il termine può presentare un certo aspetto volgare a causa del suffisso *-torium* tipico di formazioni solitamente volgari'. There is nothing necessarily Vulgar about such suffixes, however (Mazzini's examples notwithstanding), as a glance at Leumann §278 shows; relevant too are the remarks on *amaritosa* (3.7.5) in the Introduction §v(d).

The *punctorium* itself should probably be understood as a kind of dibble (*pastinum*; cf. 4.7.1n. *pastinatos*). Mazzini 55, with an eye to *sicuti olera*, remarks: 'si tratta di uno strumento semplice e comune ancora oggi, usato soprattuto dagli ortolagni per trapiantare l'insalata, i cavoli e vari altri ortaggi, il "piolo" o "foraterra"'. *sicuti olera* indeed applies to the whole phrase *subacto solo et praecedente punctorio*, showing that what G. has in mind is the intensive cultivation of the planting soil for the vegetable garden universally recommended by the agronomists.

demergere: Not a common application of the word, but used also by Columella of 'sinking' seeds or shoots 'down' into the earth, i.e. planting: see *TLL* v 1.481.60–6 (Gudeman). For G.'s other uses of *demergo*, see 1.1.9n. *demergunt*.

2.5.9 calcata plus iusto et terra densante constricta inertia ligni: Although *calco* is applied most often to earth or soil (*terra*, *solum*, vel sim.), here it must agree either with a supplied *planta* (if *inertia* is ablative) or with *inertia ligni* (if *inertia* is nominative; see 2.5.9n. *inertia ligni*). This is indicated both by the rhythm of the colon (*calcātă plūs iūsto*) and by the position of *terra*: the skeleton of the whole phrase comprises the participles *calcata . . . constricta*, filled in by the balancing ablatives *plus iusto / terra densante*. The semantic range of *calco* is wide enough to support this transferred sense: for another example of *calco* in reference to 'packing down' a plant rather than soil, cf. Plin. *HN* 19.158 *si* (sc. *apium*) *satum calcetur cylindro pedibusue*.

plus iusto: For the meaning of *iustus*, see 2.11.2n. *mole iusta*. For the *iunctura*, cf. *Med.* 21.11 (sc. *sectiuus*) *ex abortu plus iusto sanguinem profluentem detinere potus solet*.

terra densante: With *denso* we must supply an object from *constricta*. Although the verb is capable of supporting the meanings 'compress' or 'press on/crowd', I have not found a precise parallel for the present usage. The earth itself is more often said to be 'compressed' than 'compressing': cf. Liv. 10.5.11 *male densatus agger*; 32.17.16 *densati soli*; Veg. *Mil.* 4.3 *terra . . . uectibusque densatur*; also Col. 10.319 *grauibus densate cylindris*, where some word for soil must be supplied (White (1975) 3–4, Boldrer *ad loc.*). Perhaps the unusual use of the verb is licensed by a desire for *uariatio* next to *constricta*: for the collocation of the words, cf. Cassius Felix 82 p. 193.12–15 *oportet . . . ea adhibere quae ualeant* *uirtute styptica omnia soluta densando uel constringendo sanare.* A final possibility is that *densante* is used intransitively (this phenomenon is discussed later at a more secure instance: see 3.4.3n. *maturant*); though the intransitivization is not attested elsewhere, it would give an appropriate meaning ('constricted by the packed earth').

inertia ligni: I take *inertia* as a causal ablative (cf. 2.5.2 (sc. *ossa*) *obruta ulterius germine infirmo ad superna non scandent*; note also the rhythm, cretic-trochee *inērtiā līgni*), and hence supply *planta* with the preceding *calcata . . . constricta* as subject of *sentit*. But it is possible that *inertia* is nominative, used concretely so that 'the weakness of the wood' = 'the weak wood'.

maturum . . . occasum: *maturus* in the sense of 'premature' or 'too early' applied to the death of the plant (cf. 2.7.2 *maturius consenescat*) might just be punning, for *maturus* more often denotes among the agronomists ripeness of fruit (see 1.1.3n. *matura cydonea legenda sunt*).

2.6 G. records instructions for maintaining the plants in the nursery until they are ready for transplantation into the orchard after two years (2.6.3 *bima translata*). They need hoeing and weeding to eliminate vegetation that would poison them or block their access to the sun and air; they must also be supported with props until they are hardy enough to maintain themselves in the face of harsh weather.

2.6.1 Sed: See *OLD* s.v. 2c: '(in passing to a new subject, line of argument, etc.)'.

utrisque: I.e. both the *germina* from seeds and the slips.

frequens sarculi opera et desiderata medicina continuae runcationis: What is referred to here is *sartio* (or *sarculatio*, hoeing) and *runcatio* (weeding). *sartio* designates the operation of turning over the soil with a light hoe (*sarculum*) in order 'to keep down the weeds and prevent the surface soil from "caking"' (White (1967) 45; Plate 9): see White (1970) 180–1, Kolendo (1980) 122–8, Spurr (1986*b*) 58–64, Marcone (1997) 58–9. G. provides a concise description of *sartio* at 3.1.5 *quidquid in summo scrobe induruit leui sarculo relaxetur.* Frequent application of *sartio* is absolutely essential in the growing season of cereals and legumes (see the lengthy discussion of Col. 2.11) and would confer similar benefits upon young fruit trees. The 'layer of loose, dry earth obtained by frequent hoeing' both prevents water loss and, because it is a poor conductor of heat, 'helps to keep the young plants at a more suitable temperature for growth' (White (1970) 180–1); the advice would thus be especially important in the summer months of, e.g., southern and eastern Italy, where the heat and aridity might check plant growth. *runcatio* refers to the different task of grubbing up or rooting out weeds (though *sartio* certainly contributes to that end). *runcatio* was done mostly by hand: see White (1967) 92, Kolendo (1980) 88, Spurr (1986*b*) 63, Halstead (2014) 233–8. Palladius in his account replaces hoeing and grubbing with digging, also intended to kill weeds: 12.7.2 *dum tenera sunt germina, saepe herbis circumfossa liberentur.* (He makes a similar substitution in the case of the almond: see 3.1.5n. *quidquid . . . relaxetur.*)

sarculi: A type of mattock and one of the most common agricultural implements: it is mentioned more frequently in the Roman technical writers and poets than any other hoe (White (1967) 44) and could stand as a synecdoche for farming (cf. e.g. Ov. *Fast.* 4.927, Juv. 3.310–11). See further the illustrations and discussion at White (1967) 43–7.

desiderata: The verb (also 2.7.3, 4.5.1), like *amo* (2.4.2, 2.7.1), is used very commonly by the agronomists to mark the conditions of growth to which a plant responds favourably.

tenero germini noceant flatus herbarum: *herba* is used here, as often, with an accessory notion of 'undesirable', i.e. in reference to weeds (*OLD* s.v. 1b). It is hardly certain what the *flatus* are that G. speaks of as presenting a danger to the bud; if the text at 2.8.1 *inhalantibus fibris* (n.) is sound, however, a common doctrine may underlie both places, for the fact that the root fibres are *inhalantes* is given there as a reason for pruning them (cf. 2.8.1n. *fibris de radice purgatis*). It may be relevant to note that the corrupting action of quince when stored with other fruit is spoken of in some sources as due to its vitiating 'odours' (*odor*/ὀσμή) or 'exhalations' (*halitus*/ἀποπνοή) (see 1.1.15n.), and that Theophrastus speaks of plants as having exhalations (ἀποπνοή) on which their fragrances depend (cf. *Caus. pl.* 6.16.8, 6.17.9).

tenero germini: Cf 2.5.2 *germine infirmo.*

noceant flatus herbarum: Perhaps for *noceant* should be read *noceat*, which would allow us to read *flātūs hērbārum* (cretic-trochee); note however that *flatūs herbarum* (as transmitted) would still make the accentual *cursus planus*.

densitas contegentium solis atque aurae nutrimenta praecludat: In giving a fair trial to the transmitted ἅπαξ *condego* it may be said that the meaning of 'live together with' is easily derived from *cum* + *dego*: compare the verb *conuiuo*, considering especially its range of absolute uses at *TLL* IV 886.48–55 (Gudeman). *densitas* can refer to a packed crowd of people ('i. q. multitudo constipata', *TLL* V 1.544.8 (Jachmann)), and might thus aid in interpreting *condego* with the personified subject *herbae*: perhaps it is relevant to compare Ennod. *Opusc.* 3.112 p. 360.5 *ad recipiendam habitantium densitatem.* The phrase *densitas condegentium* would thus mean something like 'the density of its co-habitors', in reference to the weeds that share the plot with the sprout. *t* is very often changed in the manuscript to *d* (12x in total), however, and *contego* has much more point in the context: it is insofar as the weeds cover the sprout that they can be said to 'shut out' (*praecludat*) sun and air (*solis atque aurae nutrimenta*). *densitas* will as easily refer to the 'thickness' of the crowding weeds as it would if *condegentium* were adopted.

2.6.2 Danda etiam calamo siue uirgulto propius affixo adminicula: At least 13 names for props and frames were used by the agronomists (White (1975) 19), offering more or less precise description of the supports. The most general terms were *pedamentum* and *adminiculum*, which could be used of any kind of prop because they bring into focus not the construction but the function of the support (cf. 2.1.3n. *adminiculis*). G.'s *calamus* (more commonly called *harundo* among the agronomists) and *uirgultum* in fact are not words for props *per se*, but rather denote the material out of which a prop might be made (thus giving an idea of its relative stoutness); so compare Virgil on the four props used in preparation of an *arbustum*, of which the *calamus* and *uirgultum* (= *hastile uirgae*) are presented as the lightest: *G.* 2.358–9 *tum leuis calamos et rasae hastilia uirgae / fraxineasque aptare sudes furcasque ualentis* (with Mynors). In the present instance, *adminiculum* is used with *calamus* and *uirgultum* to specify the purpose of the objects: because *adminiculum* is a very general term for a prop, denoting simply its function rather than particular character, there is no difficulty in overlaying the terms (cf. similarly 3.2.1n. *palus . . . adminiculo*).

As for types of props themselves, among the agronomists the *palus*, *pertica* and *ridica* are especially common: the *palus* is the 'ordinary term for a stake sharpened to a point' (White (1975) 21; cf. *TLL* x 1.175.3–35 (Hodges)) and is distinguished in viticulture chiefly from *ridica*, 'a prop made from a living tree', and *pertica*, properly 'a support made from the branch of a tree' (White (1975) 22; cf. Varro, *Rust.* 1.8.4, Col. 4.26.1). Varro (*Rust.* 1.8.4) asserts that of the four props (*pedamenta*) recognized for viticulture, the *ridica* and next *palus* are best, but of course the suitability of a prop would in many cases be more closely related to the specific application than to the intrinsic superiority of one kind or the other. For further discussion, see White (1975) 19–23. The Roman agronomists are generally of the opinion that the material for stakes, which varied widely, should be supplied from the farm, since they would otherwise have to be bought *extra fundum* (Col. 4.30.1), which was thought unprofitable. They accordingly devote significant attention in their writings to the reeds (*harundo*, *calamus* = κάλαμος, *canna* = κάννα) and trees (e.g. osier, white willow, chestnut, cypress, elder, juniper) that could be used to furnish supports, mostly for vines: cf. Varro, *Rust.* 1.8.4, Col. 4.30–3, *Arb.* 16, Plin. *HN* 17.141–51, 17.174. For the range of uses for reeds on the farm, see also Bergmann (2014) 269 n. 56.

calamo siue uirgulto propius affixo: Cf. 3.2.1 *iuxta uastior palus affigitur*. If *propius* here is taken to express the same thought as *iuxta* does at 3.2.1, then the comparative will stand for the positive *prope* (see 1.1.15n. *Diligentioribus*).

quibus se tueri . . . consuescant: The verb *consuesco* appears twice in *Pom.* (also at 3.2.1 *scandere in rectum . . . consuescere*), in both cases paired with an infinitive to describe the effect of props on a young plant. The idea of habituation is appropriate in discussing supports, since the goal is for the plant to become accustomed to supporting itself; while *consuesco* is not attested in this context outside of *Pom.*, one might compare *condisco* at Plin. *HN* 17.184 (sc. *uitis*) *initio adminiculum desiderabit, dum stare condiscat et recta surgere.*

uiolenta uentorum flabra: It is hard not to suppose that Lucretius lies somewhere behind G.'s words: cf. 5.214–17 *et tamen inter dum magno quaesita labore / cum iam per terras frondent atque omnia florent / . . . / flabra . . . uentorum uiolento turbine uexant.* Although the word *flabra*, indeed found first in Lucretius, retains its associations with poetry throughout late antiquity (*TLL* VI 1.832.84–833.2 (Bacherler)), it is hardly uncommon in prose. It is much harder to see why G. would write *uiolens* (whence would come the paradosis *uiolentia*), which occurs only rarely and until the fourth and fifth centuries CE is essentially confined to poetry,[36] in place of the adjective *uiolentus*, which is regularly used of winds, storms and other natural forces (*OLD* s.v. 2a); the mistake is palaeographically easy, and the scribe may have been influenced by the common collocation *uiolentia uentorum* (where *uiolentia* is the nominative singular noun).

2.6.3 Quae sollicitudo non amplius quam unius anni spatio complectitur: For *complector* used as a true passive (rather than a deponent), attested first in the first century CE and sporadically thereafter, see *TLL* III 2081.44–70 (Jachmann). Reeve's suggestion *spatia* is metrically attractive (a resolved dicretic *spătĭă cōnplēctĭtur*).

sollicitudo: Cf. 2.2.1 = 4.3.1 *anxie atque sollicite.*

ceterum: I.e. the tree is in general hardy, except for the single year (*unius anni spatio*) in which it must be supported by a prop. See *OLD* s.v. 5c: '(in slightly adversative sense) apart from these exceptions, qualifications, etc., however that may be'.

bima translata: Cf. Pallad. 12.7.2 *bimam plantam recte transferemus scrobe breui.* As also observed in the case of the almond (3.2.2n. *Triennio exacto*), the timeline of transplantation is different in present-day peach cultivation; cf. already Waugh (1913) 53–4: 'Peach trees are planted at one year old . . . A few growers still prefer a two-year-old peach tree, but in most cases it is safe to say that such

[36] See *OLD*, L&S, s.v. *uiolens*. Unambiguous forms of the word are difficult to identify. In consulting the *Zettel* of the *TLL* I could find no certain instances of the word in prose before the fourth century, unless the fragment of Cicero in Terentianus Maurus should be thought neither poetry nor paraphrase: 655 *et Iolaus ille uiolens, crede Marco Tullio.*

preference indicates a distinct ignorance of the business. It is the opinion of nearly all fruit growers that any peach tree more than one year old is worthless for planting. The June buds are here included as one-year-old trees, but as a matter of fact they are only half a year old.' See also *EFN* 722.

nulli non repugnat iniuriae: For *repugno* of a plant's resisting adverse conditions, cf. 2.7.5 *a solis ardoribus, quibus difficulter hoc genus arboris creditur repugnare.* In *Med.* G. uses it, along with similar words such as *resisto*, to describe the beneficial therapeutic effects of plants: cf. 47.7 (sc. *folia mespili*) *in uino propinata ueneficis potionibus repugnare.*

2.7 Before moving to questions of their maintenance (2.8–9), G. offers advice relevant to the establishment and arrangement of the peach trees in the orchard; the section thus corresponds to 3.4 (*amygd.*). Textual difficulties in the section obscure some of the finer points, but the principal recommendations are certain: (1) plant the trees in fairly shallow holes; (2) establish them fairly close to one another.

On the whole, the information G. provides here presents a minimum of what we might expect to find concerning the layout of fruit trees in the orchard: there is no secure mention of the distances between trees (2.7.3 *quam scilicet binos pedes* must be thrown out (n.)), as there is at least for the almond, and in neither *persic.* nor *amygd.* is there a discussion of the planting arrangements or modified instructions for landowners who intend to practise intercultivation (cf. Jashemski (2018*b*) 438–9). For the question of the possibility of a general treatment in a separate section, see Introduction §III(a). If G. did not in fact provide a general discussion elsewhere, we can assume that his audience would already have known about various ways of planting the orchard, and that this specific information for the peach would have helped them to adapt the tree into a pre-existing pattern. For a sampling of remarks on the layout of orchards and vineyards, cf. Varro, *Rust.* 1.7.2–4 (quoted partially at 3.4.2n. *moderanda sunt*), Verg. *G.* 2.274–87, Col. 3.13.3–5, 3.15.1–2, Plin. *HN* 17.78; see also the discussion of intervals in the orchard at 3.4.2n. *uiginti . . . pronuntiant.* The typical planting patterns may be summarized briefly here. A common and noteworthy way of arranging the trees in a Roman orchard was the quincunx pattern, the 'name [of which] derives from the resemblance of the pattern to a repetition of the mark on a five-*uncia* weight or coin . . . also used, then as now, for the five on dice' (Powell on Cic. *Sen.* 59). Of course, other patterns such as a square (or a simple modification thereof) were also employed. It is instructive to compare *AOM* 38, where there are adduced four planting designs typically used for the almond in present-day orchards (Figure 3): square, hedgerow (i.e. tightly packed parallel rows), quincunx and hexagon (resembling an expanded quincunx). Of these, the most commonly used now is the square. Although it is not the most efficient in terms of trees

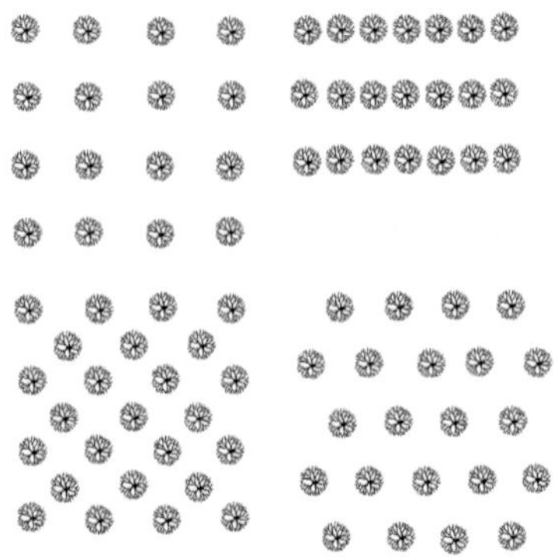

Figure 3: Common designs for planting trees in an orchard. Clockwise, from top left: square, hedgerow, hexagonal and quincunx planting designs

per acre (the quincunx, for example, is superior in that regard), it simplifies cultivation and harvest (cf. Col. 5.3.3, Pallad. 2.10.2).

2.7.1 Scrobes altos arbor persici non amat, quippe cum modica radicum concessa substantia esset; <et> tamen cito bibit: contra aliarum arborum legem de superioribus partibus mutuatur alimentum: The *quippe cum*-clause explains why the peach needs a shallow hole: because the 'substance' (2.7.1n. *radicum concessa substantia*) of its roots is 'slight' (2.7.1n. *modica*). The statement *contra aliarum*, etc., should continue the theme, for (1) the addition cannot be motivated by the following statement *Hanc maxime*, etc., which is merely parenthetical; (2) the *radices* mentioned earlier bring to mind the source of the plant's nutrition (cf. 2.7.1n. *contra . . . alimentum*), which is the explicit focus of these remarks (so *mutuatur alimentum*); and (3) the detail *de superioribus partibus* seems to be spatially apropos of the shallow holes preferred by the peach (*altos . . . non amat*). If a connection between the clauses should therefore be sought from the garbled text *essetamnecito uiuit*, Condorelli's proposal of *bibit* for *uiuit* (*b* and *u* are often exchanged in **N**[37]) is desirable, for the verb will anticipate the following remarks on the feeding of the peach tree.[38] Less certain is how to handle *essetamnecito*. Condorelli's hesi-

[37] *u* is written for *b* 5x apart from the present instance, and *b* for *u* 9x.

[38] See Condorelli xix–xx. There are parallels for *bibo* in application to a tree: cf. Plin. *HN* 13.28 *gaudet* (sc. *palma*) *riguis totoque anno bibere*; 17.74 *leniter rigatur cupressus a solis occasu in trinis diebus, ut aequaliter bibat.* Three further examples are found in Pliny and one in Apuleius: see *TLL* II 1967.56–61 (Münscher). Condorelli also compares Ov. *Met.* 14.632–3 *bibulae . . . / radicis fibras.*

tant suggestion *est et tam nec cito bibit* gives a text that is implausibly difficult to construe.[39] I prefer a mild transposition that allows us to take *tamen* from *tamne*. Because for G. *tamen* (35x in *Pom.* and *Med.*) never comes first in its clause, we might suppose that *et* has fallen out by haplography after *esset* (for *et tamen*, cf. *Med.* 13.1, 44.4). The tense of *esset* is awkward, but perhaps less so if *concessa* (for which see 2.7.1n. *radicum concessa substantia*) can be taken attributively with *substantia* (= 'insofar as the substance of the roots *with which it has been endowed* is/was slight'), as the word order suggests, and not periphrastically with *esset* as the pluperfect (= 'insofar as the peach *had been endowed with* a slight substance of roots'). An emendation to *sit et* might also be considered (cf. *Med.* 45.5 *cortici austerior natura concessa est*), but it should be noted that *esset* makes the rhythm (cretic + ditrochee *concēssă sūbstāntĭa ēssĕt*);[40] at any event we should avoid breaking *esset* into *est et*, for the ordinary mood of *quippe cum* is subjunctive (cf. Schrickx (2011) 122) and, although *quippe cum* does not appear again in G.'s writings, his other causal *cum*-clauses take the subjunctive (cf. *Pom.* 3.5.3, 3.7.4, 4.4.3, *Med.* 3.7, 41.6).

According to the interpretation offered here, *<et> tamen cito bibit* acts as a pivot, introducing a fact about the peach that is not expected from the first statement. The logic of 2.7.1–2 is thus as follows: although the tree has weak roots (cf. *quippe cum modica radicum concessa substantia esset*) and should accordingly be given a shallow hole (*Scrobes altos arbor persici non amat*), it nonetheless feeds rapidly (*<et> tamen cito bibit*); that is because (the asyndeton with *contra aliarum*, etc., is causal; see 2.4.6n. *Idem . . . excludant*) the tree has a unique mode of nutrition, being nourished from the upper parts, sc. of soil or roots (*contra aliarum arborum legem de superioribus partibus mutuatur alimentum*); indeed, its rapid growth is held responsible for the tree's premature senescence (*Hanc maxime existimant causam propter quam sola maturius consenescat*).

Scrobes altos arbor persici non amat: For the depth of the holes into which trees were to be transplanted, previous authorities (e.g. Cato, *Agr.* 43.1, Col. 5.5.2, Plin. *HN* 17.80–1 reporting the opinions also of Mago and the *Graeci auctores*) offer suggestions ranging from 18 inches to nearly 5 feet. The recommendations depend in part on the nature of the soil, with Mago, for example, recommending the deepest holes in hard, sloping clay and somewhat shallower ones in rich, black earth. Another factor, as in the present case, is the nature of the tree under consideration. Theophrastus reports that for shallow-rooted plants deeper holes are desirable in order to encourage

[39] See also below for reservations about printing *est*. The meaning of the words *et tam nec cito bibit* would presumably be 'and it drinks with such difficulty' (for *non cito* = *non facile*, see *TLL* III 1210.69–78 (Spelthahn), *OLD* s.v. 4).

[40] Regarding the tense of *esset*, cf. also 3.6.2n. *cuneum qui esset ex robore.*

downward growth of the roots: cf. *Caus. pl.* 3.4.2 ὄντων δὲ τῶν μὲν βαθυρρίζων, τῶν δ' ἐπιπολαιορρίζων, διὰ τοῦτο τοὺς γύρους οὐκ ἰσοβαθεῖς ὀρύττουσι τοῖς ἐπιπολαιορρίζοις, οἷον ἐλαίᾳ καὶ συκῇ, βουλόμενοι πιέζεσθαι καὶ ὥσπερ ἀντιταττόμενοι πρὸς τὰς φύσεις; *Hist. pl.* 2.5.1 τοὺς δὲ γυροὺς προορύττειν ὡς πλείστου χρόνου καὶ βαθυτέρους αἰεὶ καὶ τοῖς ἐπιπολαιορριζοτέροις. (Plin. *HN* 17.81 follows Theophrastus: see André *ad loc.*) G. remains vague on the precise depth for the peach, as also for the almond (3.4.1 *ne altissimo scrobe*), indicating only that the peach is averse to deep holes (this is the force of the litotes *non amat*; cf. Pallad. 12.7.2 *scrobe breui*). The relative suggestions he gives for peach and almond are probably to be related by the cultivator to the standards of practice. With the varied and somewhat loose recommendations found among the agronomists one might well compare the remarks of Wickson (1891) 149: 'The depth to which trees should be set has always been a matter of discord among planters. The safest rule is to get the tree as nearly as possible the same depth it stood in the nursery row; that is, so as to have it stand that way when the ground has settled.'

Scrobes: The agronomists' all-purpose word for a 'planting hole' (*OLD* s.v. a), although it can also be used to refer to a pit dug for other purposes (cf. 1.1.13 *scrobibus*). At least from the time of Cato, *scrobis* is usually distinguished from *sulcus* (cf. e.g. Cato, *Agr.* 40.1, Col. 4.3.9), the latter referring to a narrower trench for planting or other purposes, such as conveying water (*OLD* s.v. 2a; it also has a very common technical application to a furrow made by ploughing, *OLD* s.v. 1). Both *sulcus* and *scrobis* are more precise terms than *fossa*, *fossio*, *fossura*, etc., which can refer to digging of any shape and sort and thus generally require contextual specification: cf. e.g. Col. 3.13.4 *sulcos derigunt . . . ac per latera fossarum uitem uel malleolum disponunt*. For deverbal nouns from *fodio*, see further 3.2.2n. *ut . . . relinquatur*.

The agronomists devote significant attention to the proper way of digging planting holes for trees. Most critical were the holes into which the young saplings would be transplanted, and it is above all to these that the agronomists' special instructions apply. The instructions guiding creation of *scrobes* for fruit trees offered by Pliny and Columella are nearly identical (Plin. *HN* 17.79–80, Col. 5.10.2–4 = *Arb.* 19.1–3, Col. 5.5.1 = *Arb.* 4.3).[41] Columella's version is as follows: the planting hole should be excavated within the year

[41] Pliny attributes the advice to Mago, who should stand also behind Columella's account. The *Geoponica* would later pick up some, but not all, of Mago's instructions: 9.6.2–3 βοθρεύειν δὲ μάλιστα μὲν ἐνιαυτῷ πρότερον τῆς φυτείας, ἵνα ὑπὸ τοῦ ἡλίου, καὶ τῶν πνευμάτων, καὶ τῶν ὄμβρων, ἡ γῆ ψαθυρωτέρα γένηται, καὶ τὰ φυτὰ ῥιζοβολήσῃ. εἰ δὲ φυτεῦσαι ἐπειγόμεθα, πρὸ δύο μηνῶν, ἢ ἑνὸς τὸ ἐλάχιστον, φρύγανα καὶ κάλαμον, καὶ ὅσα εὔφλεκτα, ἐγκαταφλέξομεν τοῖς βόθροις ἐπὶ πολλὰς ἡμέρας.

prior to the intended planting time so that rain and sun can soften the soil (*sole pluuiaque macerabitur*);[42] on short notice, it can be done two months in advance, but if so the cavity should be warmed by burning straw or brushwood. Holes that are broader and more spacious (*latiores patentioresque*) in general produce more fruitful plants.[43] The hole should be shaped like an oven (*scrobis clibano similis*),[44] the cavity being wider below than above (*imus summo patentior*) so that the roots have room to spread out and are less exposed to fluctuations in temperature and so that fertile soil heaped into them is not washed away by rain.

One may note that while the agronomists universally recommend *scrobes* for arboriculture, they consider them an inferior way of digging the soil in comparison with *sulci* and the intensive trenching known as *pastinatio*: cf. Plin. *HN* 17.166 *optime in pastinato, proxime in sulco, nouissime in scrobe*; White (1970) 237. (At 5.4.2, however, Columella allows that in certain rich soils *scrobes* or *sulci* are adequate.) For the hierarchy of digging techniques and the relationship of *pastinatio* to holes (*scrobes*) and furrows (*sulci*), see more fully Col. 3.13.4–5, 5.4.2–5.5.3, *Arb.* 1.5–6, 4.2–3, Pallad. 2.10, White (1970) 237. Columella provides a lengthy overview of *pastinatio* in particular at 3.13.6–13; for *pastinatio*, see also 3.3.1n. *omnem . . . fodiunt*; 4.7.1n. *pastinatos*.

Specialized tools were developed to aid in all of this trenching and heavy digging. The one that would have been most important in excavating or overturning the soil for gardens or orchards was the heavy shovel that incorporates footrests, called a *uanga* by Pallad. 1.42.3 (cf. Adams (2016) 562–3) and still so named in some parts of Italy today (*REW* no. 9137). According to Columella (3.13.11), overseers could also use the *ciconia* (lit. 'stork'), a wooden measuring instrument, to test the depth of the excavations that the labourers toiled on (cf. White (1975) 39–40).

[42] The benefits of digging the holes well in advance of transplanting had long been known: cf. Theophr. *Hist. pl.* 2.5.1 τοὺς δὲ γυροὺς προορύττειν ὡς πλείστου χρόνου. Plin. *HN* 17.79 adds that a good sign of the maturity of the hole is the growth of a rich layer of turf in it (*pingui caespite obducantur*). One may compare the method in California almond orchards of excavating and backfilling holes several months in advance of planting (*AOM* 36–7).

[43] Cf. Pallad. 3.19.1 *scrobes autem maiores facies ut materiae prosis et fructui.*

[44] The *clibanus* was a dome-shaped portable earthenware oven, wider at the bottom than the top: for concise description, see Curtis (2001) 368–9 (cf. too the dome-shaped *furnus*, pl. 27), Dalby (2003) 101; for plates and discussion, Hilgers (1969) 148–9, Cubberly et al. (1988), Cubberly (1995). Pliny (*HN* 17.80) does not speak of the *clibanus*, but says simply that the hole should be *caminata fossura* ('an oven-shaped excavation'), with no further specification.

modica: The word here verges on *paruus* or *exiguus*, in reference to the mass or extent of the roots (if *substantia* refers to their bulk) or to their natural slightness or weakness (if *substantia* refers to their quality; see following n.). This range of meanings for *modicus* (*TLL* VIII 1229.72–1233.81 (Buchwald)) becomes increasingly common in later Latin, tracking the retreat of words such as *paruus* (not found in *Pom.* or *Med.*) that previously occupied that semantic field: see Löfstedt (1911) 71–2, Hofmann (1926) 89–90, Lundström (1948) 107, Stefenelli (1962) 81–3, Väänänen (1981) 79, Adams (2016) 547; cf. also 3.4.1n. *mediocritas*. As Adams notes, the words are replaced because frequency of use causes them to become mundane over time. The same meaning of *modicus* is found at 3.3.2 *modicis scrobibus excitatis* (n.).

radicum concessa substantia: For the use of *concedo* here, cf. *Med.* 45.5 *cortici austerior natura concessa est*, etc. (Cf. also 2.7.4n. *eiusmodi folia sortita*.) The word *substantia* refers basically to the 'material' of the roots, picking out not only their sensible bulk but also the inherent nature or quality which affects their strength and growth. In the latter sense, *substantia* is like the more common *lex* or *natura*; these nouns are used by the agronomists to create periphrases that describe in concrete terms the obscure causes for a tree's disposition to certain conditions of growth: cf. 2.7.1n. *arborum legem*, 3.4.1n. *natura . . . arborum*. For *substantia* referring to an inherent quality, cf. also *Med.* 7.1 *atriplex umidae atque frigidae substantiae holus est.*

contra aliarum arborum legem de superioribus partibus mutuatur alimentum: For *mutuor* meaning 'take in' or 'receive' without a suggestion of exchange, see *TLL* VIII 1731.61–1732.17 (Lumpe), 'vi mutui fere detersa'. *mutuatur alimentum* thus refers simply to feeding, and will provide *uariatio* for the earlier *bibit.* The preposition *de* is construed closely with *mutuatur* to indicate from where the plant receives the nourishment. As for 2.5.2 *ad superna non scandent* (n.), G. allows the context to aid in interpreting the vague directional expression *superioribus partibus*, which could conceivably refer to the 'upper parts' of either soil or roots, though G. would probably have expressed the roots in another way if that is what he meant (bare ablative, *per*, vel sim.).

In any case, what does G. mean when he says the peach is exceptional (*contra aliarum arborum legem*) in feeding from its upper parts? The explanation must have to do with roots, for, as Theophrastus says, the root was considered *per definitionem* the part of the plant by which the organism is nourished (*Hist. pl.* 1.1.9 ἔστι δὲ ῥίζα μὲν δι' οὗ τὴν τροφὴν ἐπάγεται); this fact was already theorized to a certain extent by Aristotle, who gives a number of analogies between the roots of a plant and the mouth or umbilical cord of an animal (e.g. *Gen. an.* 740b9–10, *Juv.* 468a9–11, *Part. an.* 678a11–12, 683b19–20, 686b35–687a1). Theophrastus further holds that the ultimate source of nourishment for the roots is the heat of the sun and that, accordingly, the roots will not travel below its reach (*Hist. pl.* 1.7.1 οὐδεμία <δὲ> καθήκει πλέον ἢ ὅσον ὁ ἥλιος ἐφικνεῖται· τὸ γὰρ θερμὸν τὸ γεννῶν; cf. *Caus. pl.* 1.12.7, 3.3.1). Nevertheless, the natural direction of root growth is downwards (as Aristotle emphasizes in his teleology, *Ph.* 199a28–30), and provided that the soil conditions are good nothing will prevent them from extending deeply. Indeed, deep roots seem generally to produce better nutrition and growth (cf. Theophr. *Caus. pl.* 1.18.1, *Hist. pl.* 1.7.1), and it was an appreciation of this fact that led to the agronomists' strategies for promoting their downward growth: cf. 2.7.1n. *Scrobes . . . amat* and esp. 3.2.3n. With the nutritive function of the roots in mind, it is probable that G.'s statement should be interpreted as referring to the shallow-rootedness of the peach. Other trees which the ancient agronomists maintain are shallow-rooted include the vine, olive, apple and pomegranate (Theophr. *Hist. pl.* 1.6.4, *Caus. pl.* 3.4.2, 3.8.1, Plin. *HN* 17.81). G. is the first to make this claim of the peach.

arborum legem: Columella draws out the figurative basis for this usage of *lex* to describe the principles or conditions of the plant's existence (*TLL* VII 2.1251.14–33 (Hübner); cf. *OLD* s.v. 7c, 8): 3.8.1 *igitur si rerum naturam, Publi Siluine, uelut acrioribus mentis oculis intueri uelimus, reperiamus parem legem fecunditatis eam dixisse uirentibus atque hominibus ceterisque animalibus*; cf. also 3.9.1.

2.7.2 Hanc maxime existimant causam propter quam sola maturius consenescat: *Hanc . . . causam* refers to the weak (*modica*) and shallow roots of the peach tree. The parenthetical statement, a digression typical of G.'s style, is prompted by mention of the reason for avoiding a deep planting hole.

causam propter quam . . . consenescat: For the subjunctive of attributed reason, see 2.1.3n. *siquidem . . . uideamus*. The transmitted *consenescant* requires emendation: the subject is *arbor persici*, as shown both by the preceding *mutuatur* and the following *desiderat* (cf. Mazzini 141).

2.7.3 Nec interualla maiora desiderat eiusmodi folia sortita, quorum tenuis umbra opacitatis incommodum non parat [quam scilicet binos pedes]: For fuller discussion of the considerations that influenced planting distance, see 3.4.2n. *uiginti . . . pronuntiant*. It is in general true, as G. says, that the peach requires less planting space than many other fruit trees. There is no mention in the ancient agronomists of the precise distance that should be maintained between peach trees, but we may compare modern reports: Wickson (1891) 297 gives a range from 18 feet (on triangles) to 24 feet (on squares), remarking also that the 'present tendency with the peach . . . [is] to give more room than was the custom a few years ago'; Waugh (1913) 60 offers 18 to 20 feet (and compare the results of his poll of growers, 61); and *EFN* 722 says that '[t]ree densities used to be about 6 x 6 m but most are planted

1–2 m closer now, depending on land availability, irrigation and equipment'. By any standard, the transmitted *quam scilicet binos pedes* yields impossibly small intervals for the mature trees and cannot stand in the text. Winterbottom plausibly suggests that it is a gloss, presumably on *interualla maiora*, from 2.5.2 *duobus intermissis pedibus*, which is the description of the intervals between *sulci* in the nursery, not the orchard. It is true that the lack of a precise measure here will make the following statement (2.7.4 *Sic Curtius optime positas arbitratur*) vague, but *sic* could be interpreted as referring generally to a preference for intervals that are 'not too large' (*nec . . . maiora*). An alternative is to rewrite *binos* to express a more appropriate measure, but without any ancient standard for the peach tree the emendation would be in large part guesswork. No measure is given in Palladius' short statement which summarizes *Pom.* 2.7.3–5: cf. 12.7.2 *nec a se longius statuendae sunt, ut inuicem se a calore solis excusent*; this may argue in favour of deletion rather than emendation. (Compare also Palladius' elative comparative *longius* with G.'s *interualla maiora*.)

interualla maiora: For *interuallum* in application to the space between the *ordines* of the orchard, as also at 3.4.2, see Varro, *Rust.* 1.7.2 (quoted at 3.4.2n. *simili ratione moderanda sunt*), Col. 3.15.1, Plin. *HN* 17.88. After deletion of *quam scilicet binos pedes* (see previous n.), the comparative *maiora* is either elative (as translated) or positive in sense (see 1.1.15n. *Diligentioribus*).

eiusmodi folia sortita: G. uses the word *eiusmodi* four times (also 3.3.4, 4.1.2, 4.4.3) in order to specify a plant or plant part. This striving for precision is paralleled by other features of G.'s writing: see the remarks on over-specification in the Introduction §v(b). *sortior*, like *acquiro* (2.1.3), *tribuo* (2.5.6) and *concedo* (2.7.1), belongs to the varied vocabulary that G. employs in order to describe the properties that plants 'acquire' or are 'granted' by nature or by nurture. *sortior* is also so used at *Med.* 5.1, 36.1.

tenuis umbra opacitatis incommodum: The genitive *opacitatis* specifies the source of injury (*incommodum*; cf. *TLL* VII 1.987.82–988.23 (Buchwald)): it does not merely vary *umbra*, but rather picks out what is harmful about it, namely, deprivation of light.

2.7.5 Quin immo inuenti qui densius consitas [tutas] existiment tutiores dum se inuicem protegunt a solis ardoribus, quibus difficulter hoc genus arboris creditur repugnare: G.'s additional statement reinforces and extends 2.7.3 *nec maiora interualla*, etc.: it is not just that the trees can tolerate small intervals because of their harmless shade, but, in fact, those so planted derive the extra benefit of protection from the sun. *tutiores* is the key word, picking up *densius* ('those sown *more closely* together are *safer*') and filling out the corroborative (*Quin immo*) statement to the effect that the trees actually thrive when planted nearer one another. Thus *tutas* is pointless and indeed somewhat deflationary in context (albeit not impossible); it is presumably an error of anticipation under the influence of *tutiores*. Winterbottom observes that the deletion improves the rhythm. Mazzini 141 defends *tutas* by arguing that 'in ogni caso non è fuori luogo in un contesto che presenta aspetti di oralità', but the rebuttal is weak, especially because the colloquial nature of G.'s writing has not been satisfactorily demonstrated (see Introduction §v(d)).

dum: *dum* is causal. Although the sense is not mentioned in L&S or *OLD*, the *TLL* has many examples of *dum*-causal, both where it retains an idea of time (*TLL* v 1.2221.13–50 (Hey)) and where the notion of cause alone predominates (= *quia*, *quod*, *TLL* v 1.2221.51–2222.10). For the indicative after *dum* in indirect speech, cf. Keeline (2013) 247.

se inuicem protegunt a solis ardoribus: Palaeographically the emendation *praetegunt* for the paradosis *pretegunt* is easier than *protegunt* (*e* is often written for *ae*, and vice versa, in **N**; see Mazzini 25), but *protegunt* is hardly difficult and is preferable on other grounds. *protego* is regularly construed with *a*(*b*) + ablative (*TLL* x 2.2257.8–9 (Schröder)), whereas *praetego* seems never to be (cf. *TLL* x 2.977.73–5 (Beikircher); it is found once with *ex*). Moreover *protego* is very commonly used in reference to protection offered by plants or their foliage ('adversus iniurias caeli . . . arbores, folia, locus naturalis sim.', *TLL* x 2.2258.32–43 (Schröder)), including by the shade they cast (as here).

2.8–9 These mutilated sections preserve some of G.'s recommendations for tending to and improving the productivity of the peach tree. Enough survives to show that the disposition of material here resembles the careful but implicit organization found in the corresponding sections on the almond: cf. 3.5–7n., where it is argued that the advice is structured according to different aspects of the growth and productivity of the tree. Thus in the present instance 2.8 discusses ablaqueation, root-pruning and manuring, but when the text resumes in 2.9 the subject has obviously changed to the improvement of the produce of the tree. Palladius (12.7.3–6) repeats most of the advice in these sections (observed *passim* below), and we may cautiously infer the missing contents of G.'s text by consulting his discussion: Palladius has methods for avoiding and remedying sun damage (12.7.3–4), for restoring an unproductive tree (12.7.3), for protecting the tree from frost (12.7.4), for killing parasites (12.7.5) and for remedying dropping or wrinkled fruit (12.7.5).

2.8 Palladius simplifies G.'s extant instructions, omitting both the reason for ablaqueation (in order to remove the rootlets) and the recommendation to mix the mulch with soil: 12.7.3 *ablaqueandae sunt per autumnum et suis stercorandae foliis*.

2.8.1 Ablaqueare: The meaning of *ablaqueo* is rather broader than that given by White (1970) 263, 'trench around vine-stock, exposing surface rootlets' (*ad hoc* definition of Col. 4.8.1 notwithstanding). The purely technical term is used to refer to the operation of trenching around and exposing the roots of any plant, not only the vine: in Cato for example it is used most frequently of the olive (see further *TLL* I 103.29–40 (Bannier)). It is true, however, that the word is applied often to the practice in connection with the vine – in part simply because of the enormous attention it receives in the agronomists relative to other plants – and as a consequence could be used as a specialized term in application to it. Cf. thus Pallad. 2.1 *Ianuario mense locis temperatis ablaqueandae sunt uites, quod Itali excodicare appellant, id est circa uitis codicem dolabra terram diligenter aperire et purgatis omnibus uelut lacus efficere, ut solis teporibus et imbribus prouocetur*; also Isid. *Etym.* 17.5.31. In suggesting a link between *ablaqueo* and *lacus*, Palladius follows a derivation of the word that Varro seems to favour (*Rust.* 1.29.1 *in pratis circum uites ablacuari*), that is, as though it were formed from *ab* + *lacus*. Current authorities prefer a derivation from *ab* + *laqueo* (*TLL* I 103.25–8 (Bannier), E–M 607–8, de Vaan 327).

ablaqueatio itself was practised for one or more of five reasons: (1) as here and at 3.7.1 (*ablaqueatis radicibus*), to expose the roots for the application of water, manure or other fertilizers (Cato, *Agr.* 29, 36, 37.2, Col. 4.8.3, 4.22.3, 5.9.16–17, 11.2.87, Plin. *HN* 17.127, 17.259, 17.263, Pallad. 3.25.23, 3.25.30, 4.9.29, 11.8.3); (2) also as here, for the purpose of cutting off the surface rootlets (Theophr. *Caus. pl.* 3.8.1, Cato, *Agr.* 114.1, Col. 4.8.1–2, 4.29.10, 5.5.5, Pallad. 11.5.1, 11.8.1–2); (3) to loosen the soil and allow the penetration of water, sun, etc., to the roots (Col. 4.8.3, 4.14.2, 5.5.5–6, *Arb.* 10.4, Plin. *HN* 17.140, 17.246, Pallad. 2.1); (4) to remove weeds from the surrounding soil (Cato, *Agr.* 114.1, Pallad. 2.1); and (5) for the other benefits attendant upon hoeing and turning over the soil, e.g. to provide an insulating blanket against heat (Col. 5.9.12; cf. 2.6.1n. *frequens . . . runcationis*). See further White (1970) 238, Martin on Pallad. 2.1.

nouissimo autumno: An autumn time for ablaqueation is standard (Cato, *Agr.* 27.1, Col. 11.2.79), especially for olives (Cato, *Agr.* 5.8, Col. 5.9.13, 11.2.87, Plin. *HN* 17.127, Pallad. 11.8.1–2) and vines (Col. 4.8.1, 4.8.4, 4.14.2, 4.17.2, 5.5.5, Plin. *HN* 17.189). As usual, the meaning of *nouissimus* in reference to time is 'the last phase of, the end (of a period)' (*OLD* s.v. 2); cf. e.g. Col. 2.10.23 *Ianuarii parte nouissima uel toto Februario.*

inhalantibus fibris: *inhalo* may mean either 'send out exhalations' or 'send out an odour' (*TLL* VII 1.1591.43–9 (Hofmann-Ehlers)); on the strength of various passages where it refers to the smell of a plant, Hofmann and Ehlers assign the present instance to the latter category. The paradosis has its difficulties for, as Winterbottom reasonably asks, 'why should fibre smell, and if it did what matter?'. But before we accept his suggestion *inhaerentibus*, attractive though it may be, we should reckon with the fact that G. appeals elsewhere to the harmful effects of vegetative exhalations and odours: cf. 2.6.1n. *tenero germini noceant flatus herbarum*.

fibris de radice purgatis: *purgo* is used commonly by the agronomists for a wide range of tasks that involve some kind of 'cleaning'. G.'s application of it to plants to describe pruning (so too 3.5.1 *ramos . . . falce purgare*) is rare, but found also at e.g. Pallad. 8.2.3, 12.4.2. More common are references to weeding or otherwise clearing (e.g. of stones) a field, trench, etc. (*TLL* x 2.2682.62–71 (Spoth), *OLD* s.v. 2c; cf. Cato, *Agr.* 2.4, Col. 2.21.3, 11.2.27, Plin. *HN* 18.295, Pallad. 3.6); cleaning plant parts or products (e.g. Cato, *Agr.* 65.1, Col. 11.2.92, 12.32.1, 12.36.1, Pallad. 12.7.12); or simply cleaning any space on the farm (*TLL* x 2.2682.45–62; cf. Cato, *Agr.* 2.3 *uillam purgari*). The verb *purgo* is most often joined with *ab* or the bare ablative, but in later texts *de* becomes more common. Even in early Latin, however, it may be found with *de*: cf. Cato, *Agr.* 151.3 *herbas de areis purgare*. The evidence for an increasing usage of *de* here is at best ambivalent (see 2.5.7n. *de codice auellitur*).

folia sua in uicem stercoris cum terra molliore congerere: The compound verb *congero* is often used of heaping soil (cf. Col. 5.10.4 = *Arb.* 19.2 *terra quae in eum* (sc. *scrobem*) *congesta est*; see *OLD* s.v. 3), but the prefix *con-* here and at 3.1.2 has special point: in both cases it refers to a combination of soil and manure (so also 3.1.2 *terram . . . cum stercoris admixtione*). For *in uicem stercoris*, cf. 2.1.3 *ea res* (= *poma ipsa*) *et ossibus uicem stercoris praebeat*; and for remarks on vegetative manure, see Appendix II.

cum terra molliore: The comparative assumes as an implicit standard (see 1.1.15n. *Diligentioribus*) the surrounding soil in which the tree is planted.

2.9 This section is oriented towards maintaining and improving the produce of the mature tree. There is a method for making the fruit larger, for pruning so as to induce early ripening, and, more vaguely, for protecting the tree against 'many harms' (*plurima incommoda*) with the magic of *spartum* (esparto). The several borrowings of Palladius are noted *passim*.

2.9.1 magna fieri asseuerant si, cum arbor in flore sit, lactis caprini per triduum ternis sextariis irrigetur: Cf. Pallad. 12.7.6 *magna poma persicus affert, si florenti per triduum ternos sextarios caprini lactis ingesseris.* The benefit rendered by the goat's milk is obscure, but perhaps germane is Pliny's claim that it is the next most nourishing to human milk: *HN* 28.123 *maxime autem alit quodcumque humanum, mox caprinum*.

fieri asseuerant: G. also constructs *asseuero* with a *quod*-clause (3.3.4), for which see Introduction §v(c).

cum arbor in flore sit: The expression *in flore* (also at 3.5.3, 3.6.4) is regular both in the agronomists and in other authors: see *TLL* vi 1.931.73–84 (J. Kapp).

per triduum: See 2.2.2n. *paucis diebus*.

ternis sextariis: The *sextarius* (Gk. ξεστής) is one-sixth of the Roman measure *congius* (Gk. χοῦς, 'pitcher'). A current estimate for the measure is 540 ml or 18.25 oz. See Richardson (2004) 41–4, *NP* s.v. 'sextarius' (Schulzki).

2.9.2 Omnibus persicis, maxime tamen praecoquis et armeniis: The *armenia* are apricots in general, the *praecoqua* an early-ripening variety: see 2.1.2n. *armenia*.

quotiens abundantibus pomis arbor onerata est, decerpenda sunt aliqua: sic enim reliquis celerior maturitas imperatur: The idea is that with the burdensome fruit removed the tree will redirect its energies to those that remain: cf. *Geop.* 10.13.2 τινὲς δὲ ἀφαιροῦσι τὰ πολλὰ περσικά, καὶ ὀλίγα καταλίμπανουσι τῷ δένδρῳ. οὕτω γὰρ ἔσται μείζονα, τῆς τροφῆς εἰς ὀλίγα χωρούσης; Pallad. 8.3.1 (on pear and apple) *interlegenda sunt si qua uitiosa* (sc. *poma*) *reppereris, ut arboris sucum uertamus ad meliorum nutrimenta pomorum*; also 3.25.16, 7.5.1. (G. similarly recommends culling almonds so that the rest will become better: 3.7.6 *Si uiridibus auferas tenuiora, quaeque liqueris meliora sunt.*) The practice of thinning peach trees remains absolutely vital in their production today (Wickson (1891) 303, *EFN* 722–3, Marini and Reighard (2008)): 'Most peach trees produce thousands of flowers and, if conditions are favourable, may set several thousand fruit per tree. If all these fruit are allowed to develop, the weight of the fruit will break branches and the fruit will be small and have low sugar concentrations. To avoid overcropping, the number of fruit per tree must be managed' (Marini and Reighard (2008) 289).

abundantibus pomis arbor onerata est: *onero* is often used of the abundance of a plant's parts or produce: see *TLL* ix 2.631.60–71 (Beikircher). It can also mean to 'overburden', sc. by excessive weight, as in the case of the danger presented to the young vine by its boughs: cf. Col. 5.5.10 *sed haec bracchia non statim primo anno tam procera summittuntur, ne oneretur exilitas uitis*. Here the foremost idea is obviously that of abundance, but there is also a sense in which the numerous fruit are burdensome to the tree (see previous n.).

decerpenda sunt aliqua: *decerpo* (found already in Cato, cf. *Agr.* 112.3) is very common in both the agronomists and other authors in reference to flowers, fruit and leaves: see *TLL* v 1.157.66–158.7 (Simbeck).

sunt: **N** transmits the impossible *sit*. A slightly easier fix is *sint*, but the subjunctive is grammatically and semantically unmotivated. There are to be sure some uncertainties in G.'s usage of the subjunctive (cf. e.g. Introduction §v(c) on *quod*-clauses, and 1.1.10n. *Sunt qui . . . subigunt . . . existimant*), but those uncertainties relate to his usage of the moods in structures – indirect statements, relative clauses – where the subjunctive has an explanation, not, as here, where it would be used independently. The claim of Mazzini 141 that '*sint* non è impossibile se si suppone una struttura indiretta' is therefore unconvincing; no such structure can be found here, as it can, for example, where G. is attributing the reason for a piece of advice to another author: cf. 2.1.3n. *siquidem . . . uideamus*.

sic enim reliquis celerior maturitas imperatur: *impetratur* may be considered (Reeve), but *impero* has a good pedigree among the agronomists. Mynors on Verg. *G.* 1.99 *imperat aruis* notes that *imperium* is used for the farmer's control over his fields at Cic. *Sen.* 51 (but Powell argues for the reading *impendium*), Ov. *Fast.* 2.296, Tac. *Germ.* 26.2. Of greater relevance for the present usage, he also observes for the verb the 'special use, "to force into bearing", especially of vines', citing Sen. *Dial.* 9.17.5, Col. 3.3.6, 4.24.21, 4.29.12, Plin. *HN* 17.178 (cf. Powell on Cic. *Sen.* 51, *TLL* VII 1.589.2–6, 'fere i. q. instare, vim inferre' (O. Prinz)).

2.9.3 Sparto ligandam aduersum plurima incommoda, ut amphora, existimant, et hodieque nonnulli iumentorum solias in itinere derelictas pro remedio truncis ramisue suspendunt: G. appends two further pieces of advice for protecting the produce on the tree. The recommendations present some interpretative difficulties, but a clue is offered by Palladius' version: 12.7.6 *contra uitia persici proficit spartum ligatum uel spartea suspensa de ramis.* As his condensed account shows, the element common to both recommendations is *spartum*, a fibre from the esparto plant used in cords, basketry and so forth (see 2.9.3n. *Sparto*). Shoes for cattle and other pack animals (G.'s *iumentorum solias*) were regularly made of *spartum* (cf. Col. 6.12.2 *solea spartea*; see further D–S III 2.2012, Brouquier-Reddé (1991) 44), and the practice was familiar enough that *spartea* alone could denote such a slipper (*OLD* s.v. 1c), as it does also in the quoted passage of Palladius.[45] But granting that *spartum* is the common thread, what exactly is the purpose of the rope (G. *sparto ligandam* / Pallad. *spartum ligatum*) and shoes (G. *iumentorum solias* / Pallad. *spartea suspensa*)?

The second piece of advice, *et hodieque*, etc., a peculiarly detailed stipulation to hang just those shoes that 'have been left in the road' (*in itinere derelictas*), suggests a superstitious basis for the advice. D–S III 2.2012 and Kaufman (1931) 620 observe the magical value of discarded shoes in antiquity, comparing it with that of the horseshoe in modern times (for which see Lawrence (1899) 1–139). Moreover, the advice to hang various objects from the tree – a kind of

[45] To be sure, iron shoes, if not the modern horseshoe as such, were by no means unknown: see Brouquier-Reddé (1991).

'anchoring' magic – is commonplace in Palladius and the *Geoponica*: see Ager (2010) 187–8. Note thus that, on the peach, Palladius also says that hanging a snake skin will be beneficial (12.7.4 *iuuat in ea et spolium serpentis appendi*). For other clear examples of anchoring magic found in *Pom.*, see 3.6.2–3n.

The magical nature of the second piece of advice employing esparto suggests that the first should be interpreted along the same lines. Indeed, it is difficult to find a physical purpose for tying up the fruit (if *liganda* (sc. *poma*) is read with Scotti) or tree (if *ligandam* (sc. *arborem*) with Mazzini). Columella recommends putting up a net of esparto around ripe pomegranates to protect against the forays of birds: 12.46.2 *deinde sparteis retibus arborem cludunt, ne aut coruis aut cornicibus aliisue auibus pomum laceretur*. But G.'s *sparto liganda(m)* will hardly support the meaning 'enclose with an esparto net', and in any case, the detail *ut amphora* excludes that meaning. This parenthetical remark shows that we are meant to think of a relatively close-fitting wrap, in the way that amphorae and other vessels were bound with esparto as protection against hard usage (Plate 10): cf. Cato, *Agr.* 11.2 *amphoras sparteas* with White (1975) 124. But would *ut amphora* also suggest that the esparto on the tree is meant to protect the wood or fruit from knocks or blows? That seems improbable (where would the danger of blows come from?), and similar advice is not found in other agronomists.[46] The only function that remains for the esparto knot is therefore a magical one. This conclusion is corroborated by Palladius' version *spartum ligatum uel spartea suspensa de ramis*, where the esparto knot and shoes are presented on a par – they are alternatives, thus *uel* – and there is no hint of a physical solution to possible harms. I prefer *ligandam* (supplying *arborem* from 2.9.1–2) against *liganda* (with which we would supply *poma*), for it is difficult to imagine that G. would intend the individual fruit to be wrapped *ut amphora*: the detail is more easily understood if the trunk or branches are meant, where the shoes are also hung (cf. *truncis ramisue*).

On the role of magic in Roman agriculture in general, see Diederich (2007) 77–155, Ager (2010); and on magic more broadly in (Greek and) Roman culture, see e.g. Tavenner (1916), Dickie (2001) 124–41 and *passim*, Luck (2006) 1–29, Johnston (2007*b*), Collins (2008), Gordon and Marco Simón (2010).

Sparto: '[N]om de plantes utilisées en sparterie et corderie' (*NPRA* 244). *spartum* may refer to any of three species that 'yield a similar fibre, with similar

[46] When Columella reports Mago's advice to tie picked pomegranates with flax or esparto before dipping them in boiling water (12.46.5 *Poenus quidem Mago praecipit aquam marinam uehementer calefieri et in eam mala granata, lino uel sparto ligata, paululum demitti, dum decolorentur*), the idea is rather that the fruit are on strings whereby they can be 'lowered in little by little' (*paululum demitti*) and subsequently pulled out.

commercial uses' (White (1975) 29–30): *Spartium junceum* L. (Spanish broom), *Stipa tenacissima* L., and *Lygeum spartum* L. The last two are both called 'esparto grass' (*Lygeum spartum* being 'true' esparto), and the rope produced from them may be called in Latin simply *spartum* (esparto). See further *LTBL* 298, Boscherini (1970) 105, White (1975) 29–30, *NPRA* 244.

ligandam: G. uses *existimo* with voluntative force with both the gerundive and the infinitive (see 1.1.10n. *illiniri existimant*), but if the infinitive were intended we would expect *ligari*: the easy emendation to the gerundive is therefore desirable. For *ligandam* (sc. *arborem*) rather than *liganda* (sc. *poma*), see 2.9.3n.

aduersum: For *aduersum* in the sense of '(as a remedy) against' (cf. *Med.* 9.5, 25.12, 39.13), see *OLD* s.v. 7c.

plurima incommoda: For the collocation, cf. *Med.* 48.2 (sc. *zizypha*) *habent . . . plurima incommoda.*

solias: The spelling *solia* is condemned by *App. Prob.* Heraeus 80 *solea non solia.*

pro remedio . . . suspendunt: Mai says that he believes he saw *pro remedio* in the MS, which is a far happier reading than the vague *prae medio* ('in front of the middle'). *pro remedio* is parallel to *aduersum plurima incommoda*, explaining the reason for hanging the slippers from the tree.

truncis ramisue: G.'s use of *-ue* is significant for the same reason as his use of *-que*, *atque* and *ac*, that is, insofar as it reveals the literary quality of the text (see Introduction §v(c)): *-ue* had all but disappeared as a productive coordinator in the first century BCE (H–Sz §271, Janson (1979) 103, Torrego (2009) 456–7), and its continued use was associated with higher-register Latin, whether poetry or prose.

2.10–11 These sections present three methods of grafting the peach tree (Plate 11). Grafting, which may be defined as the 'the natural or deliberate fusion of plant parts so that vascular continuity is established between them . . . and the resulting genetically composite organism functions as a single plant' (Mudge et al. (2009) 439), was one of the most important developments in ancient agriculture, allowing for the domestication of many species of tree that could not be propagated by shoots or cuttings (Mudge et al. (2009) 439, Zohary et al. (2012) 114–15; for the example of the apple, see Juniper and Mabberley (2006) 91–7). For the Roman agronomists in particular, grafting solved several problems presented by the Mediterranean climate and offered significant advantages over propagation from seed and slips. Grafting allowed the grower to avoid the juvenility of plants, which presented the problem of getting the root system of the young tree established and delayed the yield of the fruit (White (1970) 248, Mudge et al. (2009) 441–2). As the agronomists also appreciated, 'whereas seedlings tended to revert to the wild form or to deteriorate in quality, grafted stock gave a yield true to type' (White (1970) 248; cf. Theophr. *Hist. pl.* 2.2.4–5, *Caus. pl.* 1.9.1, Verg. *G.* 2.57–60, Col. 3.1.2, Plin. *HN* 17.59, Pallad. 3.25.2, 3.25.28). Grafting also enabled the grower to 'introduce new and better varieties, and to acclimatize them' (White (1970) 248, 262; cf. Plin. *HN* 15.35). Also recognized was the ability of a graft to repair or restore a tree (cf. Col. 5.9.17, *Geop.* 10.13.4; 2.10.8n. *cito recurrunt in senectutem*). On these and other benefits, see more extensively Mudge et al. (2009) 441–5, Hartmann et al. (2013) 437–42, Bilderback et al. (2014) s.v. 'Reasons for grafting and budding'. The relative indifference to the value of grafting shown by the earlier Roman agronomists – this in spite of the fact that the advantages had long been known to the Greeks (cf. Theophr. *Caus. pl.* 1.6.10) – gave way to an increasing appreciation for its superiority over other methods of propagation. Its advantages were well enough accepted by the time of the early Empire that it could be casually stated that grafted trees are, as a rule, more fruitful than ungrafted ones: *Arb.* 20.2 *arbos insita fructuosior quam quae insita non est, id est quam quae ramis aut plantis ponitur*; cf. already Verg. *G.* 2.36 *fructus . . . feros mollite colendo*; 2.47–8 *sponte sua quae se tollunt in luminis oras / infecunda quidem.*[47] (The idea is also echoed at *Geop.* 10.75.13 ἐπὶ τοσοῦτον δὲ χρήσιμός ἐστιν ὁ ἐγκεντρισμός, ὅτι κἂν αὐτά τις τὰ φυτὰ ἑαυτὰ ἐγκεντρίσῃ, τροφιμώτερα καὶ καλλίονα γίνεται.)

Roman authors speculated on the origins of grafting, often deriving the art from human observation of chance natural grafts (e.g. Lucr. 5.1361–9, Plin. *HN* 17.99–101; cf. Mudge et al. (2009) 445–9). Many techniques for grafting and budding were known (White (1970) 248, 257, Hardy and Totelin (2016) 137). For grafting proper (= *insitio*, ἐγκεντρισμός), the principal types discussed by the Romans are (1) cleft-grafting (*insitio per fissuram*), (2) bark-grafting (*insitio in cortice*) and (3) bore-grafting (*terebratio*). The last of these applies only to the vine (Cato, *Agr.* 41.3–4, Col. 4.29.13–17, Plin. *HN* 17.116, Pallad. 3.17.7; exceptionally of the olive at Col. 5.9.16). Budding (= *inoculatio*, *emplastratio*, ἐνοφθαλμισμός) was done especially through patches or chips. G. mentions cleft- and bark-grafting in 2.10 (*uel in cortice uel in ligno*) and patch-budding in 2.11 (*emplastratio*) as viable options for the peach. In cleft-grafting, the simplest method, the stock is split and the scion is trimmed and fitted into the fissure (Cato, *Agr.* 41.2, Verg. *G.* 2.78–80, Col. 4.29.8–12, 5.11.7–8 = *Arb.* 26.6, Plin. *HN* 17.102, 109, Pallad. 3.17.3, *carm. ins.* 42, *Geop.* 10.75.5). In bark-grafting, a slot

[47] Although not likely to be a significant factor for most growers, one may also note the symbolic value attached to grafting in Roman poetry and culture, which attests to great interest in it: see Ross (1980), (1987) 104–9, Thomas (1987) 244–6, 258, Bovey (1999), Gale (2000) 211–15, Clément-Tarantino (2006), Hunt (2010), Lowe (2010), Gowers (2011), esp. 113–14, Henkel (2014), esp. 58–61, von Stackelberg (2014), esp. 415–23.

is cut into the bark of the stock and the bark carefully teased back from the wood; the scion is then trimmed and fitted into the space between bark and wood (Cato, *Agr.* 40.1–4, Col. 5.11.3–6 = *Arb.* 26.2–5, Plin. *HN* 17.102–10, Pallad. 3.17.1–3, *carm. ins.* 41, *Geop.* 10.75.2–4). The method produces a better graft because the scion is thus brought into contact with the layer of cells below the bark called the cambium, which is responsible for the secondary growth of the tree (Bilderback et al. (2014)). In patch-budding, the most involved of the grafts (cf. Col. 5.11.8 = *Arb.* 26.7 *subtilissimum*), the scion-bud is carefully removed from its tree together with a precisely shaped portion of the bark; on the stock, a patch of bark of identical shape is excised and replaced with that of the bud (Cato, *Agr.* 42, Col. 5.11.8–11 = *Arb.* 26.7–9, Plin. *HN* 17.100, 118–19, Pallad. 7.5.2–4, *carm. ins.* 43–4, *Geop.* 10.77).

Further details on methods of grafting in antiquity and the details believed to affect the success of these grafts will be found *passim* below, but one aspect of special importance deserves mention here: the determination of compatible graft-pairs. It is now known that 'taxonomic affinity is the arbiter of which species can be grafted successfully onto any other' (Mudge et al. (2009) 440). This fact was not recognized as such by the ancient agronomists, but they did appreciate that a 'principle of limitation' applied in the compatibility of different grafts (Pease (1933) 66; cf. Hardy and Totelin (2016) 138–9). Thus Theophrastus says that 'like always grows easily onto like' (*Caus. pl.* 1.6.2 εὐπρόσφυτον δ' ἀεὶ τῷ ὁμοίῳ τὸ ὅμοιον), adding that similarity of bark leads most to successful grafts, and Varro emphasizes that the nature of the trees (*Rust.* 1.40.5 *qua ex arbore in quam transferatur*) is an important consideration in grafting (cf. Plin. *HN* 17.103). Nicolaus also says that grafting of like onto like is better, calling attention to the success of intra-specific grafts of vine onto vine, olive onto olive, and almond onto almond (*De plantis* 820b35–7). Columella himself makes the similarity of bark a decisive factor (5.11.1 = *Arb.* 26.1 *omnis surculus inseri potest, si non est ei cui inseritur arbori cortice dissimilis*) and reports that the *antiqui* prioritized similarity of bark, rind, and fruit (5.11.12 = *Arb.* 27.1 *eos tantum surculos posse coalescere qui sint cortice ac libro et fructu consimiles îs arboribus quibus inseruntur*). In spite of their generally sound principles, however, many fallacious grafts are reported by the agronomists; even if some are to be defended (Pease (1933) *passim*), explained for example by inarching (i.e. grafting a branch without separation from the parent stock, as at Col. 5.11.12–15 = *Arb.* 27.1–2), they cannot all be maintained. The systematic treatment of graft compatibility offered by Palladius' *carmen insitionis*, for example, contains more than two dozen impossible pairs, and fallacious grafts are widely reported from Virgil to the *Geoponica* (for details see Pease (1933), Mudge et al. (2009) 455–8). Happily, all of the stocks that G. reports for the peach (peach, almond, plum) are compatible and indeed continue to be used as rootstocks today in pure or hybridized form (Reighard and Loreti (2008) 194–5).

To illustrate the superiority of grafting as a technique for propagation, one may note the dominance in American orchards, both those today and in recent centuries, of propagation by budding and grafting. Fruit trees for commercial production are not grown from seedlings all the way to fruit trees; instead, by budding a scion cultivar onto a rootstock and promoting vigorous growth with intensive irrigation and fertilization (obviously unavailable for most Roman farmers), trees are rapidly developed for the market and made ready for transplantation within a year. For discussion of present-day grafting techniques in connection with the peach and almond, see Wickson (1891) 110–23, Waugh (1913) 43–56, *EFN* 712, 722, Loreti and Morini (2008) 237–41. As remarked at 3.1–4n., propagation by budding does not eliminate the need for growth of rootstock; a nursery is still required, and attention to planting from seed for both the peach and almond is thus not rendered entirely superfluous even by advanced grafting methods (cf. Wickson (1891) 99–109, Loreti and Morini (2008) 221–4).

2.10 We are told that the peach can be sown either by bark- or cleft-grafting (2.10.1 *uel in cortice uel in ligno praecedente fissura*) and offered a miscellany of remarks pertaining to the methods. We are told where on the stock to graft them, which seasons are best for grafting, what kind of scions to gather (not from the top of the tree), and which trees the peach will take well to. In a digression at the end of the section, G. briefly considers the viability of reciprocal grafts of plum and almond onto peach stock. His remarks are for the most part narrowly focused on the peach, and it is clear from certain omissions that he assumes more than a passing familiarity on the reader's part with the two *modi* of grafting: we find no technique for trimming the scions, splitting the stock for a cleft-graft, opening a slot in the bark to receive scions in a bark-graft, or binding and protecting the grafts, among other relevant subjects. More to the point, G. does not even clearly distinguish in the passage between those instructions applicable to bark-grafting and those to cleft-grafting or remark on the suitability of the methods in different conditions. This may in part be because the instructions could apply to both methods, but in at least one case he assumes that the reader will understand that the instructions are only pertinent to cleft-grafting (2.10.2n.). It is therefore clear that G. either was relying on the tacit knowledge of his audience, or else that he provided fuller introductory statements in a general exposition of grafting: for discussion of these possibilities, see Introduction §III(a), and cf. 3.1–4n.

2.10.1 Inseri: *insero, inserere, inseui, insitum* may refer simply to planting, but it is also the *uox propria* for grafting, especially through *surculi* (*TLL* VII 1.1876.35–1877.48 (Stiewe), *OLD* s.v. 2, *LAT* 38, White (1970) 264); other verbs (*inoculo, emplastro*, etc.) are preferred for patch-budding. The present forms of this verb are indistinguishable from those of *insero, inserere, inserui, insertum*, with which it is sometimes confused in meaning and form: for a synopsis see *TLL* VII 1.1869.7–22, 1875.35–61 (Stiewe); also Clément-Tarantino (2006) 7 n. 17.

fissura: The standard term in Columella and Pliny for the cleft in the tree (made by a *cuneus*) into which the scion is to be inserted: see *TLL* VI 1.828.45–51 (Leonhardi).

2.10.2 Sed quantum propius ad terram recisa sit, eo melius inseretur: fortior est enim pars eius ima: Although G. does not say so, these instructions obviously apply to cleft-grafting, in which the stock is cut low (cf. *recisa*) before being split. Cleft-grafting was thus recommended principally for smaller trees, for which such a procedure was feasible. The standard advice is to cut the tree close to the ground (as G. recommends): Col. 5.11.7 = *Arb.* 26.6 *si pusillam arborem inserere uoles, imam* [*Arb. iuxta terram*] *abscidito ita ut sesquipedem e terra extet*; Pallad. 3.17.5 *si arborem minorem desiderabis inserere, in qua sine dubio meliora incrementa proueniunt, circa terram secato.*

quantum propius . . . recisa sit, eo melius inseretur: *quantum* is used for *quanto* with comparatives from the time of Livy (H–Sz §84.b.γ, Oakley on Liv. 10.35.2, Adams (2016) 622–3). For *quantum* correlated with *eo* (not uncommon), cf. Liv. 3.15.2 *quantum iuniores patrum plebi se magis insinuabant, eo acrius contra tribuni tendebant*, etc. The subjunctive (especially perfect) is extremely rare in correlative *quantus*-clauses, and *erit* would be more easily understood than *sit* (note the future *inseretur*). Nevertheless, it is perhaps possible to explain it as an extension of the iterative subjunctive, which is well attested by the end of antiquity at least for the generalizing *quantuscumque* (H–Sz §302; cf. §297.1.a on the wider extension of the subjunctive in subordinate clauses). For another iterative subjunctive with *quantum* + *propius*, cf. *Paneg.* 6.18.2 (*anno* 310) *inde arreptis armis portas petiuerunt, tot dierum iter a Rheno usque ad Ararim sine ulla requie peregerunt indefessis corporibus, animis flagrantibus, crescente in dies ardore uindictae quanto propius accederent.* Despite the lack of *tanto*, *eo*, vel sim., the phrase *crescente in dies ardore* is correlated with the *quanto*-clause: 'with their zeal for vengeance daily waxing greater the nearer they approached'.

2.10.3 Quidquid in summo est, quoniam suci minus retinet, ut fragile ac putre damnatur: The *quoniam*-clause within this parenthetical statement incidentally explains the reason for the strength of the 'lowest part' of the tree (*pars eius ima*): it has more *sucus*. *sucus* refers to the nutritive element for the plant both inside and outside of it, i.e. not only its sap (*OLD* s.v. 3a) but also the raw nourishment (*alimentum*) in the soil (*OLD* s.v. 4; cf. e.g. Verg. *G.* 1.89–90, quoted at 3.1.5n. *leui sarculo relaxetur*). According to the most widespread theory of plant growth in antiquity (varying in its particulars in different authors; cf. Hardy and Totelin (2016) 143–6), vital nutrients in readily usable form are drawn out of the soil by the roots and siphoned through the trunk into the more remote parts of the plant. Columella has an admirably concise description of this process in reference to the vine: 3.10.2 *naturali enim spiritu omne alimentum uirentis quasi quaedam anima per medullam trunci ueluti siphone, quem diabeten uocant mechanici, trahitur in summum; quo cum peruenit, ibi consistit atque consumitur.* (For the roots as the part of the plant drawing the *alimentum* from the soil, see 2.7.1n. *contra . . . alimentum.*) On the basis of this respiration or siphon theory of nutrition, Columella argues that the most material for growth (*materiae uehementissimae*) is found in the 'head' (*caput*, the furthest or uppermost part) and 'shank' (*crus*, the lower part of the stalk) of the plant, the head because the *sucus* comes to rest there and the shank because all *sucus* is funnelled through it. Furthermore, the lower parts are essentially the most fertile, *quod . . . ex proximo terrae integro atque illibato suco aluntur* (3.10.3). It seems that G. would accept some broadly similar theory of consumption up to the point of the trunk, rejecting, however, at least in the case of the peach tree, the idea that the sap is channelled to and retained in the furthest reaches of the plant.

fragile ac putre: *fragile* refers not only to the brittleness of the upper parts (*OLD* s.v. 1) but also to their general weakness or lack of vitality owing to a dearth of sap. *putre* is a remarkably strong way to describe the upper branches of an otherwise healthy tree; in application to plants it generally picks out rot or decay, e.g. as a result of age (examples at *TLL* x 2.2751.59–64 (Clementi)). With *ac* it is clear at least that *putre* is climactic to *fragile*, suggesting a translation such as 'apt to fall apart' (sc. if supporting a graft?).

2.10.4 Inserendi tempus aptissimum est nouissimi autumni, si loca non sint adeo pruinosa, uel inter brumam et exortum Fauoni, id est extremo Decembri ac deinde Ianuario, ita exigente praecoquo germine, quod saepe et ante ueris aduentum florere festinat: Although there are two timeframes offered here, they are not strictly interchangeable: the first (*nouissimi autumni*) applies to the peach generally, presumably also the apricot, and the second (*inter brumam et exortum Fauoni*) to the variety of apricot called *praecoquum* (see 2.1.2n. *armenia*). The agronomists offer an extremely wide range of grafting times both for fruit trees considered generally and for particular species. The overriding consideration in each case remains the budding time of the specific tree, and the agronomists sensibly advise adjusting the grafting time in accordance with it (Theophr. *Caus. pl.* 1.6.3, Varro, *Rust.* 1.39.1–2; cf. Col. 5.11.2 = *Arb.* 26.2, Plin. *HN* 17.113). This does not, however, prevent dispute over the best time to graft: as Theophrastus observes, the arguments for the ideal season for grafting any plant are similar to those for its planting time (*Caus. pl.* 1.6.3 παραπλήσιοι δὲ καὶ οἱ λόγοι καὶ περὶ ἑκάστης οἵπερ καὶ περὶ τῆς φυτείας). There are some who recommend spring, he says, because it coincides with the time of budding, when the bark will readily grow onto and enclose the graft (ὁ φλοιὸς ἐπιφύεται καὶ περιλαμβάνει); Columella might fall into this camp, for he urges grafting as soon as the buds appear (Col. 5.11.2 = *Arb.* 26.2 *omnes arbores simulatque gemmas agere coeperint . . . inserito*; cf. Plin. *HN* 17.113). There are others, however, who counsel autumn, because this will allow the plant to have a 'more powerful start' (ἰσχυροτέρας ἀρχῆς) as it grows in the fall and winter before bursting forth in the spring (cf. also *Caus. pl.* 1.12.1–2). Theophrastus mentions as a final possibility a summer grafting time, which is found elsewhere connected with the pear and apple (Cato, *Agr.* 41.1) and fig (Varro, *Rust.* 1.41.1–2). There is truth to the viability of multiple seasons for grafting, as shown by a similar variation today and in recent history (Wickson (1891) 122–3, *EFN* 722).

G.'s suggestion of an autumn grafting time agrees with his recommendation to plant both seeds (2.3.2) and *plantae* (2.5.7) in autumn. He shows some concern about frosty conditions (*si loca non sint adeo pruinosa*), which recalls observations by earlier agronomists that the grafting time depends on local conditions (e.g. Theophr. *Caus. pl.* 1.6.9) as well as his own discussion of the dangers of cold weather when propagating the peach by seed (cf. 2.3.2n. *metus . . . experitur*, 2.3.5n.), but he does not suggest an alternative schedule for such climates. One will also note that G. revises the grafting time for the *praecoquum* apricot in order to take into account its earlier budding time: indeed, he describes this modification as one that the *praecoquuum* bud itself 'demands' (*ita exigente*) owing to its earlier bloom. The range of grafting times reported by Palladius for the peach does not seem to depend directly on *Pom.* He makes no mention of the *praecoquum* and offers months for the peach quite different from G.'s: 12.7.6 *mense Ianuario uel Februario locis frigidis, Nouembri calidis persicus inseratur.*

est . . . si . . . sint: The mixed condition is an example of 'substitution', whereby the logical apodosis is suppressed and 'in its place is substituted a statement of fact, often of a more general nature, the truth of which is unaffected by the fulfilment or non-fulfilment of the condition' (Handford (1947) 128–9): see also Nutting (1923) 156–8, (1925) 93–8. In the present case, it is presented as a fact that the end of autumn is the most suitable (*aptissimum*) time for grafting. The truth of this claim as a general statement does not depend on the climate of any particular place, notwithstanding the ideal conditions stipulated in the *si*-clause.

inter brumam et exortum Fauoni, id est extremo Decembri ac deinde Ianuario: *id est*, etc., is corrective, specifying a range of times smaller than that suggested by *inter brumam et exortum Fauoni*, for although *bruma* = *extremo Decembri*, Favonius commences blowing later than January (it begins in the spring: see 2.2.1n. *ab exortu Fauoni*).

praecoquo germine, quod saepe et ante ueris aduentum florere festinat: The *germen praecoquum* is the *germen praecoqui*; there is no difficulty with calling the bud rather than the variety *praecoquum*, because it is precisely in virtue of its swift blooming and production of fruit that the type as a whole earns the name. Indeed, the relative clause virtually spells out what is meant for the apricot to be *praecoquus*, with *festinat* in particular glossing *praecoquo*. For *festino* of the swift maturation of a tree, cf. Plin. *HN* 16.98 *festinat et platanus*; 19.7 *seritur* (sc. *linum*) *sabulosis maxime unoque sulco, nec magis festinat aliud.*

ueris aduentum: Cf. 2.3.5n. *hiemis aduentu*.

2.10.5 Meliores ad inserendum surculi habentur aliquatenus pleniores et de ea parte sublati quae propinqua arbori fuerit, uixque proueniet aut non diu durat pars summa ramorum: Cf. Pallad. 12.7.6 (sc. *persicus inseratur*) *maxime circa terram surculis plenioribus et prope arborem natis. nam cacumina uel non tenebunt uel diu durare non poterunt.* See 2.10.3n on the aetiology for the weakness of the upper or outer parts of the branches, i.e. the tips which are further from the trunk. For remarks on the desirable properties of the *surculus* taken from the tree – it should above all be vigorous and rich in buds – see Col. 5.11.3 = *Arb.* 26.2, Plin. *HN* 17.103–5, Pallad. 3.17.4, *Geop.* 10.75.7.

ad inserendum surculi: *surculus* is the agronomists' standard technical term for the scion (*OLD* s.v. 2), especially one that is to be cleft- or bark-grafted; *oculus* is favoured for the buds used in patch-budding (*emplastratio*, *inoculatio*): see *TLL* VII 1.1876.35–45 (Stiewe).

aliquatenus pleniores: *plenus* is one of the regular ways to refer to the abundant growth of a plant or its part (*TLL* X 1.2420.47–58 (Reineke)): here it evokes the stoutness of the slip, its vigour, and perhaps also the density of its foliage or buds.

uixque proueniet aut non diu durat: How to account for the variation in tenses? Mazzini passes over the future *proueniet* in his translation ('[l]e punte dei rami crescono a stento e non durano a lungo'), but if the tenses are to be assimilated it would be more likely that the present *durat* is instead used for the future, a phenomenon found already in early Latin (comedy) and increasingly widespread in a variety of contexts in later Latin: for documentation and discussion, see Löfstedt (1911) 212–13, H–Sz §172, Petersmann (1977) 167–9, Adams (1995*a*) 470, Leiwo (2010) 287–91, Adams (2013) 666–72, (2016) 277 and index s.v. 'present tense for future'. Yet the variation in tenses is not necessarily pointless: *uix proueniet* may refer to the future growth and productivity of the graft, with *aut* functioning correctively to introduce a more immediate problem, i.e. the swift demise of the graft (*non diu* stakes a strong position; for the litotes, see 1.1.7n. *non leui pondere*).

2.10.6–7 Persicus et in semet ipsam inseritur et in amygdalum et prunum . . . Sed armenia uel praecoqua peculiariter prunis insita

1 Quince (*Cydonia oblonga*).

2 Villa of Livia, Prima Porta, detail of a quince tree in a garden behind a low wall. Late first century BCE.

3 Marble relief depicting a farmer bringing his produce to the market in town. Late first century BCE/early first century CE.

4 Herculaneum (Insula IV, Casa dei Cervi), still-life of peaches on a shelf, perhaps in the kind of cool, dry environment recommended by the agronomists. The fruit have been picked *cum ramulis*. Before 79 CE.

5 Panel illustrating autumn from the seasons mosaic, Saint-Romain-en-Gal: a man perched in a tree collects fruit to hand down to a partner waiting with a basket. Note the ripe fruit that has fallen in the right foreground. Early third century CE.

6 Panel illustrating autumn from the seasons mosaic, Saint-Romain-en-Gal: with the help of a ladder a man scales a tree to gather fruit from the upper branches. Another man waits below with a basket. Early third century CE.

7 Peach (*Prunus persica*).

8 A rustic tableau from Utica, Tunisia, depicting a system of artificial irrigation using water channelled from a tower. Note the cultivated fruit trees in the foreground. Third or fourth century CE.

9 Great Palace of Constantinople, Istanbul, detail of a peasant engaged in *sartio* (hoeing) with a *bidens* (two-pronged hoe) under a fruit tree. First half of the sixth century CE.

10 St. Maximin, Trier, funerary relief, detail of tavern scene: an amphora (left foreground), wrapped tightly with rope as protection against hard use, is propped against a barrel, also wrapped with rope above and below. Second or third century CE.

11 Panel illustrating spring from the seasons mosaic, Saint-Romain-en-Gal: two men graft fruit trees, attaching the scion (slip) to the stock (tree). Early third century CE.

12 Almond (*Prunus dulcis*).

13 Unswept floor mosaic signed by Heraklitos and found near the Aurelian Wall. Detail of food debris, including almonds, walnuts, hazelnuts and other food scraps. Early second century CE.

(a)

(b)

14(a) Panel illustrating autumn from the seasons mosaic, Saint-Romain-en-Gal: intercultivation in an orchard. One man goads the oxen yoked to a ploughwhile the other prepares to scatter seed. Early third century CE. (b) Mosaic from Cherchell (Caesarea), Algeria: scenes of intercultivation among olive trees. Above, two men goad oxen ploughing the soil in preparation for sowing. Below, a man scattering seed is followed by another driving oxen to plough it in. Second or third century CE.

15 Chestnut (*Castanea sativa*).

16 Panel illustrating winter from the seasons mosaic, Saint-Romain-en-Gal: two men in hooded cloaks transport mixed material on a stretcher. The material should represent compost or manure, the stretcher a *cratis stercoraria* (manure hurdle), and the structure in the background a shelter for a *sterculinum* (compost heap or manure pit). Early third century CE.

floridiora sunt; cetera, id est duracina, sic insita in amygdalo melius assuescunt: All of the graft pairs are viable: see further 2.10–11n. Palladius remains very close to G. here: 12.7.6 (sc. *persicus*) *inseritur in se, in amygdalo, in pruno, sed armenia uel praecoqua prunis, duracina amygdalis melius adhaerescunt et tempus aetatis acquirunt.* Cf. also Pallad. 2.15.20 *pruno armenia inseremus et praecoqua*; *carm. ins.* 95–8 *ipsa suos onerat meliori germine ramos / persicus et pruno scit sociare genus / imponitque leues in stipite Phyllidis* (= *amygdali*) *umbras / et tali discit fortior esse gradu*; *Geop.* 10.17 τὸ δωρακινὸν ἐγκεντρίζεται εἰς δαμασκηνόν, καὶ εἰς ἀμυγδαλῆν (also 10.76.4). Condorelli also quotes from Clément-Mullet's translation of Ibn al-ʿĀwwām (after the *Geoponica*): 'suivant Marsial . . . le pêcher [*scil.* se greffe] sur l'amandier et sur le prunier'.

2.10.6 Persicus et in semet ipsam inseritur et in amygdalum et prunum: For G.'s use of repeated *et*, which is here emphatic and disjunctive, see 2.1.3n. *et . . . praebeat et . . . acquirat.* Special stress is placed on the peach's candidacy as a stock for itself not only by the intensifiers *-met* and *ipsam* but also through the word order.

in semet ipsam inseritur: When using the verb *insero* G. designates the stock onto which the scion is grafted with *in* + accusative (as here), the bare dative (2.10.7 *prunis insita*),[48] and *in* + ablative (2.10.7 *insita in amygdalo*). When the bare ablative is used (2.10.8 *surculo . . . insita*), it denotes rather the scion with which a stock is grafted (cf. 2.10.8n.).

amygdalum: The almond, *Prunus dulcis* (Miller) D. A. Webb, for which the older Latin name is *nux Graeca* (*LTBL* 222, *NPRA* 173); the word *amygdala* (*-um*/*-us*), from Gk. ἀμυγδαλῆ, is not found in Cato or Varro. The most common form for the tree is *amygdala, ae*, f., and for the nut, *amygdalum, i*, n. The form *amygdalus, i*, f., which refers to the tree, is late: the *TLL* (1 2029.75–80 (Vollmer)), though not taking into account the evidence of G., gives its earliest attestation as Arnobius, and thereafter Palladius. G. uses *amygdalum* for the nut and both *amygdala* and *amygdalus* for the tree: *amygdalus* is found unambiguously three times, all in the present section;[49] *amygdala* is found unambiguously only at 3.4.3 (*amygdalas*). In some other cases where the tree is referred to, such as 3.3.5 *amygdalis* (for the gender note the following *illas*), the declension cannot be decided.

prunum: Here the plum (*Prunus domestica* L.), though *prunus* (from Gk. προύμνη/προῦμνον) may also refer to other species, among them *Prunus*

[48] For the dative, less common, cf. Col. 5.11.1 = *Arb.* 26.1 *cui* (sc. *arbori*) *inseritur* (sc. *surculus*).

[49] At 3.3.4 (*nouam* <*normam*> *amygdali*) the reference is probably also to the tree, although the nut cannot be ruled out.

spinosa L. (*LTBL* 262, *NPRA* 208–9). Pliny observes that the plum was introduced only after Cato's time (*HN* 15.46 *pruna quoque omnia post Catonem coepisse manifestum erit*), but by the early Empire it was already a major crop: see *RE* (1) 19.1456.22–1461.8 s.v. 'Pflaume' (Steier), de Angelis (1995) 94–7.

magisque durabilis: The analytic comparative of *durabilis* is found elsewhere (once, at Col. 12.55.1), but from the time of Apuleius only the synthetic *durabilior* is found (*TLL* v 1.2286.54–6 (Bannier)). Maltby (2016) 353, 355, 359 observes that analytic comparisons are relatively more common in technical writing than in higher styles of prose. It is also germane to note that the distribution of analytic comparatives suggests that it was not specifically a colloquial feature (Maltby (2016) 355, 359), and in many Late and low-register texts it is scarcely attested (Bauer (2016) 338).

in ligno fortiore cinguntur: If the verb *cinguntur* is right, then with Antonelli we must understand it as *uariatio* for *inseruntur* (note his translation 'furono annestate'). There are no good parallels for the usage, but presumably the idea is that of 'girding' or 'equipping' the tree with the scions that are to be propagated (*in* + ablative is not what we would expect with *cingo*, but it may be relevant that it is one of the ways that G. constructs *insero*: cf. 2.10.7 *insita in amygdalo*). The emendation *panguntur* (Reeve) is also attractive: although the verb is not elsewhere used of grafting, it is a standard term in the agronomists for planting seeds, slips, etc. (*TLL* x 1.208.61–209.8 (Fröhlke), *OLD* s.v. 2), and the idea would be that the *surculi* are 'set down' into the wood of the stock as though into earth.

2.10.7 peculiariter: Before G. the adverb (first in Scribonius Largus in this sense) is nearly peculiar to Pliny: see *TLL* x 1.926.43–62 (Bruun).

floridiora: The comparative form *floriora* transmitted by **N**, otherwise unattested, would presumbly derive from the adjective *florus* ('light-coloured', 'fair'). Despite the late-antique reanalysis of the word as though from *flos* (*TLL* vi 1.927.41–55 (J. Kapp)), this meaning has no obvious relevance in the context. The word is simply a mistake for *floridiora* (Scotti); although the adjective *floridus* is not common in the agronomists, it has the appropriate meaning ('abounding in, covered with, or producing flowers' *OLD* s.v. 1a, *TLL* vi 1.925.21–37 (J. Kapp)).

duracina: The *duracinum* peach (cf. 2.1.2n. *armenia*) is praised by Pliny as being especially juicy (*HN* 15.109). He reports also that its flesh is not easily separated from the stone: 15.113 *duracinis adhaeret corpus et ligno auelli non quit, cum in ceteris facile separetur*. As André on Plin. *HN* 15.39 observes, peaches are now separated into 'fruits à chair molle non adhérente au noyau et en fruits à chair dure adhérente au noyau', or 'freestone' and 'clingstone' respectively (cf. *EFN* 697–8 on the plum). The *duracinum* is clearly a clingstone to judge from Pliny's comments, and as the name itself suggests (*durus* + *acinus*; see *LTBL* 121, de Vaan 184),[50] but nothing more can be said of the variety.

sic insita in amygdalo melius assuescunt: *assuesco* is used intransitively (as often; see *TLL* II 909.33–49 (v. Mess), *OLD* s.v. 2) and absolutely (although an oblique form of *amygdalus* may be easily supplied with it). The idea is that the slips adapt better to the stock when grafted onto almond than onto other trees. In the same spot Palladius has *adhaeresco* (causing Condorelli to wonder whether it should also be printed here): 12.7.6 *duracina amygdalis melius adhaerescunt.*

2.10.8 At haec ipsa prunorum atque amygdali germina difficilius <recipiunt> et cito recurrunt in senectutem; quotiensque surculo alterius generis insita sint infirmitatem suam confessa moriuntur: Winterbottom emends *Ad* to *At*; I have added *<recipiunt>*. The transmitted text *ad haec ipsa prunorum atque amygdali germina difficilius, ed* [unde *et* aut *sed*] *cito recurrunt in senectutem* is difficult. One must observe first that G. is no longer speaking of engrafting the peach slip onto another tree; rather, the mention of a successful marriage between peach slips and almond and plum stocks has led him in a brief digression to observe that the reciprocal graft (almond and plum slips onto peach stock) is not viable (cf. Antonelli, col. 1687–8). In fact, G.'s remarks are probably limited to the *duracinum* variety used as a stock: as much is indicated (1) by *difficilius*, which seems to balance in particular 2.10.7 *melius*; and (2) by *haec*, which, even if it could have general reference to the peach, more happily points back to 2.10.7 *duracina* (neuter plural).

On any interpretation, *ed* should be emended with Antonelli to *et*: we do not want the adversative force of *sed*, for the difficulty of the grafting (cf. *difficilius*) should be related causally to the plant's rapid senescence (*cito recurrunt in senectutem*).[51] There is a way to give a sense to the transmitted text if with *ad haec* we may supply *assuescunt* from the previous clause (for *assuesco* + *ad*, see *OLD* s.v. 2d), where the subject is *germina* and *haec* refers to the *duracina* peach: 'the buds of plums and almonds (*ipsa . . . germina*) for their part [adapt] to these [peaches] (*ad haec*)[52] with more difficulty, and they lapse quickly back into

[50] Plin. *HN* 14.14 has the correct etymology. Isidore (mistakenly) states that it is rather so called from its sour taste (*acor*): *Etym.* 17.7.7 *duracenum nuncupatur eo quod pomum eius in gustu acorem referat.*

[51] Nor is there a strong preference for *sed* on palaeographic grounds: haplography is common in **N**, but so is the change of *t* to *d* (cf. Mazzini 26).

[52] We would arguably want *ad has* (sc. *persicos*) since it is the stocks (trees) to which the buds are becoming accustomed, but if *haec* is meant to refer only to the clingstone peach (as suggested) then the gender is likely motivated by the neuter antecedent *duracina.*

old age'. This interpretation is perhaps already too strained, but it runs into a more serious problem with the following words *quotiensque surculo alterius generis insita sint*: if the text is left undisturbed, the subject of *insita sint* must be *germina*, and *surculo* must accordingly refer to the stock (= 'whenever they [sc. the buds of almond and plum] are grafted onto a sapling (*surculo*) of another kind').[53] But *surculus*, the agronomists' technical term for a slip or scion (cf. 2.10.5n. *ad inserendum surculi*), will not in a careful discussion of grafting bear the meaning of 'stock'.[54]

The difficulties outlined above can be avoided with two changes to the text. Winterbottom's easy suggestion of *At* for *Ad* (the substitution of *t* and *d* a very common error in **N**) will complement the contrastive and transitional *ipsa* (*OLD* s.v. 3; cf. also 2.1.3n. *poma ipsa matura integra*) and allow *haec*, still referring probably only to the clingstone variety (2.10.7 *duracina*), to be construed in the nominative. A verb such as *recipiunt* (used commonly by the agronomists of a stock 'receiving' a slip or graft; cf. *TLL* XI 2.343.21–3 (Schrickx)) may then be furnished to complete the meaning of the sentence. (Commentators have indeed long wondered about a lacuna in the first clause: cf. Antonelli, col. 1688.) On this reading, the subject of *insita sint* in the final sentence will be the peach stock, and *surculus* will be allowed its ordinary meaning 'scion'. The following statement *quotiensque*, etc. offers a general conclusion that subsumes the previous statements: owing to its weakness the *duracina* peach in fact perishes when grafted with any other species (*alterius generis*), something which holds perforce for plum and almond scions.

cito recurrunt in senectutem: As I remarked at 2.10–11n., one of the reasons for grafting (recognized also by the Romans) is to restore an unproductive plant. The prefix of *recurro* can therefore be given semantic force if the idea is that the stock will 'run back' to the old age (i.e. unproductive state) from which the engrafted scion was meant to save it. For this meaning of *recurro* ('restitutio status prioris'), see *TLL* XI 2.453.25–51 (Rey). (The phrase may also provide stylistic variation for 2.7.2 *maturius consenescat*.)

quotiensque: = *quotienscumque*, for which see *OLD* s.v., H–Sz §108.c.

2.11 G. offers limited comments on a third method of grafting, that of patch-budding. Most of the key instructions for this exacting method of grafting are omitted; as for the treatment of cleft- and bark-grafting in 2.10 (n.), G. assumes that the reader will already be familiar with the general technique

[53] This interpretation is adopted by *TLL* VII 1.1876.78–9 (Stiewe).

[54] It is never so used by the agronomists; the occasional application of the word to a 'low-growing woody plant' (*OLD* s.v. 3; cf. L&S s.v. II.B, 'a small tree') is irrelevant here.

of patch-budding. Indeed, he signals this explicitly with the words 2.11.3 *solito more*. Yet in spite of the limited scope of the section, many of the statements apply not only to the peach but also to other trees. This of course holds true for much of the information in *persic.* (and other chapters of *Pom.*); as elsewhere, such statements should likely be taken as confirmatory, that is, as stressing the applicability of more general principles to the specific tree in question.

2.11.1 emplastratio: The word ultimately originates from Gk. ἔμπλαστρος (-ον), a medical term for a plaster or salve. In agricultural language, the 'plaster' (*emplastrum*) is conceived of as the patch or ring of bark which surrounds the 'eye' (*oculus*, whence *inoculatio*) from which emerges the bud itself. Interestingly, the noun *emplastrum* takes a different course from *emplastratio* and *emplastro*: despite its technical agricultural application, *emplastrum* remains largely a medical term in Latin (cf. *TLL* v 2.531.83–5 (Kapp-Meyer)); conversely, *emplastratio* is exclusively a term of agriculture, and *emplastro* nearly so (Chiron uses it a few times to mean 'apply a plaster or salve'; see *TLL* v 2.531.46–9). See further *OLD* and *TLL* s.vv. *emplastratio*, *emplastro* and *emplastrum*. Columella straightforwardly equates *emplastratio* with *inoculatio* (5.11.1 *quam uocant agricolae emplastrationem uel, ut quidam, inoculationem*),[55] though Pliny (*HN* 17.118, 17.121) draws a distinction between them, employing *inoculatio* for excision of the bud without the bark patch. Like Columella, G. treats the terms synonymously; so cf. also Pallad. 12.7.7 *persicus inoculari potest, quod implastrari dicitur.*

olea, persicus, sicuti et nonnulla ficus et huius generis cetera quibus seponitur sucus pinguis: The inclusion of olive and fig on the list of trees for *emplastratio* is traditional: cf. Cato, *Agr.* 42.1. Columella makes the general connection between the sap content of the tree and its suitability for *emplastratio* (mentioning fig, too): 5.11.8 = *Arb.* 26.7 *tertium genus insitionis* (sc. *emplastratio*) *cum sit subtilissimum, non omni generi arborum idoneum est, sed fere* [*Arb. et fere eae*] *recipiunt talem insitionem quae umidum sucosumque et ualidum librum habent, sicut ficus, nam et lactis plurimum mittit et corticem robustum habet. optime itaque ea inseritur scarifici* [*Arb. tali*] *ratione*. Palladius lifts G.'s statement out of the chapter on peaches and places it in his own general exposition of patch-budding: 7.5.2 (sc. *emplastratio*) *solis arboribus conuenit quibus pinguis sucus in cortice est, ut ficis et oleis ac similibus, ut Martialis dicit, et persico.*

nonnulla ficus: *ficus* (Gk. συκῆ) by itself refers especially to the common fig (*Ficus carica* L.). There were, however, a vast number of fig varieties and species known to the Greeks and Romans (*RE* (1) 12.2119.50–2125.51 (Olck), *LTBL* 136–8, *NPRA* 104), hence *nonnulla*, restrictive here: 'some [types of] fig'.

[55] Note however that the words *uel, ut quidam, inoculationem* are omitted at *Arb.* 26.1.

quibus seponitur sucus pinguis: As for 2.4.6 *excludant* (but see 2.4.6n. *sicuti . . . excludant* for the conjecture *exsudant*), the semantic range of *sepono* encompassing 'separate', 'shut out', and 'exclude' is exploited to mean 'secrete' or 'exude'.

rite inoculantur prius trunco reciso: Cf. Pallad. 12.7.7 *inoculari potest, quod implastrari dicitur, praeciso super trunco*, etc. The advice to remove the stock's branches and stem above the site of *emplastratio* is standard: cf. Col. 5.11.11 *arboris autem insitae subolem et ramos superioris praecidito, ne quid sit quo possit auocari, aut ne cui magis quam insito seruiat*; *Arb.* 26.9 *materiam quam inseueris, si subolem uel supra ramum habebit, omnia praecidito, ne quid sit quo possit auocari aut cui magis quam insito seruiat*; Plin. *HN* 17.118 *ergo amputatis omnibus ramis, ne sucum auocent*, etc.; Pallad. 7.5.4 *ramos superiores eius arboris secabis ac stirpes*; *Geop.* 10.77.6 γενομένης δὲ συμφύσεως, εὐθὺς τὸ ὑπεράνω τῆς ἁρμογῆς ἀπόκοπτε, ἵνα μὴ εἰς τοῦτο ἡ τροφὴ χωρήσῃ, ἀλλ' εἰς τὸ ἁρμοσθέν. As all the authors save G. and Palladius spell out, the step is intended to redirect nutrition (*sucus*/τροφή) to the bud. The same principle is met in a different form at 2.9.2 and 3.7.6, where culling some peaches and almonds will improve the remainder on the tree by a transfer of nutrients.

2.11.2 Aduerti: One of only two first-person singular verbs in the text (or even the only one if 2.12.5 *deprehendi* has rather been truncated by the lacuna), though G. speaks in his own voice also at 1.1.7 (*nos non leui pondere affirmamus*). Cf. also 1.1.1 *uerum de fructibus pauca nobis dicenda sunt.*

mole iusta: *moles* here has a notion of large size or bulk, but, as *iusta* shows, not of massiveness (compare the remarks on *uastus* at 3.2.1n. *uastior palus*). *iustus* simply means 'fitting' or 'appropriate', sc. with respect to some implied standard. This is a regular meaning of the word when it is used without reference to legal (*ius*) or ceremonial propriety (*TLL* VII 2.720.33–721.65 (Baer)), and as the *OLD* s.v. 6b emphasizes, it is especially common of the size, amount or degree of a thing: cf. thus also 2.5.9 *calcata plus iusto*; 3.4.2 *interualla . . . iusta*; 4.2.2n. *iusta maturitas . . . adeo ut*; 4.7.3 *mensura non satis iusta . . . nisi.* A good comparandum for *iustus* in reference to the size of a plant (as here) is Col. 4.19.3 *at si cuncta competunt uoto, iusta est altitudo uineae pedum quinque.*

in glarea argilla: If the text is correct, *glarea argilla* is asyndeton bimembre (for which see 4.6.2n. *calidas . . . credit*). I construe *glarea* as an adjectivized noun (for which cf. H–Sz §92, Adams (2016) 104), which together with *argilla* will mean 'gravelly clay'. The asyndeton can perhaps be defended as formulaic, for the soil type that was a mixture of *glarea* and *argilla* was apparently so well known that it had its own name: cf. Plin. *HN* 33.72 *est namque terra ex quodam argillae genere glarea mixta – gangadiam uocant.* (Note also Pallad. 3.24.5 *argillam et glaream timent* (sc. *caules*).) It may therefore be supposed that the expression would not have afforded great difficulty to G.'s expert readers; but if the asyndeton cannot be tolerated, it is better to accept a connector such as *atque* (or simply *et*) into the text or even to emend to *glareosa* than to adopt Scotti's *aut*, which would give the wrong relationship between *glarea* and *argilla*.

2.11.3 Et quidem solito more possunt emplastrari: G.'s *solito more* is more ambiguous than Palladius' version: 12.7.7 *implastratis pluribus gemmis more quo dictum est* [= 7.5.2–4]. G. could be appealing to such an earlier section, but there is nothing preventing a more economical assumption that he refers only to the touchstone of practical experience.

quod agere in fine prodest mensis Aprilis, sed initio Maii satius proueniet: For general considerations affecting grafting times, see 2.10.4n. G. is followed in the present advice by Palladius and the *Geoponica*: Pallad. 12.7.7 *mense Aprili uel Maio locis calidis, in Italia uero utroque exeunte uel Iunio persicus inoculari potest*; *Geop.* 10.77.1 ὁ ἐνοφθαλμισμὸς εὐκαίρως γίνεται πρὸ τροπῶν τῶν θερινῶν. ἐγὼ δὲ καὶ περὶ ἰσημερίαν ἐαρινὴν ἐνοφθαλμίσας ἐν ἡμέρᾳ εὐδιεινῇ, ὅταν ἐν ἀρχῇ τῆς βλαστήσεως ᾗ τὰ δένδρα, σφόδρα ἐπέτυχον. Palladius' statement that in Italy the season for budding is later should be understood with reference to its temperate regions.

satius proueniet: It is at least possible that the subject of *proueniet* is the bud, for when *prouenio* appears elsewhere in *Pom.* (6x) it always describes the growth of the plant (see 1.1.2n. *melius proueniunt*); yet the subject more likely remains *quod* (i.e. *emplastratio*), with *prouenio* meaning 'to have a specified course or issue' or 'turn out' (*OLD* s.v. 5b, c).

2.12 In this final, truncated section G. turns immediately to preservation of the peach (see Plate 4). He is pessimistic about the longevity of the fruit, opening with a negative conditional (*nisi*) allowing that the only way to keep the fruit until the next summer is by pickling – and even this, he says, is done with difficulty (*uix . . . peruenit*). G. attributes the poor longevity of peaches to their *cutis exilitas* and *carnis squalor* (2.12.1); the meaning of the last is not well understood (n. *ad loc.*), but G.'s attempt to explain their rapid degradation on the basis of the nature of the fruit is not misguided: '[p]each fruits ripen rapidly and have a short postharvest life, usually limited to 3–4 weeks depending on storage conditions' (Ramina et al. (2008) 550). Add that ripening would have proceeded even more quickly in antiquity without the sophisticated temperature and humidity controls now available. While the refrigerated and stable environment offered by the 'little Suessian jars' (2.12.4n. *Suessatibus ollulis*) would certainly not have hastened the spoiling of the fruit, the benefits of the method would have been limited, hence G.'s proviso *hoc modo saepe in breue tempus flaccescunt*. If not pickled, drying the fruit (2.12.2) is a more promising way to keep them; peaches are still dried in some quantity today (*EFN* 719), and appropriate methods for dehydrating have long been the subject of attention (cf. Wickson (1891) 544–5). Palladius largely takes over G.'s methods, omitting some details as usual (everything from *uixque* to *uetustatis*), but he also adds a means of conserving them not found in G. (phrased as his own observation): Pallad. 12.7.8 *item saepe uidi detractis ossibus duracina melle condiri et saporis esse iocundi.*

2.12.1 Persici pomum, nisi conditum muria et oxymelle, asseruari diu non potest: Cf. Pallad. 12.7.8 *duracina seruantur condita muria et oxymelle.* The connector *et* is not disjunctive, for the pickle is prepared in most recipes with a mixture of brine (*muria*) and some other ingredient, usually vinegar (*acetum*), but also *sapa, defrutum*, honey, *passum* (raisin wine) and occasionally more specialized ingredients such as *oxymel*; herbs were often added for flavour. A common ratio for a pickle for herbs, vegetables, etc., is two parts vinegar to one part brine (Col. 12.7.2, 12.9.2; cf. 12.49.8), but as one might expect, the procedures and ratios vary widely depending on what is being preserved (cf. Thurmond (2006) 168–9, 171, 184–5). Apicius has a recipe for pickling the peach similar to that of G., but omitting *oxymel* and adding savory: 1.26 *duracina persica ut diu durent: elige optima, et mitte in muriam. postera die exime, et spongiabis diligenter, et collocabis in uas. fundes salem, acetum, satureiam.*

muria: Brine, i.e. salt and pure water. It is intermediate in strength between *aqua salsa* (salted water) and *muria dura* (hard brine) (Thurmond (2006) 170–1). Cato (*Agr.* 105.1) gives a simple recipe for *muria* that is to be used in the production of 'Greek wine'. Columella (12.6) has more detailed instructions for the concentrated *muria dura*; the basic idea is to progressively add salt to the water until it no longer dissolves, that is, the solution is saturated. It is best to use rainwater (*aqua caelestis*) or, failing that, sweet spring water (*fontana dulcissimi saporis*). Very different is the production of a brine to substitute for seawater (*aqua marina*), in the course of which the water is allowed to putrefy before being strained, boiled down, and salted (Col. 12.25.2).

oxymelle: Honeyed vinegar (Gk. ὀξύμελι). The mixture was held to have important medical properties and was very widely prescribed both simple and compounded from the fifth and fourth centuries BCE on. Pliny has the following recipe (identical to Dsc. 5.14 save for Pliny's specification of rainwater): *HN* 14.114 *quin et acetum melle temperabatur . . . oxymeli hoc uocarunt, mellis decem libris, aceti ueteris heminis quinque, salis marini libra, aquae pluuiae sextariis quinque sufferuefactis decies, mox elutriatis atque ita inueteratis.* See further *RE* (1) 15.1301.67–1302.25, 1307.40–60 s.v. 'Met' (Maur-Schuster). It is much less widely recommended for preserving foods, but cf. e.g. Col. 12.58.2 (*siser*), Plin. *HN* 19.128 (lettuce), *Geop.* 9.29 (olive).

uixque hoc modo peruenit in alteram aestatem: cutis exilitas et carnis squalor consentanee prodibunt ad parsimoniam uetustatis: *uix* throws emphasis on the unsuitability of the fruit for preserving: pickling (*hoc modo*) is the only way to keep them, but even then they last 'barely' or 'with difficulty' (*uix*) to the next year (lit. 'to the next summer'). The sentence *cutis exilitas*, etc., could be taken as an independent statement about the fruit, but I prefer to take it more closely with what precedes as causal asyndeton (see 2.4.6n.): it explains the difficulty of preserving the fruit even when pickled (*uix . . . peruenit*) by appealing to an intrinsic property of the peach.

cutis exilitas et carnis squalor consentanee prodibunt ad parsimoniam uetustatis: For the stylistic qualities of this line, see Introduction §v(a).

cutis . . . carnis: Exactly as we would speak of the 'skin' and 'flesh' of a peach (or other pome), although the metaphorical extension of these words to fruits is not as common in Latin as in English. Before G. only Pliny uses *caro* of fruit or any other part of a plant (*TLL* III 484.10–30 (Meister)), though he does so very often; it is later found also in Palladius. Columella uses *cutis* once of the grape (3.2.15), but that word too is largely restricted to Pliny in its application to plants until after G.'s time (*TLL* IV 1578.84–1579.16 (Gudeman)). G. uses *cutis* also of the melon at *Med.* 15.4 *detracta cute ac semine abiecto.*

carnis squalor: The meaning of *squalor* (*-ll-* **N**) is very uncertain, but there is no reason to suspect the text. Antonelli translates *ad hoc* 'la polpa troppo molle', and Mazzini follows, 'la mollezza della polpa'. Of course, *squalor* and related words do not mean 'soft', and it should be kept in mind that not all peaches have 'melting' flesh, although these are the ones preferred for consumption (*EFN* 721). Yet it is not clear how peach flesh, even if firm, could be described as 'harsh' or 'rough' (*OLD* s.v. *squalor* 1; cf. s.vv. *squaleo, squalidus*) – might it refer to a fibrous or stringy quality? My translation 'roughness' is merely a stopgap.

consentanee prodibunt: The use of *prodeo* of a thing, rather than a person, that progresses to an end or result is late and relatively uncommon, found first in *Pom*: see *TLL* X 2.1600.71–8 (Vestergaard). Regarding the adverb *consentanee*, Antonelli translates 'al pari', and Mazzini follows ('congiuntamente'). That is not, however, what *consentaneus* means. The adjective, like *conueniens*, means 'together with' in the sense of 'consistent with', 'appropriate to', vel sim., not 'together' in the sense of accompaniment. Here the adverb refers rather to the natural properties of the fruit that conduce to early degradation in a way 'consistent with' or 'appropriate to' those same properties (i.e. *cutis exilitas et carnis squalor*). The adjective *consentaneus* itself is old (on its formation see Leumann (1940), Leumann §272.3), but the adverb is first attested here.

parsimoniam uetustatis: *parsimonia uetustatis* is equivalent to *parca uetustas*, 'short old age': for this meaning of *parcus* ('respicitur status rerum non abundans, non amplus, fere i. q. exiguus, modicus sim.'), see *TLL* x 1.344.84–345.6 (Paśkiewicz), and in its application to extent of space or time, 345.50–70. In spite of the well-established meaning of the adjective *parcus*, the noun *parsimonia* is only very rarely used in the same sense. The *TLL* offers a handful of instances under the heading 'pertinet ad statum rerum' (x 1.490.74–82 (Korteweg)); in the lemma 'de quantitate exigua, parua' are included the present usage and Ambr. *Epist.* 63.26 *namque ut redundantia ciborum mitigatur ac temulentia etiam maturior aetas calescit, ita mitigatur iuuentutis ferocitas epularum parsimonia* (note the crucial opposition to *redundantia ciborum*).

2.12.2 Plerumque temptarunt ossibus detractis more ficorum in sole siccare atque in frigido seruanda suspendere: Cf. Pallad. 12.7.8 *uel detractis ossibus ficorum more in sole siccantur ac pendent.* How far *more ficorum* refers to elaborate procedures for drying figs (cf. Col. 12.15, Pallad. 4.10.34–5), such as the provision of tents to protect the fruit from dew and rain (Col. 12.15.1), is uncertain. It could mean nothing more than the bare fact of drying them in the sun, but there is a detail recorded by both Columella (12.15.5) and Palladius (4.10.35) that may be plausibly assumed for the peach as well, that of splitting and spreading open the fruit before placing them in the sun to dry. This detail would fit well with the instruction to remove the stones (*ossibus detractis*), for the peaches must have been cut in half for this purpose.

Plerumque temptarunt: Lit. 'they have very often tried', but the subject of *temptarunt* is so vague that it is virtually an impersonal expression. For the syncopated form *temptarunt*, see Introduction §v(a).

ossibus detractis: At *Med.* 15.4 (quoted 2.12.1n. *cutis . . . carnis*), G. uses *detraho* rather of removing the skin of a melon and *abicio* of the seeds.

in frigido seruanda suspendere: There is no precedent for the substantive use of *rigidus* (as transmitted by **N**) to refer to the cold or a cold place, whereas *frigidus* is often so used: see *TLL* vi 1.1327.69–1328.24, 1330.76–15 (Rubenbauer). The adjective *frigidus* is very common in *Pom.* (1.1.10, 2.3.1, 2.3.2, 2.4.2, 3.6.4, 4.3.3, 4.6.3, 4.6.6), and G. uses it of the storeroom for fruit at 1.1.10 *in loco sicco et frigido reponantur.* Relevant too for the establishment of the text here are Col. 12.46.5 (pomegranate) *loco frigido suspendi*; 12.46.5 (pomegranate) *frigido loco suspendere*; 12.46.7 (quince) *frigido loco suspendunt.*

2.12.3 Vix in his aliquid suci inest, cuius admonere inane uidetur: This observation may give additional point to the claim in 2.5.6 that transplanting the young peach trees from baskets will yield *suci amplioris poma.*

2.12.4 Ex ueteribus multi Suessatibus ollulis impicato singula poma clausere umbilico atque in sapam purissimam demissa merserunt, sed hoc modo saepe in breue tempus flaccescunt: Palladius transmits the advice minus G.'s caution: 12.7.8 *item bene seruantur, si umbilicum pomi gutta picis calentis obpleueris, ut sic sapae innatare cogantur uase concluso.* It is uncertain who G.'s *ueteres multi* are, for these instructions are not found in any of the extant agronomists before him. The basic principle of this advice is the same as that of 1.1.6 (n.): *Aliqui putant et in dolium uini olla noua conclusa, ut in eo obtecta natent, esse mittenda* (see also 1.1n.).

Suessatibus ollulis . . . singula poma clausere: The diminutive *ollula* is used presumably because the jars (see 1.1.6n. *olla noua conclusa*, adding Hilgers (1969) 117) are large enough to fit no more than one peach each (*singula poma*). Pliny mentions this 'single-serve' technique for the preservation of pears and apples: *HN* 15.65 *alii singulis malis pirisque singula uasa fictilia assignant.* Cf. also Col. 12.44.7 (bunches of grapes) *uasa fictilia . . . quae singulas uuas laxe recipiant*; Pallad. 3.25.17 (apples) *singula in uasculis fictilibus*; perhaps also Col. 12.46.2 (pomegranates still on the tree) *nonnulli uascula fictilia dependentibus malis aptant.* The idea may be to avoid contamination if some one fruit should spoil. The connection of the pots with Suessa Aurunca (now Sessa) or the area around it is extremely obscure: it is not certain whether the 'single-serve' pot is a particular Suessian style, whether the potter's clay is sourced from there, or something else. On the colony see generally *RE* (2) 7.584.60–586.2 s.v. 'Suessa' (Philipp). Cato (*Agr.* 22.3, 135.1, 135.3) mentions Suessa three times as a place where farm equipment may be purchased. The adjective *Suessates* itself (*Suesates* **N**) is not attested elsewhere (*Suessanus* is used), but the formation is regular: see Leumann §309.

impicato singula poma clausere umbilico: The word order is unusual, perhaps a stylistically ambitious attempt to form a 'word picture' in which the pitched-over stem (*impicato . . . umbilico*) covers and seals the *singula poma.*

impicato . . . umbilico: The *umbilicus* is the peduncle, the stem that connects the fruit to the tree. The standard word for it is *pediculus* (*TLL* x 1.997.49–66 (Stewart)); before G. *umbilicus* in reference to the 'navel' of a plant part is a largely Plinian word (*OLD* s.v. 2d). The verb *impico* is found only here and at Col. 12.29, 12.44.1. At Col. 12.44.1 it is also used of sealing stems with pitch: *pediculos earum* (sc. *uuarum*) *impicato dura pice.* The advice to pitch over fruit stems is found additionally for pomegranates (Plin. *HN* 15.60), sorb apples and pears (Plin. *HN* 15.62) and especially grapes (Col. 12.44.1–2, 12.44.5, 12.45.1, Plin. *HN* 14.17, Pallad. 10.17). Varro (*Rust.* 1.59.3), too, probably has pitch in mind when he recommends 'sealing over' (*obleris*) the end of a branch picked with green pomegranates on it; he says this is done *ne extrinsecus spiritus afflet,* and promises that the fruit thus protected will not only remain uncorrupted but become even larger than if they had remained on the tree (*non modo integra eximi, sed etiam maiora quam in arbore umquam pependerint*). Pitch, the product of boiling down vegetable resin, is used much more commonly in the context of preserving for making the storage vessels air- or water-tight: see the abundant examples at *TLL* x 1.2077.4–21, 46–67 (Sauer-Gaertner), x 1.2249.60–7 (Pieroni). On pitch, see also Meiggs (1982) 467–71.

in sapam purissimam demissa merserunt: For *sapa*, see the helpful note of André on Col. 12.19: 'La *sapa* et la *defrutum* sont des vins cuits différant seulement par le degré de réduction. L'élimination de l'eau, la concentration du sucre, la réduction de la fermentation permettaient la conservation des moûts faibles après vinification. Les termes latins désignant les vins cuits n'ont pas la même signification suivant les auteurs, et sans doute suivant les époques et les régions: la *sapa* est obtenue par réduction du moût des 2/3 (Pline, *nat.* 14,80; Pallad. 11, 18), de moitié (Varron *ap.* Non. 551, 18; Col.), ou seulement d'1/3 ou d'un 1/4 comme ici. – Le *defrutum* par réduction des 2/3 (Varron, *ibid.*; Col. 12, 21, 1) ou de moitié (Pline, *nat.* 14, 80)'.

in breue tempus: We would expect *breui tempore*, vel sim.: either *in* stands loosely for *intra* (cf. 2.3.5n. *in ceterum autumni*) or the accusative is used for the ablative (cf. 3.6.2n. *in locum taedae*).

flaccescunt: I.e. as a result of their decay. G. connects degradation of the plant with 'withering' also at 3.5.3 *nubilo . . . marcidae* (n.).

<III DE AMYGDALIS>: Fr. 10 Condorelli. Ch. III Mai. Ch. II Scotti.

3.1–4 The text of *amygd.* picks up in the middle of a discussion of different ways of preparing a *seminarium*, i.e. a nursery for raising young plants. As for the peach, cultivation of the almond (Plate 12) requires preparation of the seeds, soil and planting holes, as well as the proper method for planting the seeds and caring for them during growth. Transplantation from a nursery is recommended as the best way to grow them (cf. Pallad. 2.15.6 *in hoc genere arboris* (sc. *amygdali*) *nihil utilius est quam seminarium facere*). 3.1–2 detail the recommendations principally of Mago and Celsus, 3.3 those of the Quintilii and the alternative giant fennel method of Mago and Diophanes. 3.4 deals with certain general questions about the layout of the orchard and the position of the trees. All of the methods of cultivation reported in these sections treat growing the almond from the seed. In fact, *amygd.* surprisingly omits mention of two other crucial methods employed in the cultivation of fruit trees in antiquity, both of which are discussed in connection with peaches (2.1–7n., 2.10–11n.): (1) grafting, and (2) propagation from a living portion of a tree (i.e. a cutting, slip, sucker, layer, etc.). Some brief remarks on these omissions will be useful, in the spirit of *Geop.* 10.3.2 ἀναγκαῖον οὖν σαφηνίσαι τοὺς τρόπους, καθ' οὓς χρὴ ἕκαστον δένδρον φυτεύεσθαι.

Of the omissions in methods of propagation, the more unusual is that of grafting.[56] It is strange to find no mention at all of grafting in connection with the almond, not only because G. treats the principal techniques (cleft-grafting, bark-grafting and patch-budding) in reasonable detail for the peach – a context in which the almond is mentioned as a rootstock (2.10.6–7) – , but more generally because the advantages of grafting over growing from seed were well known in antiquity (2.10–11n.). It is true that the almond, in contrast to many other fruit trees, can produce attractive fruit when grown from seed and that it was regularly so cultivated in antiquity (Zohary et al. (2012) 147, 149): Pliny thus includes almonds 'among those trees which are grown from seed' (*HN* 17.63 *inter ea quae semine seruntur*), and he is followed in this respect by Palladius (2.15.6) and the *Geoponica* (10.57.5) (both quoted below). Nonetheless, an explanation for G.'s complete silence on grafting in *amygd.* is wanted, and it may be possible to explain it by consideration of generic precedents. Consider that Columella, who touts the advantages of grafting for fruit trees, includes no specific discussion of it in his section on almonds; nor does he for the fig (5.10.9–11 = *Arb.* 21), pomegranate (5.10.15–16 = *Arb.* 23), pear (5.10.17–19 = *Arb.* 24), peach and so forth (cf. 5.10.20 = *Arb.* 25). This is not because he thinks that those trees could not or ought not to be grafted – far from it – but rather because he defers a careful discussion of grafting to a separate section that immediately follows the advice on cultivating specific trees (5.11 = *Arb.* 26–7) and further shifts recommendations regarding the timing of grafts for specific plants to the calendar of book 11: the almond is mentioned there at 11.2.11 and 11.2.96 (cf. also 11.2.37). Columella himself signals the intentionality of this organization at the conclusion of his discussion of the olive: 5.9.17 *quemadmodum autem olea inserenda sit postmodo dicemus*.

Even more tellingly, at the beginning of his section dedicated to grafting (5.11) Columella boldly declares that *omnis surculus inseri potest*, where *omnis* emphasizes that the remarks on grafting are applicable to *all* of the trees in the foregoing discussion. It is implied that, since the grafting techniques do not significantly differ among species, general instruction will suffice; indeed, Columella only mentions a specific tree where the process of grafting it differs in some important respect from other trees (so note the contrast at 5.11.2 *omnes arbores, simulatque gemmas agere coeperint, luna crescente inserito, oliuam autem circa aequinoctium uernum usque in Idus Aprilis*). The same organization of material is found in Pliny: a reader who considered only the remarks specific to almonds

[56] A discussion of grafting for the almond is not likely to have been lost in the lacunae, for our extant sections run from planting (3.1) to preservation (3.8); taking *persic.* as a guide, we would have expected to find grafting handled somewhere within these sections, most likely after 3.7, if such a discussion had existed. See also Introduction §III(a).

(*HN* 17.63) would find no indication, if they were otherwise inexperienced in arboriculture, that the tree was a candidate for grafting; but it would be ridiculous to assume on that basis that the tree could not or should not be grafted. Like Columella, Pliny has elected to treat the complicated rules for grafting in a separate discussion (*HN* 17.101–14, 118–21) that follows the recommendations for planting specific trees; and, like Columella, Pliny intends his remarks to apply to all of the fruit trees he has mentioned (cf. esp. *HN* 17.103, 120–1). Finally, he too adds notes specific to one sort of tree or another as needed (e.g. *HN* 17.121 on the fig and pomegranate), and occasionally gives lengthier treatment to those trees for which special instructions apply, such as the olive (*HN* 17.125–7, quoting Cato). This system of organization, remarkably durable because logical and economical, is found also in Palladius, who overviews general techniques for grafting at 3.17 and 7.5.2–4 while providing a wealth of observations on individual trees *passim* in the particular chapters and in the concluding *carmen insitionis.*

How far can these observations inform our understanding of *Pom.*? The basic difficulty in answering this question remains, of course, our poor knowledge of the overall shape of G.'s original text and of his sources (as stressed in the Introduction §III(a)). When the standard organization of material in the other agronomists is taken into account, it seems *a priori* possible or even probable that G. followed traditional precedents in discussing grafting at a higher level of generality in a separate part of *Pom.*, which has not been transmitted. Yet it must be acknowledged that the evidence elsewhere in *Pom.* for the existence of such a section is ambiguous. The discussion of grafting peaches in 2.10–11 obviously presupposes some familiarity with the basic techniques (cf. 2.10n., 2.11n.), but G. could merely be assuming practical experience on the part of the reader. A detail such as 2.11.3 *solito more* (n.) is suggestive but not decisive, since it could refer equally well to tacit knowledge on the part of the audience or to a previous chapter. In the other direction, the generalizing character of some of G.'s advice in those sections[57] is not a sure argument against a separate, high-level treatment of grafting, for those remarks could simply be intended to stress the applicability of standard techniques to the tree under discussion (and, we add, would not be contrary to G.'s amplifying and digressive manner). It cannot therefore be decided with confidence whether G. had a separate treatment of grafting, but a more modest conclusion can be advanced, namely, that G. *either* had such a section *or* assumed a knowledgeable audience with prior experience of grafting (cf. Introduction §III(a), §IV). Significantly, in either case G.'s silence on grafting the almond would not be proof of his rejection of the possibility; indeed, we must surely

[57] Cf. e.g. 2.11.1 *olea, persicus, sicuti et nonnulla ficus et huius generis cetera quibus seponitur sucus pinguis, rite inoculantur prius trunco reciso.*

go further to deny that G. was unaware that the almond could be grafted, especially because he discusses its reciprocal function as a rootstock (cf. 2.10.6 *inseritur . . . in amygdalum*; 2.10.7 *insita in amygdalo*).

As regards the connection between grafting the almond and planting it, it should be added that those who wished to graft it would require at all events the instructions that G. includes here for the *seminarium*, since the rootstocks would have to be grown from seed (*EFN* 713–14). G.'s extensive discussion of cultivation from seed does not therefore rule out grafting any more than do the admissions of Pliny, Palladius or the *Geoponica* that the plant can be grown from seed. We briefly set out the remarks on grafting the almond found in Palladius and the *Geoponica*: Pallad. 2.15.12 *mense Decembri uel Ianuario circa Idus amygdalus inseritur, locis uero frigidis et Februario, si tamen surculos condias antequam germinent. utiles sunt qui de summitate sumuntur. inseritur et sub cortice et in trunco. inseruntur in se, in persico*; *Geop.* 10.57.3–4 καὶ ἐγκεντρίζειν δὲ χρὴ τὴν ἀμυγδαλῆν ἐν τῷ μετοπώρῳ, ἐπειδὴ τῷ ἔαρι σχεδὸν ὑποφαίνει. δεῖ δὲ τὸν μέλλοντα ἐγκεντρίζεσθαι ἐπίπηγμα τῆς ἀμυγδαλῆς μὴ ἐξ ἄκρου τοῦ δένδρου, ἀλλ' ἐκ τοῦ μέσου λαμβάνειν.[58] (The instructions of the *Geoponica* contradict Palladius' account in some important respects, but nevertheless offer another perspective on grafting the almond tree.)

The second group of planting methods missing from G.'s collection for the almond is that of planting from a live portion of another tree (e.g. *planta, surculus, propago, uiuiradix*, etc.; cf. 2.5.7n.).[59] As for grafting, these techniques are not mentioned by Columella or Pliny with specific reference to the almond, but there are again abundant indications that their advantages were well known and that a variety of methods were routinely employed, especially for those trees that established their root systems slowly. Varro, for example, touts the benefits of planting from cuttings (*surculi*) for the fig: *Rust.* 1.41.4–5 *fici enim semen naturale intus in ea fico, quam edimus, quae sunt minuta grana, e quibus paruis quod enasci coliculi uix queunt – omnia enim minuta et arida ad crescendum tarda, ea quae laxiora, et fecundiora, ut femina quam mas et pro portione in uirgultis item . . . quare ex*

[58] These instructions are repeated at 10.62.1, where they are attributed to one Paxamos. The mention of air-layering almonds that follows the advice for grafting is noteworthy: 10.57.6 τινὲς δὲ ἀπὸ ἄκρας τῆς κορυφῆς τοῦ δένδρου εὐθαλέστατον κλάδον λαβόντες ἐφύτευσαν, καὶ ἐπέτυχον σφόδρα.

[59] Zohary et al. (2012) 149 observes that the 'almond does not lend itself to vegetative propagation from suckers, cuttings, or rooting', stating even more forcefully elsewhere that '[a]lmonds cannot be propagated by rooting' (147). Does G.'s omission therefore merely reflect agronomic reality? Perhaps, but at least some of the ancients thought that the almond could be grown from suckers and cuttings: see Palladius and the *Geoponica* quoted below.

terra potius in seminariis surculos de ficeto quam grana de fico expedit obruere, to which he astutely adds, *praeter si aliter nequeas, ut siquando quis trans mare semina mittere aut inde petere uult*. Theophrastus (*Caus. pl.* 1.8.1) had already noticed the quicker growth of slips than of seeds, and the point is artfully stressed by Virgil (*G.* 2.57–8 with Spurr (1986*a*) 166). Thus, Columella also presents a variety of techniques for propagating (e.g. 4.33.4 layering chestnut, 5.9.1–5 branches/buds of olive, 5.10.8–9 = *Arb.* 20.3, 5.10.21–2 = *Arb.* 25.2 cuttings/branches of fruit trees), stressing their applicability to all trees (so 5.10.21 = *Arb.* 25.2 *omnium . . . generum ramos*), and affirms the benefits of planting fruit trees from suckers or quicksets (5.10.6 = *Arb.* 20.1 *si cum radice plantam posueris, incrementum ei maius futurum quam ceteris senties*).[60] Pliny too treats various techniques in detail (*HN* 17.66–8 suckers, 17.96–8 layers, 17.123 branches, 17.124–6 cuttings, 17.154–5 fig slips). As with grafting, Columella's and Pliny's omission of these methods in their discussion of almonds (and often other fruit trees) should be explained on the basis of the authors' organization of material rather than any ignorance of or opposition to their application for particular species. In general, one is therefore faced with the same difficulties in explaining the absence of these techniques in *amygd.* as encountered in the case of grafting. It should be noted, however, that Palladius and the *Geoponica* depart from the organization of Columella and Pliny on the subject, being more inclined to report specific instructions for propagation from slips in the sections on individual trees. For the almond, cf. Pallad. 2.15.6 *amygdala seritur . . . semine et plantis quae de maioris radice tolluntur*; *Geop.* 10.57.5 φυτεύονται δὲ αἱ ἀμυγδαλαῖ καὶ ἀπὸ σπέρματος, καὶ ἐξ αὐτορρίζων, καὶ παρασπάδων ἢ μοσχευμάτων. The later evidence of Palladius and the *Geoponica* may in part reflect a greater diversity of and improvement in cultivation methods of fruit trees but could also simply represent a shift away from the agenda of discussion found in Columella and Pliny.

3.1 It is difficult to estimate the extent of the lacuna at the beginning of this section, but this and the following nn. contain some speculation. Unlike *persic.* or *cast.* (*cydon.* is too lacunose to be helpful here), *amygd.* as transmitted does not begin with any discussion of the ideal planting time for the almond tree. G. is not likely to have passed over this point, not only because planting time was of the highest importance for ancient agronomists, but also because he dedicates a good deal of space to surveying the authorities' opinions on planting times in *persic.* (2.1–3) and *cast.* (4.1–3) (consider the 'typical' structure of the chapter as presented in the Introduction §III(a)).

[60] Palladius curiously misreports Columella's stance on these methods of propagation: 3.19.3 *sed, ut Columella dicit, feraciores sunt quae seminibus, hoc est nucibus suis, quam quae plantis ponuntur aut ramis.*

A range of dates is found for the almond in earlier authorities: Columella recommends the beginning of February (5.10.12 *nucem Graecam serito circa Kalendas Februarias*), with *Arb.* allowing up to the third week of February (22.1 *nucem Graecam serito Arcturi signo uel circa Kalendas Februarias*, with Goujard), but Pliny offers a very different range, from the beginning of November to the winter solstice (*HN* 17.131 *ab occasu Arcturi ad brumam*[61]). Palladius toes a middle line, recommending January and February in general, but October and November in hot climates (2.15.6 *amygdalus seritur Ianuario et Februario, item locis calidis Octobri et Nouembri*). Both Columella (*ibid.*) and Pliny (*HN* 16.103) observe by way of explanation for their recommended dates that the almond blooms first among the fruit trees, which is true: it is the 'earliest temperate tree crop to bloom in spring' (*EFN* 709), something that Theophrastus attributes to its ἰσχὺς καὶ πολυτροφία (*Caus. pl.* 1.10.2). The *Geoponica* follows Pliny's advice closely, again attributing the early planting time to its early bloom: 10.57.1 τῷ μετοπώρῳ ἄμεινον τὰς ἀμυγδαλᾶς φυτεύειν, ἕως τροπῶν χειμερινῶν. τῷ γὰρ ἔαρι φυτευθῆναι ταύτας οὐ ῥάδιον, ὅτι τὸ φυτὸν τοῦτο πρωϊμώτατα βλαστάναει.

Also missing from *amygd.* are comments on the appropriate climate and soil for the almond, which are found for the peach (2.3–4, 2.11) and chestnut (4.5–6). A warm, dry climate was preferred for the almond, as also for the soil: see 3.1.2n. *terram . . . minutissimam*.

3.1.1 *<in> fimum bubulum transferunt:** Something like *nonnulli amygdala in* (and perhaps more) is missing at the beginning of this sentence: cf. Pallad. 2.15.8 *alii prius fimo liquido per triduum nuces eas macerant.* Palladius adds that one may alternatively soak the almonds in honey-water: 2.15.7 *quae* (sc. *amygdala*) *antequam ponimus pridie mulsa aquata nimis maceremus, ne germen extinguat ex multo melle mordacitas.* This recommendation may have been lost in the lacuna, but it is not strictly necessary to suppose that G. reported both techniques, especially if he thought the manure technique superior. The opposite situation is found in Columella, who mentions only honey-water: 5.10.12 = *Arb.* 22.1 *antequam nucem deponas, in aqua mulsea nec nimis dulci macerato; ita iucundioris saporis fructum, cum adoleuerit, praebebit et interim melius atque celerius frondebit* [*Arb. nascetur*]. The two traditions of preparing the almond seeds by soaking them

[61] *ab occasu Arcturi* refers to the evening setting of Arcturus, which at Rome would have occurred around the fourth day of November: cf. Col. 11.2.78 with *DGRA* I 228–9, André on Plin. *HN* 17.131, Goujard on *Arb.* 22.1 (the Loeb's 'rising of Arcturus' is a mistake). For the parapegmata relating to Arcturus, see Lehoux (2007) 522. *ad brumam* refers to the winter solstice: see *DGRA* I 234, Plin. *HN* 18.221. On the general importance of the stars and planets in determining the agricultural calendar, see Taub (2003) 38–9, 176–9, Lehoux (2007) 30–5, 52–3, Ager (2010) 52–151.

in manure or honey-water probably go all the way back to Mago, for Pliny has both of the methods and reports them in the context of Mago's writings on the almond: *HN* 17.63 *fimo . . . diluto maceratas* (sc. *nuces*) *per triduum aut pridie quam serantur aqua mulsa*.[62] The attempt to affect the produce through soaking in honey-water seems to be at least partly magical in nature (Ager (2010) 200–2; cf. Rose (1933) 68–9, 87): 'sweetening' the seeds is supposed to sweeten the produce. But the principle is not incompatible with a naturalistic explanation: cf. Theophr. *Caus. pl.* 5.6.12 πέφυκε γὰρ οἷον ἂν σπαρῇ τοιοῦτον καὶ γεννᾶν. Soaking in manure seems intended primarily for the nutritive benefits, although the dividing line between magical and conventional farming techniques in Roman agriculture is often fuzzy: see Introduction §IV, and cf. 3.6.2–3n. For other recommendations to soak seeds in honey-water, cf. Pallad. 4.9.7 (cucumber), 4.10.17 (citron).

Palladius further includes an exhortation to choose the almonds healthiest in appearance for sowing (2.15.7 *sed ipsa amygdala ad ponendum et noua legamus et grandia*), and it is possible that this too may have been in the lacuna. In this vein, Pliny (*HN* 17.63) reports Mago as preferring those seeds for planting that are 'sickle-shaped' (*serendas quam maxime falcatas*).

fimum bubulum: As Adams (1982) 234–7 has shown, *fimus* (also *-um*) is originally a more specialized term than *stercus*: the former refers to ox or cow dung, the latter to excrement of all creatures (including human, birds and quadrupeds) and occasionally vegetative ordure (for the last, cf. Cato, *Agr.* 37.2).[63] In Cato (*Agr.* 40.2, 46.2), Varro (*Rust.* 3.16.16) and Livy (38.18.4), *fimus* is nearly always flanked by the adjective *bubulus*, and the single exception at Cato, *Agr.* 28.2 suggests that the word is used elliptically in the same sense as the full expression (Adams (1982) 236). The evidence indicates that the distinction between the two terms was fading by the time of the Imperial agronomists (Adams (1982) 234, 236), but there are signs that they retained some differences: *bubulus* is the only specifying adjective of its type that Columella applies to *fimus* and is never applied to *stercus*; and at Col. 4.8.3 (*aliquid fimi uel, si est commodius, columbini stercoris*), the contrast requires that *fimus* refer to ox or cow dung in order to make

[62] The *Geoponica* follows Pliny and Palladius in reporting both techniques: 10.57.7 ὅταν μὲν οὖν ἀπὸ σπέρματος τὴν ἀμυγδαλῆν φυτεύωμεν, δεῖ τὰ σπέρματα πρόσφατα λαμβάνειν, καὶ ταῦτα προβρέχειν κόπρῳ δεδευμένῃ μεθ' ὕδατος. ἔνιοι δὲ εἰς ὑδρόμελι νύκτα μίαν προβρέχουσι τὰ σπέρματα.

[63] There may have been a similar distinction in Greek between ὄνθος and κόπρος: see Bain (2007) 46–8. His n. 29 (p. 48) shows that the seemingly anomalous cases of *fimus* at Verg. *Aen.* 5.333, 358 (where it has been suspected that it refers to mud or filth: cf. Adams (1982) 237) can still be explained as referring to cow dung.

sense of it. G.'s usage of *stercus* and *fimus* provides evidence that the original distinction between them was still felt long after Columella's time. In addition to the present, telling collocation, *fimus* appears twice elsewhere in *Pom.* (2.5.5, 2.5.8), and, in at least one of those cases (*fimo oblitam*), almost certainly stands for *fimus bubulus*: see 2.5.8n. *fimo oblitam*. Even in Palladius, *bubulus* is still paired most commonly with *fimus* (6x), and only rarely with *stercus* (3.25.15, 4.15.4) or other words denoting manure (e.g. *laetamen*, 3.25.4). G. uses *fimus* of chicken manure at *Med.* 8.2 *cum fimo gallinaceo tunsum* (sc. *lapathum*) *paniculas rumpit*.

Besides its use as manure (Appendix II), cow dung had a variety of specialized applications on the farm: (1) as a coating for grafts or transplants (cf. 2.5.8n. *fimo oblitam*); (2) as a source of salutary smoke for bees (e.g. Varro, *Rust.* 3.16.16, Col. 9.14.1, Pallad. 4.15.4);[64] and (3), especially in Palladius, as a means of remedying the ailments of trees (e.g. 3.25.4, 5.4.3).

transferunt: Although *transfero* here only means 'transfer' or 'put into', in the agronomists the word more often bears the technical sense 'transplant': see 2.5.6n. *transferant*.

nec . . . nisi cum: *nisi* very commonly = 'except' when it follows negations or negative concepts (H–Sz §367.d) and is regular in *nisi* + conjunction (*nisi qui, cum, ubi*, etc.): cf. thus also 4.7.3 *non satis iusta sit nisi quae*, etc. Because *cum* here specifies a temporal condition that must be fulfilled before the main verb, 'not . . . except when' can virtually be taken as 'not . . . until' or 'only when' (cf. similarly *Med.* 27.11 *nec nisi post*, etc.).

nuces: On the word *nux* in general, see concisely André on Plin. *HN* 17.63: '*nuces*: terme général qui désigne tous les fruits secs à amande (cf. Serv. *ad B.*, 2, 52 *nuces generaliter dicuntur omnia tecta corio duriore*): châtaigne . . . amande . . . pomme de pin . . . pistache . . . noisette, noix; le mot est ordinairement précisé par un épithète etc.' The last part of André's statement is exemplified by 4.1.2 *Graecam nucem*. Otherwise in *Pom.* the identity of the nut is always inferred from the section and *nux* refers to the physical 'nut' or seed of the species under discussion.

cum hiascere coeperunt: This phrase refers to the same process of seed-split previously described by 2.3.3 *cum extra hiare coeperint*, which it stylistically varies. *hiasco* is substituted for *hio* (see following n.), the (superfluous)

[64] The supposed *cognatio* between cattle and bees that explains this advice stems from the ancient belief in the generation of bees from dead cattle, 'bougonia': see e.g. Verg. *G.* 4.218–314, 531–58, Varro, *Rust.* 2.5.5 with Flach, Harrison (1913) 338 n. 1; also *Geop.* 15.2 περὶ μελισσῶν, καὶ πῶς ἂν ἐκ βοὸς γένοιτο, ὃ καλεῖται βουγονή, with Lelli.

adverb *extra* is dropped, and the generalizing perfect *coeperunt* replaces the 'sequential' use of the future perfect *coeperint* (for which see 1.1.7n. *ubi tempus erit . . . pollicentur*).

hiascere coeperunt: *hiasco* is found only here and at Cato, *Agr.* 17.2 *eae* (sc. *nuces*) *ubi primum incipiunt hiascere, tum legi oportet.* Like the far more common *hisco*, it is derived from *hio*: see *TLL* VI 3.2831.2–3 (Kroeker), E–M 526, de Vaan 285. The *-sco* suffix is not pleonastic with *coeperunt*, for *-sco* verbs usually do not bear an ingressive (or inchoative) sense when paired with verbs that mean 'to begin' or 'to start'; they are rather inceptive, and denote 'achievement', that is, 'entrance into an event or process' (Haverling (2000) 120; see generally 115–21). For the semantic distinction between *hio* and *hiasco*, see Haverling (2000) 155, although in *Pom. hio* adopts not only a 'non-dynamic' function ('be open'), but also a dynamic one ('open'). Consider thus that the action of 'opening' mentioned here is described by *hio* in the parallel expression at 2.3.3 (*cum extra hiare coeperint*). G.'s *hiasco* is probably *uariatio* for *hio* at 2.3.3, but it is semantically entirely appropriate on its own terms and does not require emendation.

3.1.2 scrobes . . . effundunt: Editors since Marx have printed *effodiunt*, which may be right: *effodio*, although it does not appear elsewhere in *Pom.*, is regularly applied to the excavation of *scrobes* (e.g. *Arb.* 4.3, and see *OLD* s.v. 4), and the simplex is found in the same sense (*Arb.* 19.1 *scrobes fodito*). But it may be that we can get Antonelli's 'scavano fosse' from *effundunt*, if the verb means to 'empty out' the hole (sc. of soil). No parallels can be adduced for this usage from the agricultural writers, but cf. Hor. *Sat.* 2.3.149 *effundi saccos nummorum*; Sil. *Pun.* 2.389 *laxos effundit amictus* (in these cases, 'i. q. evacuare', *TLL* V 2.218.31–3 (Leumann)).

in eos terram congerunt quam minutissimam cum stercoris admixtione: N transmits *in eos tercorum congerunt quam minutissima cum stercoris admixtione.* The easiest fix for *tercorum* is *stercorum*, presumably to be construed as partitive with the neuter plural *minutissima* (taken substantively), but that reading would be pointless next to *cum stercoris admixtione.* It is best to read *terram* and emend *minutissima* to fit: this collocation is found elsewhere in reference to the finely sifted soil used for young plants, as e.g. at Col. 5.10.14 = *Arb.* 22.3 *in eo* (sc. *scrobe*) *terram minutam in modum semipedis ponito.* Cf. also 2.5.6n. *cribrata terra* and the following n.

terram . . . minutissimam: The adjective *minutissimam* refers not to the amount of soil, as the following specification *eatenus ut pars dimidia compleatur* shows, but rather to its granularity, as at Col. 5.10.14 = *Arb.* 22.3 (quoted in previous n.). It denotes not so much the basic texture or structure of the planting soil (i.e. its composition in terms of clays and sands, its moisture content, richness, etc.) as the necessity of breaking up clods to aid in the passage and retention of oxygen and water.[65] (Both Columella and Pliny use the word in this same application: see *TLL* VIII 1040.48–9 (Rubenbauer).) G. does not in fact specify the general soil characteristics preferred for the almond, although he does so for the peach (2.4, 2.11) and the chestnut (4.5). Columella (5.10.12 = *Arb.* 22.1), Pliny (*HN* 17.63) and Palladius (2.15.7) all report that the almond thrives in soil that is hot (*calidum*), dry (*siccum*) and hard (*durum*), with Palladius adding the description *calculosum*.[66] The agronomists' emphasis on *calidum/siccum* very likely refers not only to the tree's adaptation to hot and dry climates (cf. *Geop.* 10.57.2 χαίρει δὲ τὸ φυτὸν τοῦτο θερμοτέροις τόποις, *EFN* 709), but also to its susceptibility to fungal and bacterial disease in more humid settings (*EFN* 709, 714–16, Rosengarten (1984) 11): hence Columella's claim that the almond rots in other climes (5.10.12 *in locis diuersis si deposueris, plerumque putrescit* = *Arb.* 22.1 *in locis diuersis eiusmodi natura si posueris nucem, protinus putrescet*), or Pliny's that it 'perishes or grows barren in rich or moist soil' (*HN* 17.63 *in pingui aut umida* (sc. *terra*) *mori aut sterilescere*) – something of an overstatement, to be sure.

eatenus ut pars dimidia compleatur: Based on the earlier statement *altos latosque trinis pedibus*, this will be a depth of 1½ Roman feet of the soil–manure mixture.

3.1.3 Ibi tria amygdala . . . triangulari lege disponunt: G.'s advice to set the three almonds in a triangle is traditional: cf. Col. 5.10.13 = *Arb.* 22.2 *ternas nuces in trigonum statuito*; Plin. *HN* 17.63 *ternas* (sc. *nuces*) *simul serendas, triangula ratione palmo inter se distantes*. Three seeds are recommended in case not all of them germinate, which would be a danger especially in hot, dry soils. Consider thus Palladius' remarks for planting the pistachio: 11.12.3 *pistacia terna constituet, ut ex omnibus germen quodcumque procedat.*

quo inter se non palmo, ut Columellae uidetur, sed pedis siue dodrantis spatio separentur: The *palmus* is one quarter of a *pes*, the *dodrans* three-quarters (for the *pes*, see 2.5.2n. *pedibus*). G. is not especially charitable in his report, for Columella actually says *at least* one palm's width apart:

[65] Therefore it is absent from White's list ((1970) 97–102) of descriptive terms for soil used by the Roman agronomists.

[66] Compare Wickson (1891) 513 on the soil for the almond grown in California: 'the almond prefers a loose, light, warm soil, and heavy, poorly-drained soils should be avoided'. Although almonds are grown well in a wide variety of conditions (cf. Duke (1989) 251), the best soils are now considered deep and well-drained loamy soils, i.e. those comprising a mix of fine-textured clayey and coarse-textured sandy soils: the coarser-textured soils improve drainage and aeration, while the clayey mix aids in moisture retention (*AOM* 30–5).

5.10.13 = *Arb.* 22.2 *nux a nuce minime palmo absit.* Probably the advice of a palm's separation between the nuts can be traced to Mago, since Plin. *HN* 17.63 (following Mago) has simply *palmo inter se distantes.* In the present case, it is not certain whether G. is correcting Columella (and Pliny and Mago) on the basis of his own experience or through an intermediate source. By the time we come to Palladius, the recommended distance has grown significantly to 2 feet: 2.15.6 *spatio inter se binorum pedum separata.*

quo . . . separentur: This use of *quo* without an accompanying comparative adjective or adverb is not exclusively post-classical (*pace* K–S II §186, H–Sz §375; it is found in Cicero and Caesar), although it does become much more common in the Imperial period: see Oakley on Liv. 9.10.9. For *quo* with the comparative in *Pom.*, cf. 3.3.2 *quo . . . diutius immoretur.*

3.1.4 Acuta pars nucis ab utrisque prima defigitur: *Arb.*, Pliny, Palladius and the *Geoponica* all agree on this recommendation, which goes back to Mago (cf. *ab utrisque*, referring to Celsus and Mago). The terms used differ quite significantly, however: cf. *Arb.* 22.2 *parsque acutior inferior sit*; Plin. *HN* 17.63 *mucrone defigi*; Pallad. 2.15.6 *ita ut cacumina figamus*; *Geop.* 10.57.8 δεῖ δὲ τὸ φυτευόμενον σπέρμα ὀρθὸν τιθέναι, τὸ μείουρον πρὸς τὴν γῆν ἄγοντας, τὸ δὲ ξυλῶδες καὶ λεπτὸν ἄνω (also 10.63.4). All of these descriptions clearly refer to the almond's tapering point, which is most conspicuous in the seed but still evident from a glance at the stone or fruit (see Plates 12, 13). Of the various terms used by the agronomists to describe the narrower end of the seed, *cacumen* is perhaps the least marked; although never appearing in *Pom.*, it is so used commonly in the agronomists to refer to the extremities of various plant parts: see the examples at *TLL* III 11.3–21 (Spelthahn). Pliny's *mucro*, on the other hand, is quite singular, for elsewhere in the agronomists the word describes only agricultural implements (*LAT* 51; cf. *TLL* VIII 1556.8–20 (van den Hout-Buchwald)); in conjunction with *acies* (*HN* 17.63), it gives a vivid *translatio* of the nut as a sword or other pointed blade. As is suggested below (3.1.4n. *aciem . . . Aquiloni*), Pliny's metaphor is probably owed to Mago. G. opts for the simple but unambiguous *acuta pars.*

radices exitum moliuntur: Marx adds *inde* before *radices* (Reitzenstein (1884) 55 n. 77 had already suggested *qua*), a judgement with which Winterbottom agrees, comparing 2.5.3 *unde radicis exordium est*, which the present phrase varies, and citing Mazzini's translation ('donde ha principio la radice'). Mazzini himself defends the reading of the palimpsest, arguing that while the addition of *unde* clarifies the sentence, its omission 'non rende la comprensione impossibile' (142). That is true, for the asyndeton could be explanatory, i.e. implying a causal relationship between the clauses (see 2.4.6n.): 'they set the nut point-down, [because] the roots force an exit [there]'. The reader will supply *ab acuta parte*, vel sim., from the preceding statement. Although the parallel 2.5.3 *unde radicis exordium est* is instructive, it is not of decisive value in establishing the text here, since the asyndeton may be intended to vary the hypotaxis of the earlier phrase: see generally the remarks on *uariatio* in the Introduction §v(a).

Besides G., only *Arb.* 22.2 spells out the reason for burying the almond nut point first: *quia inde radices mittit.*

in posteriores dissentiunt: No emendation is necessary if *partes* (sc. *nucis*) is understood with *in posteriores*, supplied from the preceding *acuta pars.*[67] *posterior* will mean the 'later' or 'next' part of the nut not in a physical or temporal sense, but in the logical sense of enumeration (*TLL* x 2.208.54–81 (Buchwald)), for the edge of the nut (*acies lateris*) is handled 'later' (i.e. after the tapering end) in the instructions given by Mago and Celsus. Nor is the construction *dissentio* + *in* + accusative impossible: although *dissentio* + *in* ordinarily takes the ablative (*OLD* s.v. 1a), the accusative can be explained on the analogy of *consentire* + *in* + accusative (so Antonelli, Winterbottom; cf. 4.6.5n. *in eadem fere uerba consentiunt*), and there is evidence elsewhere in *Pom.* for the confusion of accusative and ablative after *in* (very common in later Latin: see 3.6.2n. *in locum taedae*).

aciem lateris in Aquilonem Mago inuertit, Celsus ad Fauonium dirigit, Columella cum dicat, *anceps* nucis *ad Fauonium spectet*, [et] sine dubio et ipse unum latus opponit Aquiloni: For the stylistic features of this sentence, see Introduction §v(a). Both Mago and Celsus (cf. *ab utrisque*) believe that the almond ought to be set downward in the hole (3.1.4n.), so that it is standing perpendicular to the soil. When so planted, the almond will show four parts: its two broad sides or halves and the two tapering 'seams' or edges that join the sides. **N**'s *faciem lateris* should be emended to *aciem lateris*, following Pliny's report of Mago on the same matter: *HN* 17.63 (sc. *Mago iubet nucem*) *mucrone defigi, aciem lateris in Aquilonem spectare.*[68] The *acies lateris*, lit. 'sharp edge of the side', can refer to nothing else but one of the two 'seams' of the nut (see Plate 12).

Bearing this in mind, it can be appreciated that the issue at hand is the proper way of orienting the seed in the planting hole once it has been set point down: should it be rotated so that its edges (*acies*) are aligned on a north-south or east-west axis? Mago recommends turning the edge to the north and Celsus to the west. G. next quotes Columella's words *anceps nucis ad Fauonium*

[67] So Condorelli (*app. crit.*): '*meo iudicio, nisi* in partes posteriores *Gargilius scripsit, lectio* in posteriores *seruari potest, subaud.* partes *ex superiore . . . uoce* pars'.

[68] The metaphorical *acies* – not elsewhere used in this way (although *TLL* I 400.21–2 (Kempf) is no help here) – could easily have confused the scribes, especially because G. substituted the bland *acuta pars* for *mucro*, the other half of the nut-as-blade metaphor (see below and 3.1.4n.).

spectet (5.10.13 = *Arb.* 22.2; *nucis*, absent from Col., is G.'s clarification and should not be deleted) and adds a gloss that will help the reader understand what position Columella takes on the issue. (The words *sine dubio* thus mark an intervention on the part of G. that is intended to reconcile the varying reports of the authorities, as also at 4.6.6 (n.).[69]) This gloss is necessary because Columella's *anceps* (sc. *nucis*) is an unusual expression, and it is not immediately obvious what part of the nut is intended.[70] Nonetheless, *anceps* should refer to the edge of the almond; Columella's peculiar word choice can be explained if we recall his and Pliny's Magonian source. As we have seen, when Pliny adapts Mago he describes the nut metaphorically as a blade, with a 'point' or 'tip' (*mucro*) and 'edge' (*acies lateris*). Now, Columella is also following Mago in his passage, and if we suppose that the figure is not Pliny's but Mago's (G. after all attributes it directly to him), so that Columella could also have read it, we can explain why he would have written *anceps*: the word is used regularly to describe the cutting double edge of a weapon (*TLL* II 23.46–61 (Hey), *OLD* s.v. 1b) and therefore appeals to the same metaphor as Mago's/Pliny's *acies* does, and with the same meaning in reference to the edge of the nut. But G. apparently foresees the confusion that Columella's *anceps* might provoke and attempts to forestall it with the explanation *unum latus opponit Aquiloni*: if G.s' *unum latus* = Columella's *anceps*, then Columella's advice will amount to the same as Mago's (i.e. turn the edge to *Aquilo*) – just what we would expect, if Mago is Columella's source.

The emphasis on the question of the proper orientation of the almond may seem strange, and we speculate briefly on two possible origins for it. First, it could have been extrapolated from the extreme importance of aspect in viticulture (e.g. Col. 3.12.4–6, Plin. *HN* 17.19–24, White (1970) 230–1) and, to a lesser extent, in arboriculture (e.g. Plin. *HN* 17.10). Thus Pliny states (*ibid.*) that trees *Aquilone maxime gaudent, densiores ab afflatu eius laetioresque et materie firmiores*, and Palladius (2.15.6) says specifically of almonds that *ita statuendae sunt arbores ut ad meridiem spectent*.[71] Still, it is not clear how rotating the seed in any particular

[69] For *uariatio* in expressions within this semantic field (e.g. *non dubitanter*, *non dubie*, *indubitate*), see Maire on *Med* 4.1 (n. 2).

[70] It puzzles Forster and Heffner (on *Arb.* 22.2): 'The reading here is doubtful: the MS reading *anceps* gives no sort of sense.' They emend the text to *apex ad Fauonium spectet* and translate 'let one apex of the triangle face towards the West'. G.'s quotation argues for *anceps* (cf. Goujard on *Arb.* 22.2), if not decisively (he might have had a corrupt text of Columella).

[71] The separate issue of ensuring that the tree transplanted into the orchard face the same direction as the plant did in the nursery was always thought very important (cf. Col. 5.10.7–8 = *Arb.* 20.2, Plin. *HN* 17.83–4), and still is today, but that is not what is at stake here.

direction should be thought to influence the orientation of the grown tree. (A further argument against a relationship between the two strands of the advice is the fact that Palladius' *ad meridiem* for the direction of the tree is nowhere found recommended for the seed.) A second possibility is that the trick of orienting the nut is a bit of magic. Spells and rituals enacted upon the seeds to ensure the fertility of the plant, affect produce or keep pests away are found in all the agronomists from Cato to the *Geoponica*: cf. 3.1.1n., and for magic in Roman agriculture, 2.9.3n. (with general literature), 3.6.2–3n.

***spectet*, [et] sine dubio et ipse:** *et* is almost certainly dittography after *spectet*: for other instances of syllabic dittography in *Pom.*, see 2.4.6n. *contrarium* [*tum*] *crediderunt*. Antonelli, col. 1691 suspects its redundancy, and Winterbottom concurs that the text can be improved by its deletion. Mazzini 143 claims to have discovered several other cases of *et* 'pleonastica' (citing 2.10.6, 4.1.8, 4.3.2), but none of them is apropos.[72] If one wished to defend *et* here, a better explanation would be to construe it as 'apodotic', that is, as breaking the hypotactic connection between the clauses (for a precise definition, see Galdi (2014) 63). The distribution and significance of apodotic *et* are in fact complicated questions, recently and helpfully reassessed by Galdi (2014). As regards Christian and late Latin (where the phenomenon is most frequent), an older hypothesis was that apodotic *et* should be explained as a kind of anacoluthon in which the author 'lost, so to speak, the syntactic control over the sentence, slipping into the more "colloquial" parataxis' (Galdi (2014) 86): see Baehrens (1912) 426–33, Sorrento (1949) 30–91 (with Wehr (1984) 157–8); cf. also H–Sz Stilistik §256.g. The frequency of apodotic *et* in late Latin and the contexts of its occurrence do not support this hypothesis, however, and it has been argued that most of these instances of *et* are used intentionally as a focus-marker to throw emphasis on what follows: see Wehr (1984) 159–69, (2008) 183–4, Galdi (2014) 85–7. Nevertheless, as Galdi (2014) 87 n. 88 remarks, anacoluthon is a viable explanation especially in 'cases in which the subordinate and main clause are separated by several words'. That is at least possible in the present instance, where a direct quotation from Columella intervenes between *cum dicat* and the clause *sine dubio*, etc. But neither would an intentional focusing function be entirely out of place for *et* if it is retained, for the following clause (*sine dubio*, etc.) represents G.'s own intervention into the dispute among authorities (see 3.1.4n. *aciem* . . . *Aquiloni*). Cf. also 2.4.5n. *ubi* . . . *induci*.

3.1.5 Supra nuces terra sesquipedis ingeritur: I.e. another 1 ½ Roman feet of soil, bringing the total to 3 feet, the depth of the hole.

[72] His examples are simply of emphatic double *et*, for which see 2.1.3n. *et* . . . *praebeat et* . . . *acquirat*.

si nec ullus imber fuerit, ter in mense seminibus expositis umoris alimenta sumministrentur: Cf. Pallad. 2.15.8 *cum in seminario amygdala disponimus, si siccitas intercessit, ter in mense rigemus.* Pliny reports from Mago only that the almonds ought to be watered for ten days after planting until they swell: *HN* 17.63 *denis diebus adaquari donec grandescant.* Although Columella does not have any advice on watering specific to the almond, he does offer an ingenious way of supplying water to the seeds of fruit trees after planting: 5.10.8 = *Arb.* 20.3 *cum semina depones, dextra sinistraque usque in imum scrobem fasciculos sarmentorum bracchii crassitudinis* [*Arb. humani crassitudine*] *demittito ita ut supra terram paulum* [*Arb. paululum*] *extent, per quos aestate paruo labore aquam radicibus sumministres.* For the irrigation of almonds and other fruit trees, see 3.3.2n. *quo . . . immoretur.*

nec ullus: *nec* = *non* (i.e. it has no coordinating meaning): see Löfstedt (1911) 88–9, *Syntactica* I 338–42, H–Sz §241.B.a, Väänänen (1981) 151, (1987) 96, Orlandini and Poccetti (2008) 1–3, 9, Pinkster 692–3, Adams (2016) 374–5. Though an early Latin usage, it is 'noch in spätlateinischer Zeit lebendig, wenn auch nicht sehr häufig' (Löfstedt (1911) 88). It is especially common in legal and juridical writings (Pascucci (1968) 22–9, Pinkster 692). *nec ullus* for *nullus* in particular is found from Plautus onwards.

ter in mense: The preposition *in* is distributive (*TLL* VII 1.789.76–790.10 (Bulhart-Hofmann)). Although it is possible to use the bare ablative (cf. Adams (2016) 90–1), the pattern of the present type is established very early (cf. Cato, *Agr.* 43.2 *semel in mense*) and found continually up to and beyond G.'s time (for Pelagonius, see Adams (1995*a*) 166). For the construction with *mensis* in particular, cf. Celsus, *Med.* 1.3.24 *bis in mense*; Scrib. Larg. 58, 159 *ter in mense*; Plin. *HN* 29.109 *semel in mense*; 30.23 *bis in mense.*

expositis: One might think at first blush that *expositis* means 'exposed' or 'uncovered', but such advice would be futile and, in all likelihood, harmful to the delicate seeds. Here *expositis* means rather 'planted out' into the nursery (as at 2.5.5, 4.2.4); see also the discussion of *pono* in the Introduction §v(d).

umoris alimenta sumministrentur: The locution is altogether Columellan: for *alimentum umoris*, cf. 2.2.24 *ut* (sc. *radices inferiores*) *facilius capiant umoris alimentum*; and for *sumministro*, cf. 3.1.8 *radicibus umorem sumministret*; Col. 5.10.8 = *Arb.* 20.3 *aquam radicibus sumministres* (quoted more fully at 3.1.5n. *si . . . sumministrentur*).

quidquid in summo scrobe induruit leui sarculo relaxetur: This refers to *sartio* (see Plate 9): see further 2.6.1n. *frequens . . . runcationis.* Palladius describes an operation for the almond that is different (digging rather than hoeing) but has a similar purpose of loosening the soil and removing weeds that could choke growth: 2.15.8 *herbis nascentibus circumfodiendo saepe purgemus.*

summo scrobe: The hole (*scrobes*) has of course long since been filled in, so this refers by metonymy to the planting spot.

leui sarculo relaxetur: This metaphorical extension of *relaxare* to tillage or other loosening of the soil is very uncommon among the agronomists, occurring only twice elsewhere: Varro, *Rust.* 1.27.2 (sc. *glaebas*) *ad opus faciliores relaxatas*; Col. 11.3.46 *relaxata humo plenioris crassitudinis fiat radix.* The occurrence in Varro may have inspired Virgil's usage at *G.* 1.89–90, where he describes the benefits of burning the stubble after the grain harvest: *seu* (sc. *inde*, from burning) *pluris calor ille uias et caeca relaxat / spiramenta, nouas ueniat qua sucus in herbas.*[73]

3.2 The planting recommendations of Mago and Celsus are continued, as can be inferred from the plural active forms *docent*, *adobruunt* and *numerantes*. Instructions are provided for supporting the young plants while in the nursery and for transplanting them into their holes in the orchard.

3.2.1 Vbi primum . . . ostenderint . . . affigitur: For the tense of *ostenderint*, see Introduction §v(b).

stilus . . . flexuosus: *flexuosus* here means 'flexible' according to *TLL* vi 1.908.29–35 (Klee), an uncommon meaning for which the first attestation provided is the present one; but the word could also have its more usual meaning 'curving' or 'bending' (*TLL* vi 1.907.35–62, *OLD* s.v.), which would equally well explain the subsequent *scandere in rectum.*

iuxta . . . palus affigitur: For the *iunctura*, cf. Plin. *HN* 17.173 *adminiculo iuxta affixo.* It is important that the support be placed close to the plant (*iuxta*) in order to aid it in the face of damaging weather: cf. Col. 4.16.3 (on the vine) *nam quanto propius truncum ridica statuitur, etiam leuiter defixa stabilior est, quoniam constringens uitem mutua uice sustinetur et sustinet*; see also the following n. Columella also cautions against establishing it too near, however, which might injure the plant or inhibit root growth and prevent the cultivator from digging around the seeds: 4.16.2 *nam siue prope truncum deponitur palus, pedali tamen spatio recedendum est, ne aut premat aut radicem uulneret et ut fossor tamen ab omni parte semina circumfodiat.*

uastior palus: *uastior* is in sense like *uehemens* at Col. 4.12.1 (*uehementem palum aut ridicam*), with a view to the sturdiness of the prop (cf. also Col. 4.16.2 *uehementioribus statuminibus*). For *uastus* with this meaning of 'thick' or 'stout' (but also capable of referring to height or length), see Adams (1995*a*) 583–5. The weak application of the adjective without any hint of 'immense size in absolute terms' had developed long before G.'s time (Adams (1995*a*) 583), although

[73] Incidentally, Plin. *HN* 18.300 rejects the reasons Virgil offers for burning the stubble and argues instead that removal of the weeds is the most important goal: see Doody (2007) 194.

it is not properly recognized by L&S or *OLD*. Among the agronomists, a clear example can already be found e.g. at Col. 11.3.49 *idque de media parte cucurbitae inuerso cacumine ponito, ut fiat incrementi uastioris*. The same degree of size is conveyed by the noun *uastitas* at 2.4.4 *enormi uastitate* (n.). The reason for the stoutness of the prop is that it offers the young plants protection against adverse climatic conditions: on the vine, cf. Varro, *Rust.* 1.25 *in omni uinea diligenter obseruant ut ridica uitis ad septemtrionem uersus tegatur*; Col. 4.16.3 *isque palus sic ponendus est ut frigorum et Aquilonum excipiat uiolentiam uitemque protegat*. The benefits of staking young plants are of course still appreciated today: see *EFN* 712.

palus . . . adminiculo: *adminiculum* is used after *palus* to specify the function of the stake, an implement that could be used for countless purposes in an agricultural setting. Because *adminiculum* is a less, not more, specific term for a prop, there is no difficulty in substituting it for *palus*: cf. Col. 4.12.1 *notatum est enim a me plerumque teneram uineam melius adminiculo modico quam uehementi palo acquiescere* (the contrast resides not in the type of prop but in its size, thus *modico quam uehementi*). For the *palus* and other props used by the agronomists, and another example of the interchange of *adminiculum* with a more specific word describing the prop (*calamus*), see 2.6.2n.

quo rami adminiculo et in ualido requiescant: *et* complicates this clause. We cannot easily take it to mean 'and', joining two separate things on which the branches rest, for two reasons: (1) *requiesco* does not govern an ablative on which something rests, so *adminiculo* could not be syntactically parallel to *in ualido*; and (2) even if *requiesco* could so govern *adminiculo* (perhaps on the analogy of *acquiesco*; cf. Col. 11.3.7 *quo . . . quasi adminiculo . . . acquiescant*), the join of *adminiculo* and the vague *in ualido* (what would it refer to?) seems harsh. There is a case for deleting it: once excised, we might take *quo . . . adminiculo* as an instrumental ablative (the relative clause is final, with antecedent *adminiculo* incorporated) and let *requiescant* govern *in ualido* (*requiesco* + *in* being the standard construction; see *OLD* s.v. 4, and cf. Col. 5.2.10 *in se requiescit*). The translation would thus be: 'with which as a prop (*quo . . . adminiculo*) the branches might rest on something strong (*in ualido*)'. (The intrusion of *et* could be explained by the following *et scandere*.) Yet the difficulties in the transmitted text are not insuperable. There is precedent in *Pom.* for coordinating the bare ablative of location with *in* + ablative (cf. 1.1.13n. *et scrobibus et in doleis*), and *in ualido* might be joined epexegetically by *et* to *quo . . . adminiculo* (*TLL* v 2.874.34–81 (Hofmann), *OLD* s.v. 11), restating and specifying it: 'that is, on something strong', or 'indeed, on a strong [prop]'.

scandere . . . consuescere: In order to describe the effects of the prop on a plant G. pairs *consuesco* with an infinitive also at 2.6.2 *quibus* (sc. *adminiculis*) *se tueri . . . consuescant* (n.). For *scando* applied to a plant climbing a prop, cf. Plin. *HN* 17.215 *uiti adminicula addenda, quae scandat apprehensa*. It is not a common word in the agronomists but is found in this sense (most often of the vine) since Cato (*OLD* s.v. 3b): cf. *Agr.* 50.2 *ficos interputato et in uinea ficos surradito alte, ne eas uitis scandat.*

in rectum †conpletum†: The prepositional phrase *in rectum* means 'vertically, straight up' (*OLD* s.v. 6a): cf. also Scrib. Larg. 227 *multorum ramorum in rectum surgentium.* For sentiments similar to the present one with only slightly different language, cf. Col. 4.4.2 *meminerimus ergo usque ab imo scrobis solo rectum adminiculo sarmentum applicare et ita in summum perducere*; Plin. *HN* 17.184 *initio adminiculum desiderabit, dum stare condiscat et recta surgere.* The addition of *completum* is more difficult, however. Antonelli asserts that the whole phrase is equivalent to *in rectum perfectum*, and both he and Mazzini translate 'perfettamente'. While it is true that *completum* can mean *perfectum* (*TLL* III 2098.29–33 (Wulff)), it is doubtful whether the adjective could be so joined to the adverbial *in rectum*, and I have not found a comparable case among adverbial expressions formed from *in* + an accusative singular adjective (cf. *TLL* VII 1.746.78–747.15 (Bulhart-Hofmann)). Mai recalls seeing *conplexu* in his examination of **N**, which is suggestive – it would presumably mean 'by [the plant's] grasping [the prop]' – but does not admit of proof.

docent: The only possible subject for *docent* is Mago and Celsus, but the abrupt shift from *affigitur* and *rami . . . requiescant* is harsh: possibly something has fallen from the text in this sentence (or is concealed by *conpletum*) or else the reading *docentur* might be considered, perhaps transposed after *consuescere* for rhythm (a resolved cretic-trochee *consuēscĕrĕ dŏcēntur*). For the sense of *doceo* here, see Hus (1965) 59–60, 132–3.

3.2.2 Triennio exacto unum germen in loco permanet, reliqua eodem mense quo sata sunt ad scrobes transferuntur: As the trees grow, some must be relocated into the *ordines* of the orchard in order to give them space enough to flourish; one tree remains in each plot, and the *seminarium* is thus evidently converted into a part of the *pomarium*. Cf. Pallad. 2.15.7 *cum* (sc. *amygdali*) *in seminario adoleuerint, relictis ibi quae spatio sufficiant plantis, alias transferemus mense Februario.* G. will go on to discuss the appropriate distance and placement of the *ordines* in the orchard in 3.4: for more on their spacing, see 3.4.2n. *uiginti . . . pronuntiant.*

Triennio exacto: Because present-day methods of intensive almond cultivation do not generally begin from seed, the timeline is very different from that suggested here.[74] In any case, the almond typically begins to produce crops in its third or fourth year (*AOM* 15, Rosengarten (1984) 11).

[74] Note however that in India the trees are still raised from seedlings, where they are typically transplanted after one year (Duke (1989) 251).

ut pars aliqua fossurae conuexa relinquatur: *fossura*, like *fossio* (3.5.3), most often denotes *actus fodiendi* among the agronomists (on *fossura* see *TLL* VI 1.1215.43–51 (Rubenbauer); on *fossio*, *TLL* VI 1.1213.63–82 (*id.*), *LAT* 42); but whereas *fossio* can bear only that sense, *fossura* can also be employed in a concrete sense to refer to a dug area, as here (*TLL* VI 1.1215.51–9, Adams (2016) 297). Although *relinquo* is used in a normal way (*OLD* s.v. 9) and the general sense of the statement is clear, the phrase is deceptively hard to render. Strictly speaking, it is not logical to say that a part of a 'pit' (*fossura*) is 'concave' (for this meaning of *conuexus*, see *OLD* s.v. 2), since the pit is itself a concavity by excavation of soil. What is meant instead is that the pit is only partially filled, so that a small depression remains at the top with respect to the level of the surrounding earth.

3.2.3 Hanc paulatim adobruunt singula arboris incrementa numerantes, tantumque radicibus cumuli largiuntur quantum altitudinis truncus adiecerit: This piece of advice seems to go back to Mago, for we find the same instructions attributed to him, along with an explanation, in Columella's discussion of planting vines: 5.5.4 *unum tamen huic consitioni Mago Carthaginiensis adicit, ut semina ita deponantur ne protinus totus scrobis terra compleatur sed dimidia fere pars eius sequente biennio paulatim adaequetur; sic enim putat uitem cogi deorsum agere radices.*[75] (The essence of the advice is in fact found already in Theophrastus: cf. *Caus. pl.* 3.4.2 καὶ τοὺς γύρους οὐκ εὐθὺς συμπληροῦσιν, ὅπως ῥιζωθῇ τὰ κάτω πρότερον· εἰ δὲ μή, φέρονται πρὸς τὸ ἄνω· τρέφει γὰρ καὶ αὔξει πάνθ᾽ ὁ ἥλιος καὶ ὁ ἀήρ.) Columella goes on to revise Mago's advice (5.5.5): he agrees that in dry locations (he may be thinking of Mago's North Africa) the practice is useful, but warns that in wetter climes it is inadvisable because the method of slowly filling the holes leaves the roots exposed, making the plants vulnerable to drowning if an excess of water should fill the holes. He recommends instead that the holes be refilled immediately after planting to protect the roots and then dug again to expose them (*ablaqueari*) only after the autumn equinox. So, Columella claims, the plant will receive a double benefit: 5.5.5 *sic enim utrumque incommodum uitabitur, ut nec radices in superiorem partem euocentur neque immodicis pluuiis parum ualida uexentur semina.*

adobruunt: The word is a purely technical term introduced by Columella (19x): 'terram iuxta et leuiter supra semen radicem aggerare' (*TLL* I 792.73–86 (Vollmer), White (1970) 263), *Anglice* 'to earth up'. *adobruo* here is thus not an example of *compositum pro simplici* for *obruo*, *pace* Mazzini 52. For the benefits of

[75] Palladius (3.10.3) repeats the instructions with the same attribution to Mago, but the version of *Arb.* 4.5 does not mention him: *triennio deinde proximo paulatim scrobem uel sulcum usque in summum completo; sic uites consuescent radices deorsum agere.*

the process, which include moisture retention, see Foxhall (1996) 56–7. It was frequently practised in wheat farming in order to promote tillering, i.e. the growth of side-shoots from the wheat plants (= *fruticare*, *OLD* s.v. 1): cf. Col. 2.11.8 *ut potius adobruantur cumulisque exaggerentur, ut latius se frutex culmi diffundat*; see further Westerath (1938) 37–8, White (1963) 209, (1970) 134, 180–1. Outside of Columella's writings the word *adobruo* appears only here and at Pallad. 1.35.10. Perhaps Columella introduced it primarily with an eye to distinguishing it from *obruo*, which had been employed, among other things, for sowing or planting from the time of Varro, and which was commonly so used by Columella himself (see 2.1.3n. *obruenda*). Pliny uses *accumulo* synonymously (*HN* 17.139, 18.230, and *passim*), and Columella (11.2.46, *Arb.* 28.3) and Palladius (12.9) also *aggero* (*aggerare*).

incrementa: A multi-purpose word among the agronomists, most commonly used in one of two technical senses: (1) the amount or kind of a plant's growth (*TLL* VII 1.1045.14–29 (Bulhart), *OLD* s.v. 3) and (2) the process of growth (*TLL* VII 1.1045.67–75, *OLD* s.v. 1b). Semantic exactness is difficult in the present instance: the usage shades between the former of the above senses (amount of growth) and, under the influence of *singula . . . numerantes*, simply 'addition', i.e. that which has been added to something (*OLD* s.v. 2a).

altitudinis: For *altitudo* used as a marker of the maturity or readiness of a plant for different sorts of cultivation, cf. Col. 11.2.10 *sed et faba eandem culturam exigit, si iam coliculus eius in quattuor digitos altitudinis creuit.*

3.3 G. now offers two alternative methods for planting almonds in a nursery. The method of the Quintilii is very similar to that of Mago and Diophanes presented in 3.1–2; the principal differences are the Quintilii's recommendation to trench the entire field (*pastinatio* or a similar operation; see the following n.), the careful provision of irrigation furrows, and a shallower planting depth. The giant fennel method of Mago and Diophanes is not so heterodox as it first appears, since the trees are still basically grown from seed in a fertile container, here the pith of the fennel rather than manured soil (see 3.3.4–5n.). It thus also bears comparison to the recommendation to plant peach stones along with their fruit, which serve in place of manure: 2.1.3 *ea res* (= *poma ipsa*) *et ossibus uicem stercoris praebeat.*

3.3.1 omnem agrum in seminario . . . ad sesquipedis altitudinem fodiunt: What is described here is the intensive method of trenching known as (*re*)*pastinatio*, an operation referred to sometimes by the specialized verb (*re*) *pastino* (see 4.7.1n. *pastinatos*) and sometimes by *fodio* (as here), the ordinary word for digging, or one of its compounds (*infodio*, *defodio*, etc.). The purpose of this expensive method of trenching, which required many man-hours of labour, was to overturn and mix the field thoroughly, *ut mota aequaliter humus nouelli* *seminis radicibus, quamcumque in partem prorepserint, molliter cedat nec incrementa duritia sua reuerberet, sed tenero ueluti <ac> nutricio sinu recipiat, et caelestes admittat imbres eosque alendis seminibus dispenset, ac suis omnibus partibus ad educandam prolem nouam conspiret* (Col. 3.13.6). Columella is preceded in this artistic praise of trenching by Cicero, who has Cato say (*Sen.* 53), *quid fossiones agri repastinationesque proferam quibus fit multo terra fecundior?* The technique of *pastinatio* is most fully discussed by the agronomists in reference to vines, for which it was considered the best way of preparing the soil (for *pastinatio* in comparison with other methods of soil preparation, see 2.7.1n. *Scrobes*), but the agronomists constantly allude to the practice for many other plants too. The advice to trench the field deeply before planting is frequently connected with preparing a *seminarium* for trees: cf. Col. 4.33.2 (chestnut), 5.6.6 (elm and ash), 11.2.16–18, Plin. *HN* 17.69, *Pom.* 4.7.4 (chestnut), Pallad. 2.15.6 (almond).

The chief point of variation among the authors is indeed not whether to trench, but the depth to which the land ought to be dug. Consider the range of figures found in some of the principal statements on trenching:

- Col. 3.13.8 *campestris locus alte duos pedes et semissem infodiendus est, accliuis regio tris, praeruptior uero collis uel in quattuor pedes uertendus.*
- Col. 11.2.17–18 *pastinatur autem terreni iugerum ita ut solum in altitudinem trium pedum defodiatur operis octoginta uel in altitudinem dupondii semissis operis quinquaginta uel ad bipalium,*[76] *quae est altitudo duorum pedum, operis quadraginta. haec tamen in agro sicco surculis conserendis minima pastinationis mensura est; nam holeribus deponendis possit uel sesquipedalis altitudo satisfacere, quae plerumque in singula iugera triginta operis conficitur.*

[76] The word *bipalium* (found outside of the ablative singular only in this passage of Columella) has been the subject of much discussion, being taken variously to refer to an agricultural tool (probably the heavy footrest shovel, Palladius' *uanga*), to an agricultural technique or to an approximate measure of depth. Most interpreters take *bipalium* to refer to a tool, including Fabretti (1702) 574 n. 60, *TLL* II 1999.66 (Ihm), L&S, *OLD*, *RE* (1) 3.487.61–488.27 s.v. 'bipalium' (Olck), White (1967) 20–4, Goujard (1984) 217–18 (with further literature). On the side of the *bipalium* as a unit of measurement, Schneider (1794) II 674 asserted that 'bipalium esse mensuram aliquam, uel modum fossionis, non instrumentum aliquod, fossioni aptum'. Kolendo (1980) 206–12 argues that *bipalium* originally named a technique of deep trenching but that by metonymy it came to refer to a measure of depth. The last word on the issue is that of De Angelis (1997*b*), who, in a thorough review of the evidence, discards Kolendo's hypothesis that it refers to a technique, but tentatively argues for taking the *bipalium* to refer to an approximate unit of measurement between 2 and 2 ½ feet.

- Plin. *HN* 17.159 *in seminario siue in uinea bidente pastinari debet, ternos pedes bipalio alto*, etc.
- Pallad. 2.10.4 *pastino uero quod omne uersabitur trium uel duorum semis pedum altitudine terra uniuersa fodiatur.*

In this tally (representative but not exhaustive), the median and mean trenching depths are both 2½ feet, with the shallowest recommendation 1½ feet (Columella, for vegetables) and the deepest 4 feet (Columella, on steep slopes). The most commonly recommended depths are 2½ and 3 feet. By and large the figures for trenching that the agronomists give for specific plants agree with the general guidelines set out above: in *Pom.* 4.7.1–3, for example, trenching depths for the chestnut are reported as 2 feet, 2½ feet (cf. Col. 4.33.2), and 3 feet. There are some cases, however, in which the soil is trenched with good reason to a lesser depth, as with the Quintilii's advice here (repeated by Pallad. 2.15.6 *fodiemus ergo altam pede uno semis aream*): as G. makes clear at 4.7.4 (*potest qualiscumque mensura*, etc.), a shallower trenching depth is perfectly suitable for temporary locations such as nurseries (so cf. 4.7.4 *seminarium cogitant*); trenching any deeper would indeed be superfluous in view of both the shallow planting depth of the seeds (for the almond only 'four fingers', 3.3.2 *digitorum quattuor*) and the limited time for root growth.

omnem agrum in seminario amygdalis destinatum: Although Antonelli and Mazzini translate 'il terreno destinato al semenzaio' (thus taking *in seminario* as the complement of *destinatum*), I prefer to construe the prepositional phrase *in seminario* as an attribute of *omnem agrum* (Pinkster 1029–33) and join the dative *amygdalis* with *destino*. The verb *destino* is found only once with *in* + ablative (*Rhet. Her.* 1.25 *quid oporteat . . . in causa destinari*), and there is special precedent for taking it with the dative: as Önnerfors (1993) 264 n. 68 notes, '[d]ie Junktur *alqd alci rei destinare* ("wählen", "reservieren") ist landwirtschaftlicher *terminus technicus*, z. B. Colum. II 10,5 *fabae pinguissimus locus uel stercoratus destinetur*; III 11, 3 *oliuetum, melius tamen uetus oliuetum . . . uineis desinatur*; vgl. noch II 19,2; III 1,7'; add *Arb.* 1.5, among other possible examples.

ad sesquipedis altitudinem: *altitudo* is very often used in formulaic prepositional expressions by both Columella and Pliny to specify depth: see *TLL* I 1768.39–1769.29 (v. Mess). The most frequent of these formulas is *in* + *altitudinem* (cf. 4.7.1 *in duorum pedum altitudinem subleuetur*); *ad* + *altitudinem*, as here, is much less common, but cf. Col. 11.3.5 *resolutaque humus . . . usque ad mediam sulcorum altitudinem reponitur*. Also found among the agronomists are ablatival (e.g. Col. 11.3.48 *sesquipedali altitudine fossa facienda est*) and genitival (e.g. Col. 11.3.44 *sulci . . . non amplius dodrantalis altitudinis*) constructions (Pinkster 825–6).

3.3.2 Hunc in areas diuidunt: *area* is technical for a garden plot (*OLD* s.v. 3): for this *iunctura*, cf. Col. 11.3.13 *humus refossa in areas diuiditur*.

quo canalibus umor infusus diutius immoretur: The furrow irrigation system described here has always been ubiquitous, and Wickson (1891) 212–13 mentions nearly identical ditch and furrow systems used in California in the nineteenth century: trenches (≈ *canalibus*) are dug near the trees, and water is poured (≈ *infusus*) through them in order to saturate the soil around the roots. Many examples of furrow irrigation have been preserved at Pompeii: see Jashemski (1979–93) I 190–92, 238, 253, 285–6, 339, (2018*b*) 446–7. A furrow system may also be implied by 2.4.5 *ubi umor manu temperatus nocens ita auerti potest ut et possit induci* (n.). Although the almond has high drought tolerance (Duke (1989) 251, *EFN* 712–13), irrigation is especially important for the young plant until it can establish its root system.

Elsewhere in *amygd.*, G. mentions irrigation for the almond in connection with adverse climatic conditions, such as excessive heat or drought (3.1.5). In *persic.* and *cast.* his only remarks on irrigation are to report in passing some authorities who prefer well-watered land (2.4.2–3, 2.4.5, 4.1.2, 4.6.6), and to encourage the irrigation of peach slips (2.5.7). In the overall fairly modest emphasis he places on irrigation in *Pom.*, G. is in agreement with the other agronomists, who tend to recommend watering only in hot summer weather or periods of drought: cf. Col. 5.6.8, 5.10.8 = *Arb.* 20.3, Plin. *HN* 17.249, Pallad. 2.15.8, *Geop.* 9.11.3.[77] Pliny (*ibid.*), in fact, goes so far as to discourage the farmer from watering trees at any time of the year but summer. (He also asserts that young trees require less water than mature ones, but of course the opposite is the case.) The lack of urgency attached to artificial irrigation was no doubt a result of the climatic situation, for many fruit trees would not have required much watering in Italy to survive (cf. White (1970) 154–5, Kron (2012) 14) – virtually none, in fact, assuming typical climatic conditions (cf. Rosengarten (1984) 10 on almonds grown in Spain today). It is therefore not surprising that the agronomists, taking into consideration the relative hardiness of the trees and their ability to thrive on natural moisture, would insist on irrigating only during periods of drought, since at other times available water could have been used for more urgent purposes; fruit trees might at times have been an attractive crop for a farmer precisely because they would not put additional strain on limited water resources.

In any case, irrigation would have been feasible only for those farmers whose property was situated so as to provide access to natural or artificial water sources, such as a river, stream, well or aqueduct, and who had the technology necessary to harness those sources (for which see 2.4.5n. *ubi . . . induci*).

[77] An increased attention to the benefits of irrigating young fruit trees is observed *passim* in Palladius: cf. 3.25.20 (quince), 3.25.27 (carob), 4.10.2 (pomegranate), 4.10.11 (citron), 4.10.19 (medlar), 11.12.1 (date palm), 11.12.3 (pistachio).

Connected with the former point is the agronomists' emphasis on locating the farm near a good water supply (Cato, *Agr.* 1.3 *bonum . . . aquarium*): cf. Varro, *Rust.* 1.11.2, Vitr. *De arch.* 8.4, Col. 1.5.1–4, 11.3.8–10, Pallad. 1.4 with Martin, 1.16–17 with Bartoldus (2012) 85–7, 9.8–11. As suggested above, farmers who did not have the means to irrigate their orchards simply relied on rain (cf. Spurr (1986*b*) 23–7). This practice was the norm, not the exception, for Roman farmers: 'irrigation was applied for the most part to vegetable and pleasure gardens . . . grain fields, vineyards, orchards, and pastures depend on precipitation' (Oleson (2000*a*) 212).

The light emphasis on irrigation in ancient arboriculture may be further illuminated by cautious comparison with modern discussions of the same matter. In the nineteenth century, Californian farmers were undecided on the question of the proper level of irrigation for trees, with some holding that any amount was injurious to the crop. Wickson (1891) 79–80 tells of the great abandonment of irrigation among Californian fruit growers, who proudly displayed 'no irrigation' placards at fruit fairs, and he opens his own discussion of irrigation by addressing first the question whether, not how, farmers should irrigate their trees.[78] As irrigation was abandoned in California, it was replaced with the practice of frequent cultivation of the soil. Wickson (1891) himself more pragmatically espoused moderate irrigation practices (various kinds discussed 212–26), warning against only those methods taken to excess. There is no evidence for a similar controversy over irrigation in antiquity, but the modern parallel has this conclusion: the light treatment of irrigation in the ancient sources should not be cause for any great wonder when we consider that the cultivation of fruit trees is such as to permit practices that are explicitly opposed to its use.

Present-day methods of irrigation stand in contrast to the ancient situation insofar as they are much more intensive both in the nursery, where the trees are pushed to rapid development, and in the orchards: the drought in California in 2014–15 brought almonds back into the headlines in America for their water consumption (and prompted discussion of ethical water management in times of severe drought). Although the tree can tolerate arid conditions well (as mentioned above), intensive irrigation promotes quicker development and more abundant fruit.

modicis scrobibus excitatis: *modicus* here means 'small' or even 'shallow' (see 2.7.1n. *modica*), as revealed by the statement that the nuts planted within

[78] 'The general question of whether irrigation shall be employed in the growth of fruit in California has been discussed for more than thirty years, and difference of opinion still exists. There is no hope of definite settlement of the general proposition, because it is by its nature capable only of conditional answer' (Wickson (1891) 207).

in it are sunk to a depth of only 4 *digiti* (*ad digitorum quattuor altitudinem nuces mergunt*). For the meaning of *excitatis*, possibly providing *uariatio* for a form of *fodio*, see *TLL* v 2.1257.73–1258.5 (Rehm), 'terram sim. i. q. euertere, proscindere'.

ad digitorum quattuor altitudinem: Cf. Pallad. 2.15.6 *obruemus amygdala non amplius quattuor digitis*. Planting the seeds at a depth of 4 *digiti* would be about one-eighth the depth of the trenched soil (a *sesquipes* being 24 *digiti*; see 2.5.2n. *pedibus*). Everybody knows why the soil should be trenched deeper than the level of planting, Columella says: 4.1.2 *dicet aliquis, ut sit inferior tenera subiacens terra, quae non arceat nec duritie sua repellat nouas irrepentis radiculas*. (On the point see more fully Col. 3.13.6, quoted at 3.3.1n. *omnem . . . fodiunt*.)

nuces mergunt: *mergunt* provides *uariatio* for 3.1.3 *disponunt*. For *mergo* used of planting, cf. 2.3.2n. *ossa merguntur*. It is not clear whether the Quintilii share the other agronomists' concern about the orientation of the planted seed, although G.'s omission of the matter suggests not. We might assume that they would have advised planting the tip (*cacumen*/*acuta pars*) of the nut downwards, to judge from the inclusion of that detail by Palladius (2.15.6) and the *Geoponica*.

3.3.3 Stercora in terra obruunt: Cf. 3.1.2 *terram congerunt . . . cum stercoris admixtione*.

parique qua ceteri cultura lene nutritas: The *cultura* includes, but is probably not limited to, watering the young trees in times of drought and applying frequent hoeing (so 3.1.5).

lene nutritas: The neuter singular of *lenis* used as an adverb is restricted to poetry before the first century CE, from which time it begins to appear occasionally in prose: see *TLL* vii 2.1146.43–8 (Ravenna).

in scrobes . . . transferunt: Mazzini defends the transmitted *in scrobe* (*app. crit. ad loc.*): '*nulla emendatione opus esse mihi uidetur si* scrobe*, ut sensui conuenit, cum* nutritas *coniungi potest*'. This is perhaps possible, but not convincing. Not only would *in scrobe* be utterly otiose under such an interpretation, but one might also point to the rhythm of *lēnĕ nūtrītas* (cretic-spondee) that shows the colon end (so Winterbottom). Moreover, the only parallel that can be adduced for *nutrio* + *in* + ablative (I am not sure whether Mazzini has the passage in mind) is Col. 10.187 *Cypros item Paphio quam pinguis nutrit in aruo*, which has little authority here. *transfero* + *in* + accusative, on the other hand, is the standard way of speaking about the transplantation of trees: cf. 2.5.6 *in seminarium . . . transferant*; also, e.g., Col. 11.2.16 *maturae plantae in scrobes transferendae*. The present expression *in scrobes . . . transferunt* provides *uariatio* for 3.2.2 *ad scrobes . . . transferuntur*.

post triennium: *Variatio* for 3.2.2 *Triennio exacto*. The more common expression in Columella is *post triennium*.

3.3.4–5 Mago primus et post eum Diophanes nouam <normam> amygdali suadent, ut ferulam primo seramus eamque post annum recisam et in media parte diffissam nuce amygdali qua medullae sucus mollis est compleamus; nec desinunt asseuerare quod arbor eiusmodi et robore et fructu pulcriore prouenit. Sic et Columella describit ferularem; alio modo scribit, nam hoc tantum praestare amygdalis ferulas ut illas faciant Tarentinas: This 'novel' (*nouam*) method of planting the almond is also transmitted by Columella, who does not mention Mago or Diophanes. The substance of the advice offered by Mago, Diophanes and Columella was evidently very similar (see discussion below), but G. understood Mago and Diophanes, on the one hand, and Columella, on the other, to be making different claims regarding its purpose: Mago/Diophanes say that planting the almond in the fennel will cause the tree to flourish *robore et fructu pulcriore*, whereas Columella writes about it *alio modo* (as G. reports), and says 'only' (*tantum*) that it will make them 'Tarentine' (reading *Tarentinas* rather than *roncinias*; see below and also 3.3.5n. *Tarentinas*). What does G. mean when he says that Columella wrote about the fennel method *alio modo*? What is the difference between the tree growing *robore et fructu pulcriore* and its becoming Tarentine? I attempt to answer these questions below and suggest that although the methods of Columella and Mago/Diophanes appear broadly similar, they hint at two distinct traditions of advice for planting the almond (and other nuts). This two-tradition hypothesis not only has the advantage of explaining G.'s claim that Mago/Diophanes and Columella disagree on the purpose of planting the almond in the fennel (cf. *Columella scribit alio modo*), but also helps in turn to elucidate other agronomists' references to 'Tarentine' nuts.

We introduce first the evidence of Columella: 5.10.14 = *Arb.* 22.2–3 *nucem Graecam et auellanam Tarentinam facere hoc modo poteris: in quo scrobe destinaueris nuces serere, in eo* [*om. Arb.*] *terram minutam in modum semipedis ponito ibique semen ferulae pangito* [*Arb.*: *iacito*]. *cum ferula fuerit enata, eam findito et in medullam eius sine putamine nucem Graecam uel auellanam abscondito et ita adobruito.* Columella's method is on the whole very close to that of Mago/Diophanes as reported by G., but differs in certain respects: besides including advice on planting the fennel, Columella adds that the nut should be placed in the pith *sine putamine* and that one should earth up around it after planting (*adobruito*; see 3.2.3n. *adobruunt*). Otherwise, the reported method of Mago/Diophanes diverges from Columella's only in a few trivial elements: in place of Columella's *cum ferula fuerit enata* G. writes *post annum*, and for *in medullam* there is *qua medullae sucus mollis est*.

Although most of the differences between the accounts of Columella and Mago/Diophanes are relatively insignificant, Columella's *sine putamine* ('without its shell') is more important, for it cannot be independently assumed or inferred from the method of Mago/Diophanes. Moreover, we are alerted to the significance of this curious detail by its recurrence elsewhere: when Palladius provides instructions for making a Tarentine walnut, he stipulates that what makes the walnut 'Tarentine' is the fact that it is planted without its shell (*solam nucis carnem*) – there is no mention of the fennel at all: 2.15.18 *si Terentinam facere uolueris, solam nucis carnem lana propter formicas obuolutam in seminario debebis obruere.* Both Palladius and Columella thus offer a Tarentine method, and the common thread is not the fennel but planting the seed without its shell. But if the absence of the shell is the key element in the Tarentine method, why is the fennel recommended? There is in the first place the moist and nourishing environment that it would offer as it decomposed (3.3n.). Furthermore, if the fennel answers at all to the role of *lana* (wool) in Palladius' account, then it might also act as a protective barrier against ants and other insects. The interpretation may seem oversubtle, but it is supported by several further pieces of evidence that suggest that it is *sine putamine* and not the fennel of Columella's account that is the heart of the Tarentine method.

First, the *Geoponica* has an entire section describing how to make the shell of a nut tender by planting the seed without its shell: 10.66.1–2 γυμνὰ κάρυα δίχα κελύφων ποιήσεις, ἐὰν κατεάξας τὸ ὄστρακον τοῦ καρύου, καὶ συντηρήσας τὸ ἐντὸς ἀσινές, περιβάλῃς τουτὶ ἐρέᾳ, ἢ φύλλοις νεαροῖς ἀμπέλου ἢ πλατάνου, ἵνα μὴ γυμνὸς ὢν ὁ καρπὸς ὑπὸ μυρμήκων βρωθῇ, καὶ οὕτω φυτεύσῃς. τὸ αὐτό φησι Φλωρεντῖνος συμβαίνειν καὶ ἐπὶ ἀμυγδαλῆς, καὶ τῶν λοιπῶν ἀκροδρύων ('hard-shelled nuts') τῶν ἐχόντων ἔξωθεν ἔλυτρον ('husk' or 'shell'), τοῦτον τὸν τρόπον φυτευομένων. Although the nuts so produced are not called 'Tarentine', the section clearly follows the Tarentine method of Palladius: the *Geoponica* emphasizes, like Palladius, that it is the removal of the seed's shell (κατεάξας τὸ ὄστρακον) that has an effect on the mature fruit; and, like Palladius, stipulates the need for some sort of defence, either wool (ἐρέα) or leaves, against ants. Here is another indication that Columella's *sine putamine* may be closer to the centre than to the periphery of his advice. But the *Geoponica*'s account is valuable for another reason: neither Columella nor Palladius describes the properties of the mature Tarentine nut, but if the *Geoponica* bears on this question (as seems probable), then we should understand 'Tarentine' nuts more specifically as those which lack a hard shell when mature (γυμνὰ κάρυα δίχα κελύφων) owing to their special, shell-less planting. The connection is corroborated by a report on almond varieties at Plin. *HN* 15.90 *nunc . . . celebrantur et Tarentinarum* (sc. *amygdalum*) *duo genera, fragili putamine ac duro, quae sunt amplissimae ac minime rotundae.* Although he does not say how the nuts are so made, the detail *fragili putamine* shows that at least one kind of the Tarentine nut was known precisely for its 'tender shell', which is not very far wide of the *Geoponica*'s γυμνὰ κάρυα δίχα κελύφων. Cf. thus also Macrob. *Sat.* 3.18.13 *nux terentina dicitur quae ita mollis est ut uix attrectata frangatur* (quoted more fully at 3.3.5n. *Tarentinas*).

Although the evidence is patchwork, the important common thread in Columella, Pliny, Macrobius, Palladius and the *Geoponica* is the emphasis on the shell of the nut, either at the time it is planted or once it matures. There is, we further note, a good explanation for the attempt to affect the mature nut's shell by modifying the seed, since this practice is found widely in Roman agriculture (Ager (2010) 200–1; cf. 3.1.1n.). Finally, it is telling that when the *Geoponica* discusses the fennel method as an entirely separate technique, there is no mention at all of planting the almond *sine putamine*: 10.57.9 ἔνιοι δέ φασιν εὐαυξότερον γίνεσθαι τὸ φυτὸν νάρθηκος (= *narthex*, giant fennel) εἰς τὸν βόθρον παρεμβληθέντος. At least for the *Geoponica*, there is a clear distinction between the Tarentine (i.e. shell-less-nut) and fennel methods.[79]

To conclude: in 3.3.4–5 we are dealing with what were probably two distinct methods recommended for improving the almond tree: (1) a method of planting the almond in a fennel so as to improve the growth of fruit and foliage (*Pom.* 3.3.4 *et robore et fructu pulcriore prouenit*; *Geop.* 10.57.9 εὐαυξότερον γίνεσθαι); (2) a method of planting the almond and other nuts without their shells so as to make the produce more tender (Col., Plin., Macr., Pallad., *Geop.*). In Columella, the fennel and shell-less plantings are reported without distinction, but in all other places they are held distinct, and the only element common to every mention of the Tarentine method is the shell of the nut (i.e. whether it is planted shell-less, or whether the tree produces tender/shell-less produce). Palladius and the *Geoponica* are careful to recommend protection for the nut planted without its shell against marauding insects: wool (*lana*) in Palladius, wool (ἐρέα) or leaves (φύλλα νεαρὰ ἀμπέλου ἢ πλατάνου) in the *Geoponica*. It is the apparent replacement of this protective container with the *ferula* in Columella's case that would induce us (as it has previous interpreters) to see the Tarentine and fennel methods as one and the same. But if it is appreciated that the fennel in the advice of Mago/Diophanes and Columella is properly extraneous to the Tarentine method, we should ask again how and why they are conflated in Columella. The question is difficult to answer with confidence – it could as easily be attributed to his sources as to any conscious decision, but one might also counter with an entirely different question: why should Columella *not* have conflated the methods, if each promised to bring its own benefits to the fruit? Unifying the practices could have seemed a novel and convenient way to enjoy the advantages of both techniques.

In any case, it is curious to observe that Pliny does not mention the giant fennel method; nor for that matter does Palladius, an interesting omission given his dependence on *amygd.* in other respects for information about the almond. See also 3.3.4–5n. *nec . . . Tarentinas*; 3.3.5n. *Tarentinas*.

[79] Also meriting comparison with the fennel method is Palladius' recommendation (4.10.25) to plant the fig in a squill bulb.

3.3.4 post eum Diophanes: Presumably in his epitome of Mago (Introduction §II). But it is not certain what Diophanes contributed to the method that Mago had invented (thus *Mago primus*): did he revise or augment it, provide independent confirmation or simply transmit the technique? Any of these could justify G.'s including him with Mago as the subject of *suadent*. G. may have included the endorsement of the epitomizer Diophanes because he valued Diophanes' judgement (or thought his audience did) and believed that Diophanes' decision to copy the advice into his six-book epitome of Mago was important; or perhaps he wanted to put his own erudition on display by showing his control of the history of Mago's instruction; or perhaps he wanted to present the technique as having the approval of more than one authority. This last suggestion seems especially probable if, as argued in the previous n., G. understood Mago/Diophanes and Columella to hold different opinions about the purpose of the fennel in incubating the almond seed: citing Diophanes along with Mago is a way for G. to foreground that difference and legitimize it by showing that a number of authorities were involved.

nouam <normam> amygdali suadent: The transmitted *nouam* lacks anything to agree with, but editors disagree on how deep the corruption lies. Scotti proposes the simple change *nouum*. There is something to be said for it: the corruption could easily be explained by the following *am-*, and there is a regular substantive use of *nouus* in the neuter singular meaning 'a new thing', 'a novelty' (L&S s.v. I.4, *OLD* s.v. 6). Despite this, *nouum* is not happily joined with *amygdali* (in its substantive use it is mostly found in partitive phrases, e.g. *aliquid noui*), and without further specification (e.g. the addition of *hoc*, vel sim.) it is very vague as an anticipation of the noun clause. Antonelli, col. 1692–3 offers a different solution, attempting to explain the corruption of an original *scrobe* into *nouam*. It is not persuasive, but even if it were accepted, what would *scrobe amygdali* mean? *in scrobe amygdali sationi destinata*, says Antonelli, but the ellipse is improbable. Both Condorelli xx and Mazzini more reasonably suppose that a noun has fallen from the text. Mazzini supplies *sationem*, a common noun in *Pom.* and used at 2.3.4 *Huic sationi* (n.) in the requisite sense of 'method of sowing'. Condorelli gives *normam*, which, although it is not used by any of the other agronomists, is found at 4.1.8 (n.) in the sense of a rule, standard or practice. Either conjecture improves the text, but I print Condorelli's because there is a more transparent *ratio corruptelae*: 'lapsum litterarum similitudo efficere potuit' (Condorelli xx). Condorelli further emends *amygdali* to *amygdalis*, but the change is unnecessary, even if it would be easy to explain the omission of the final *-s* by haplography before *suadent*.

ferulam: The giant fennel (*Ferula communis* L.), called elsewhere the *ferula gallica*, *narthex*, *narthecia* and *ros maior* (*LTBL* 135, *NPRA* 103). It should not be confused with culinary fennel (*Foeniculum vulgare* Miller), to which it is related. A hardy perennial, the giant fennel is now mostly grown for its foliage and yellow-green flowers. When full grown it reaches 6 to 8 feet in height, 2 to 3 feet in spread. See further Huxley and Griffiths (1992) II 292–3.

in media parte diffissam: Similar locutions occur in discussions of (wedge-)grafting: cf. Cato, *Agr.* 41.2 *eam* (sc. *uitem*) *mediam diffindito per medullam*; Varro, *Rust.* 1.40.6 *surculum . . . in eius ramo praeciso ac diffisso implicat.*

nuce: 'Le regole di costruzione, il contesto, ed il confronto col passo di Columella [quoted at 3.3.4–5n.] mostrano chiaramente che nel codice fu scritto *nuci* per *nuce*' (Antonelli). But if we do choose to keep *nuci*, we would explain it as a dative of purpose: the fennel is split 'for the almond'.

qua medullae sucus mollis est: *medulla* (Gk. μυελός) refers to the vital stuff in the bones of animals and the stems of plants that animates them (*TLL* VIII 598.84–6 (Buchwald); for plants, see 599.69–600.20): in animals this takes the form of the marrow, and in plants it is imagined concretely as a channel for nutriments (cf. Col. 3.10.2, quoted at 2.10.3n.), practically what we would call the xylem. The *medulla* was thus the natural place for a kind of transfer of productive capacity from the *ferula* to the almond: so also Cato's insistence (*Agr.* 41.2, 41.3) on binding *medulla cum medulla* when grafting. When the *medulla* is imagined as an actual channel rather than a vague life force, *sucus* is the nutritive matter (sap) that flows through it: see 2.10.3n.

3.3.4–5 nec desinunt asseuerare quod arbor eiusmodi et robore et fructu pulcriore prouenit. Sic et Columella describit ferularem; alio modo scribit, nam hoc tantum praestare amygdalis ferulas ut illas faciant Tarentinas: The punctuation and interpretation of these lines have differed among editors, but the general sense should be clear: Mago and Diophanes make one claim for the fennel method, namely, that it produces a finer plant and fruit (*robore et fructu pulcriore*); Columella similarly (*sic et*) writes about the tree that is grown in the fennel, but (adversative asyndeton) he makes a different claim (*alio modo*), for (*nam*) he says that (indirect discourse after *scribit*) giant fennel only (*tantum*) renders this benefit to the almonds (*hoc . . . praestare . . . ut*), namely, that they become Tarentine.

3.3.4 asseuerare quod . . . prouenit: For the *quod*-clause in place of accusative-and-infinitive syntax, see Introduction §v(c).

3.3.5 Sic et Columella describit ferularem: *sicut* (as transmitted by **N**) does not establish the correct logical relationship between *Columella describit ferularem* and what precedes: it is not that Columella endorses Mago's and Diophanes' claim that fennel causes the almond to 'come forth with finer wood and fruit', for he says rather that the only (*tantum*) benefit it renders it is to make it Tarentine; the similarity is that he, too, writes of the method of sowing in the fennel. See also Mazzini 143–4.

describit . . . scribit: For the iteration of compound by simplex verb, cf. 2.1.2n. *disponi . . . ponenda*. The verb *scribo* is used in such iteration also at 4.1.4 *conscribere . . . scribit.*

ferularem: This word appears only here in Latin literature. The very productive denominative suffix *-lis* (dissimilating to *-ris* after an *-l-* in the stem, as here) forms adjectives of 'belonging' (Leumann §313.3). It is difficult to know whether G. has coined the word in order to vary *eiusmodi*, as seems possible, or has taken it from some other source. Although Mazzini 55 (following Antonelli) maintains that we must understand *sationem* with the adjective, *arborem* seems easier to supply from the preceding lines and would better accord with *arbor eiusmodi*.

alio modo: Antonelli adds *at* to the text before *alio* in order to clarify the relationship between the clauses. While it is true that the omission of *at* might be explained by the following *al-*, it is unnecessary to suppose that anything has fallen from the text: the asyndeton is adversative (see 2.4.6n.).

praestare: The verb is extremely common in *Med.* (19x) of the effects, usually medicinal benefits, that a plant produces (cf. *CG* s.v.).

Tarentinas: The word *roncinias* of the palimpsest is retained in the editions of Condorelli and Mazzini; the latter (55–6) defends it as a rusticity. But since G. is explicitly referring to what Columella has written on the matter (5.10.14 = *Arb.* 22.2–3), it is hard to see why he would arbitrarily change *Tarentinas* to *roncinias*. As Winterbottom 143 n. 2 observes, *Tarentinas* 'is not impossibly corrupted to *roncinias* . . . once *ta-* disappeared after *-ant*' (nor does the emendation spoil the rhythm).

For the significance of the adjective *Tarentinus* applied to nut trees, see 3.3.4–5n. Macrobius proffers (erroneously; see Svennung (1935) 109–10) the following etymology for the word: 3.18.13 *nux terentina dicitur quae ita mollis est ut uix attrectata frangatur. de qua in libro Fauorini sic reperitur: 'item quod quidam Tarentinas oues uel nuces dicunt, quae sunt terentinae a tereno, quod est Sabinorum lingua molle, unde Terentios quoque dictos putat Varro ad Libonem primo'. quam in culpam etiam Horatius* [*Sat.* 2.4.34] *potest uideri incidere, qui ait et 'molle Tarentum'.* Svennung (1935) 108–15 argues that, in Palladius' case at least, we ought to print *Terent-* for *Tarent-*. He also cites evidence (110–11) suggesting that the form may have had currency as early as Galen. If so, it is at any rate possible that G. wrote *Terentina*, but we cannot know how far the weakening of *ă* to *ĕ* had progressed in contemporary pronunciation.

3.4 As in *persic.* (2.7n.), G. discusses the transplantation and arrangement of the trees in the orchard before he turns to questions of their maintenance

(3.5–7n.). He stresses that the almond's natural disposition and its slenderness (3.4.1n. *mediocritas*) recommend shallow planting holes, and adds that the arrangement of the trees in the orchard ought to be guided by consideration of the same. He mentions a variety of opinions on the proper distance between the trees, and reports with hearty approval the advice of Diophanes (3.4.3n.): since the almonds do not ripen all at once, the farmer incurs the risk that the nuts will be stolen while he awaits the maturation of the rest; but by planting the trees near a crosspath on the farm (3.4.3n. *in transitu*), the almond trees will be kept near at hand, allowing the owner to deter thieves and gather them once they ripen.

3.4.1 Et natura et mediocritas arborum suggerit ne altissimo scrobe recondant: Cf. 2.7.1 *Scrobes altos arbor persici non amat.* Whereas G. provides a fairly substantial explanation there (2.7.1n.), here he only gives *natura* and *mediocritas* as reasons for the advice. Hendiadys of the nouns is possible (thus = *mediocris natura*), but it is safer to assume that they are distinct, *natura* being a vague reference to the tree's properties (see following n.) and *mediocritas* (n.) a concrete observation about its slenderness or small size. Both the almond's habits of growth and its physical size are thus what recommend against a deep hole.

natura . . . arborum: Svennung (1935) 520–1 classifies the pattern of *natura* + genitive noun as a 'beliebt' subclass of the 'abstraktes Substantivum mit einem genitiv des Substantives' (519). We discuss the question of abstract nouns (i.e. nominalizations) from a somewhat different perspective in the Introduction §v(b). Such periphrases with the word *natura* in particular are very common in the agricultural authors (for their effect, cf. 2.7.1n. *radicum concessa substantia*; 2.7.1n. *arborum legem*): Svennung (1935) 520 collects many examples found in Columella and Palladius; see also *TLL* IX 1.177.76–178.24 (Hajdú).

mediocritas: The abstract noun has three common meanings: (1) moderateness of size, amount or degree (*OLD* s.v. 1, 2); (2) moderation (*OLD* s.v. 3, in this sense translating Gk. ἡ μεσότης; cf. Cic. *Brut.* 149.5); and (3) mediocrity or inferiority of ability or power (*OLD* s.v. 4). From the first century CE, however, there is a rarer use of the word in reference to the small physical size of a thing ('de paruitate spatii localis', *TLL* VIII 569.25–35 (Buchwald)): cf. Col. 9.8.11 *sin . . . ipsius trunci . . . mediocritas patitur, acutissima serra, quo celerius id fiat, praeciditur primum superior pars.* (For the diachronic change in meaning, see 2.7.1n. *modica.*) *mediocritas* here probably bears this sense (as Buchwald also supposes), referring to the small size or slenderness of the tree at the time of transplanting.

recondant: Although the word is not uncommon in the agronomists for storing various agricultural products (e.g. Col. 2.9.18, 2.20.6, 7.5.8, and very frequently in book 12; also Pallad. 11.17.3, 11.21), the present application to planting is rare and paralleled only by Pallad. 3.25.33 (*nuces quoque iuglandes etiam nunc seminariis recondi*), where it provides *uariatio* for the many instances of *serere*, *ponere* and *obruere* in the preceding lines. *recondo* is perhaps less felicitous there than here, however, where it takes its point from the depth of the hole (*altissimo scrobe*).

The person and voice of *recondant* are perhaps surprising, but the easiest solution is to assume a vague subject such as 'practitioners' (as with the following *pronuntiant*, and on the whole common in *Pom.*), treating *recondant* as an absolute use of the transitive verb. It is not impossible however that the subject is *arbores* and that *recondant* is intransitive, roughly equivalent in meaning here to the passive *recondantur* ('be set down'). (Discussion of intransitivization in general is deferred to 3.4.3n. *maturant*, a more secure instance of the phenomenon.) An intransitive use of *recondo* (or of the simplex) is not found elsewhere (*abscondo*, however, at Feltenius (1977) 54, 73), but that is not a sure argument against its possibility. There is no compelling reason to emend to *recondantur*.

3.4.2 Illa interualla etiam simili ratione moderanda sunt: uiginti et quinque nonnulli, quidam et uiginti pedum iusta pronuntiant: For the resultative asyndeton, see 2.4.6n.

Illa: Mai was unhappy with *illa* and wondered whether *illas* (sc. *arbores*) should be printed, as direct object of *recondant*, and a period placed after it. It is possible however to explain *illa* as the anaphoric use described by Adams (2013) 482–527, esp. 487–8, 500–4, 507 as 'associative' and by the *TLL* as referring 'ad rem non antea memoratam, sed ex contextu vel aliunde notam' (VII 1.358.8–59 (Bulhart)). That is, *illa* is prompted by G.'s mention of the planting holes in the orchard (cf. 3.4.1 *scrobe*), which implies some particular intervals separating them.

simili ratione moderanda sunt: That is, consideration of both the *natura* and *mediocritas* of the plant ought also to guide the arrangement of the *ordines*. There may be intended a suggestion that the *ratio* consists precisely in 'moderation', since both *mediocritas* and *modero(r)* can denote regulation according to a mean and are thus sometimes conceptually linked: note the jingle (albeit in a philosophical context) at Cic. *Off.* 2.60 *ratio . . . mediocritate moderanda est* (also *Mur.* 63). *modero(r)* is not a common verb in the agronomists, who prefer to use *modicus/a* or *modice* + verb. Compare Stolo's comments on the arrangement of the orchard, where the sentiment is parallel to G.'s: Varro, *Rust.* 1.7.2 *de formae cultura hoc dico* (sc. *Stolo*), *quae specie fiant uenustiora, sequi ut maiore quoque fructu sint, ut qui habent arbusta, si sata sunt in quincuncem, propter ordines atque interualla modica*.

uiginti et quinque nonnulli, quidam et uiginti pedum iusta pronuntiant: The proper intervals for planting trees in the orchard were a subject of great importance for the ancient agronomists. Making the intervals between trees smaller has the obvious benefit of allowing a greater number to be planted in the same area; so long as they flourish, the farmer can expect better returns on the same amount of land. On the other hand, smaller intervals may be disadvantageous if the trees do not have sufficient room to grow to full size, if shade from other trees stunts growth or productivity, or if the narrowness prevents cultivation between the rows and hinders harvesting of the trees.[80] To this list of more or less obvious drawbacks, we must also reckon with groundless suspicions that the shade of certain trees was especially harmful (cf. Plin. *HN* 17.89) and that the water droplets falling from some species' leaves were noxious (cf. Plin. *HN* 17.92 *stilla saeua*; Pallad. 2.15.17, 3.25.21). It was also sometimes asserted without explanation that dense plantations would cause the trees to become sterile (Theophr. *Caus. pl.* 2.10.1 διὰ πυκνότητα . . . ἄκαρπα γίνεται; Plin. *HN* 16.111 *facit et densitas sterilitatem*).

Given the fundamental status of the question, it is unsurprising that we find discussion of it in nearly all of the ancient agronomists; it is perhaps equally unsurprising that their recommendations vary greatly. Theophrastus (*Caus. pl.* 3.7.1–2) cautiously argues that it is necessary to take a host of contingent factors into consideration when deciding on the size of intervals, such as the given species' tolerance of shade, the habits of root growth of neighbouring trees, the quality of soil and slope of the land. When reporting the agronomists' opinions, his advice is more concrete: *Hist. pl.* 2.5.6 φυτεύειν δὲ ῥόας μὲν καὶ μυρρίνους καὶ δάφνας πυκνὰς κελεύουσι, μὴ πλέον διεστώσας ἢ ἐννέα πόδας, μηλέας δὲ μικρῷ μακρότερον, ἀπίους δέ – καὶ ὄχνας – ἔτι μᾶλλον, ἀμυγδαλᾶς δὲ καὶ συκᾶς πολλῷ πλέον, ὡσαύτως δε καὶ τὴν ἐλαίαν.[81] He adds that the slope must be taken into consideration when determining the intervals: ποιεῖσθαι δὲ καὶ πρὸς τόπον τὰς ἀπόστασεις· ἐν γὰρ τοῖς ὀρεινοῖς ἐλάττους ἢ ἐν τοῖς πεδεινοῖς. Pliny's own discussion of the spacing of trees at *HN* 17.88–94 is heavily indebted to Theophrastus, both in his exhortation to pay attention to the characteristics of the tree and the land, and in certain places where he reports the specific intervals: *HN* 17.88 *huius loci pars est ad interualla pertinens: quidam punicas et myrtos et lauros densiores seri iusserunt, in pedibus tamen nouenis, malos amplius paulo, uel magis etiam piros magisque amygdalas et ficos.* We have again the

[80] For the related topic of orchard planting patterns, see 2.7n.

[81] For an explanation for this ordering of the trees, compare his claim that the fig, olive and almond are the most troublesome for neighbouring plants: *Caus. pl.* 3.10.6 χαλεπώτατα δὲ καὶ ἀμπέλῳ καὶ τοῖς ἄλλοις συκῆ καὶ ἐλάια· καὶ γὰρ τροφὴν πολλὴν ἀμφότερα λαμβάνει καὶ σκιὰν παρέχει πλείστην. χαλεπὸν δὲ καὶ ἡ ἀμυγδαλῆ, διά τε τὴν ἰσχὺν καὶ διὰ τὴν πολυρριζίαν.

initial recommendation of 9 feet (ἐννέα πόδας/*in pedibus nouenis*), with the distance growing by indefinite amounts for the apple, then the pear and finally the almond and fig (although Pliny omits the olive). Especially in view of Theophrastus' πολλῷ πλέον, it seems reasonable to calculate a distance of at least 20 feet for the intervals, and perhaps much more, by the time that we have come to the almond, fig and olive.

The relative measurements based on expanding intervals that we find in Theophrastus and after him Pliny are not employed by the other Roman agronomists. They show rather a preference for generous spacing and a readiness to offer concrete figures (see below) for different varieties of tree, always in increments of 5 or 10 feet. In some cases, increasing attention to intercultivation is responsible for the wide intervals that they recommend (Plate 14). For example, Cato's remarks on the intervals for the olive trees are limited to a simple declaration that 25 to 30 feet is best (*Agr.* 6.1); when Columella employs a similar figure (25 feet) in his own discussion of olives, however, he adds the qualification that such intervals are appropriate only if the soil is *macro neque idoneo segetibus* (5.9.7): if the soil can support wheat, then a distance of 60 feet between trees is best (5.9.7; cf. *Arb.* 17.3, Pallad. 3.18.5). Columella similarly recommends a minimum of 20 feet for vine-bearing ashes or elms if the *arbustum* will not be intercultivated, but if the soil can bear intercultivation, 40 or more (cf. Col. 5.6.11, 5.7.3, *Arb.* 16.2, Plin. *HN* 17.202, quoted in following n.).

We turn now to the Romans' evidence for the spacing of fruit trees in particular. We saw that Theophrastus, with Pliny following, recommends intervals for fruit trees ranging from a minimum of 9 feet (for the pomegranate) to probably 20 or more (for the almond, fig and olive). While Columella does not provide figures for specific trees, he offers the following valuable remarks on the subject of spacing out trees in the orchard: 5.10.5 = *Arb.* 19.3 *arbores* (sc. *pomiferos*) *raris interuallis serito, ut cum creuerint, spatium habeant quo ramos extendant. nam si spisse posueris, nec infra serere quid* [*Arb. neque infra quid serere*] *poteris nec ipsae fructuosae erunt, nisi interuolseris. itaque inter ordines quadragenos pedes minimumque tricenos relinquere conuenit.* Although motivated in part by thought of intercultivation (cf. *infra serere quid poteris*), he is mostly concerned about the hindrance that tightly packed trees could pose to one another's flourishing; consequently, a minimum of 30 feet is recommended, 40 being better. Palladius also adopts a general figure of 30 feet in his remarks on the layout of the orchard: 3.19.1 *si pomarium facies, inter ordines tricenos pedes relinques.* Therefore it is in spite of Columella's and Palladius' 30-foot minimum for fruit trees, and Theophrastus' inclusion of the almond among trees requiring the most space, that the recommended intervals for the almond in particular underwent a steady contraction, first to the distance of 20 or 25 feet in *Pom.* (*uiginti et quinque, uiginti pedum*), and then to the distance of only 15 or 20 feet in Palladius

(2.15.8 *spatia inter arbores uiginti aut quindecim pedibus dedisse sufficiat*).[82] Perhaps as the agronomists showed less interest in the intercultivation characteristic of Columella's mixed-farming method, there was a tendency to reduce the intervals as much as the trees could tolerate in order to maximize the yields of the orchards.

We can obtain some perspective on the ancient situation if we compare the intervals in orchards of today and of recent history. Wickson (1891) 514 writes: 'The almond makes a comparatively large tree and should have plenty of room – not less than twenty-four feet apart (though some plantations are made at twenty feet) and thirty feet is better' (so precisely Rosengarten (1984) 9). Similarly, *AOM* 37 offers an ideal distance of 24 by 24 feet, but notes that it may be adjusted on the basis of the soil's fertility (in any case no less than 20 feet); *EFN* 712 reports at least 23 feet, and as many as 50 (7–15 metres). In view of these figures, G.'s advice would be about right for soil that was not of the highest quality; Palladius' further reduction to a minimum of 15 feet could only be justified for planting on the most meagre soils.

iusta: For the meaning of *iustus*, see 2.11.2n. *mole iusta*; and for the application to *interuallum*, cf. Plin. *HN* 17.202 *interuallum iustum arborum, si aretur solum, quadrageni pedes in terga frontemque, in latera uiceni; si non aretur, hoc in omnes partes.*

3.4.3 Bene Diophanes ait amygdalas in transitu esse ponendas, quoniam non semel maturant et per partes se fructus eius accommodant, nec possint ita fures dispoliare ut non aliquid – uel quia crudum uidetur – domino relinquatur: Maturation of the almond is marked by four processes (*AOM* 50–1, *EFN* 711–12): (1) dehiscence of the hull, (2) hull split (cf. 3.8.1 *cum corticem remiserunt*), (3) fruit abscission, i.e. separation from the tree (see Plate 5),[83] and (4) dehydration of the hull and nut. Gradziel (*EFN*) notes that the entire process of maturation and drying may require as long as four to six weeks to complete. It is true that some of the fruits will finish maturing more quickly than others (even if not by very much). Diophanes and G. imagine a danger that, if the more quickly ripened almonds are left on the tree for the rest to mature, thieves could snatch whatever nuts are furthest along, leaving only the unripe ones behind. By situating the trees near a cross-

[82] Compare with the reduced intervals for the almond Palladius' recommendations of 30 feet for the pear (3.25.3) and 'fairly wide' spacing (*interualla maiora*) for mulberry and figs (3.25.29, 4.10.25), which are more in line with the general figures he offers for a *pomarium.*

[83] Compare Cicero's exquisite simile comparing abscission and old age: *Sen.* 71 *quasi poma ex arboribus, cruda si sunt, uix euelluntur, si matura et cocta, decidunt, sic uitam adulescentibus uis aufert, senibus maturitas.*

path on the farm, however, they will be near at hand, where the cultivator can ward off thieves and perhaps even gather the nuts individually as they ripen.

in transitu: *transitus* here must refer to a passage or lane (*OLD* s.v. 3a) through the farm. *transitus* only rarely has its concrete meaning in the phrase *in transitu*, which usually means 'in the crossing' or 'in the passage' (e.g. Liv. 21.40.7 *in transitu Alpium*; cf. Plin. *HN* 3.134, Suet. *Iul.* 56; examples could be multiplied), or else is used as an adverbial expression in transition in speech from one idea to another ('in passing' or 'by the way', *OLD* s.v. 5, L&S s.v. II.B.2.b). The only instance of *in transitu* I find that offers a possible parallel for G.'s usage is Front. *Aq.* 115.3 *has* (sc. *fistulas*) *comperi per eum qui appellabatur a punctis passim conuulneratas omnibus in transitu negotiationibus praebuisse peculiaribus fistulis aquam. quo efficiebatur ut exiguus modus ad usus publicos perueniret.* Frontinus' *omnibus in transitu negotiationibus* evidently means 'to all of the businesses along their path [i.e. of the *fistularum*]', with both *negotiatio* and *transitus* employed in unusual concrete applications (see Rodgers *ad loc.*).

semel: 'All at once', 'at one and the same time' (*OLD* s.v. 5), like *simul*, with which it is sometimes confused in the MSS in this sense. Strictly speaking, the following *per partes* does not therefore add any new information after *non semel*, but helps to clarify its meaning: see 3.4.3n. *se . . . accommodant.*

maturant: At least through Columella's time, the verb *maturo* is rarely used in order to denote the process of a fruit's maturation: the intransitive *maturesco* sometimes serves this purpose (cf. also *adolesco*), but also frequently employed are phrases with a predicate adjective, such as *maturus esse* (e.g. Cato, *Agr.* 31.2, Plin. *HN* 19.173; cf. *Pom.* 3.8.1 *incipiunt esse matura*), or phrases with a noun, such as *ad maturitatem (per)uenire/fieri/(per)duci/* etc. (e.g. Col. 2.10.11, 2.11.10, Plin. *HN* 19.143). The verb *maturo* here is best understood in its intransitive active usage (though it is not inconceivable that it could be construed with a supplied object such as *amygdala* or *fructus*), which appears first probably in Pliny and with certainty in the Vulgate: see for details *TLL* VIII 495.68–73, 496.49–52 (Brandt). 'Intransitivization' of verbs, that is, the intransitive use of an originally transitive verb, is not rare in Latin of any period (it is found already in Plautus), but becomes especially common in later Latin, and is in some instances connected with non-literary or technical language: see Svennung (1935) 441–5, H–Sz §42.2, §165, Feltenius (1977) 9–23, Cennamo (1998) 90–1, Pinkster 280–2, Adams (2016) 540–1 (on *moueo*). Pinkster 280–1 distinguishes two kinds of intransitivization: an 'autocausative' (or 'middle') use of the verb, in which 'the subject is to some extent actively engaged in the process in which it is involved' (Pinkster 260); and a 'decausative' (or 'ergative') use, in which the object of the transitive verb is simply used as the subject of the intransitive verb. An example of the autocausative active use is *lauo* 'I wash (myself)'; an example of the decausative use is Varro, *Rust.* 3.17.7 *cum mare turbaret ac per* *tempestatem macellum piscinarum [mare] obsonium praeberet.* G.'s intransitive use of *maturo* is apparently autocausative, to judge from the following *se . . . accommodant* (n.), also in reference to ripening. Note that at *Med.* 49.4, G. uses *maturo* transitively: (sc. *fici*) *siccae tritae et impositae uehementes tumores aut maturant aut discutiunt*, etc. For other possible intransitivizations in *Pom.*, cf. 2.5.9n. *terra densante*; 3.4.1n. *recondant.*

per partes: For *per partes* in this distributive sense ('i. q. particulatim, singillatim'), see *TLL* x 1.458.31–49 (Teßmer). The usage is well illustrated by the antithesis at Sen. *QNat.* 2.54.2, where *per partes* is opposed to *uniuersum* and joined with *minutatim*: *hoc* (sc. *aer*) *quoque, si inclusum est, aeque fugam quaerit et cum sono euadit, ac modo uniuersum eruptionem facit eoque uehementius intonat, modo per partes et minutatim.*

se . . . accommodant: The reflexive construction is middle (anti- or autocausative), not passive, which would require the notion of an external agent. Reflexive constructions with inanimate subjects in particular have prompted much discussion, and the issue has been reevaluated by Adams (2013) 674–724 in the context of the decline of the synthetic passive in general. It has been claimed with varying degrees of subtlety that the employment of such reflexive constructions (without personification of the inanimate subject) is characteristic of colloquial, late or Vulgar Latin: with the synopsis of Adams (2013) 683–6, cf. Svennung (1935) 461–4, Wistrand (1941) 60–2, H–Sz §164, Flobert (1975) 386–90, Cennamo (1998) 88–9. As Adams shows, these conclusions are untenable as generalizations and must be qualified. For our purposes, the pertinent point from his analysis is that throughout the entire history of Latin the reflexive middle construction could be used in reference to 'spontaneous events manifesting the forces of nature' (Adams (2013) 690; cf. 691, 693–5): this usage occurred 'at all educational levels of the language, and it would be unjustifiable to suggest that there was a late flowering in vulgar texts' (708–9). See also Pinkster 273–6. G.'s use of the reflexive middle to describe the maturation of fruit thus has a long pedigree and is not by itself a marker of the colloquial, late or Vulgar nature of the text; on the contrary, it is congruent with the spontaneity of the natural process described.

The verb *accommodo* itself is a very singular way of expressing maturation, but the expression is easier to understand if it is appreciated that (as in many cases of the absolute use of *accommodo*) we must mentally supply a complement (examples at *TLL* i 331.67–80 (Prinz)): in the present context *ad legendum* or something similar would suffice (thus *se accommodant* (sc. *ad legendum*) = 'become fit [for picking]'). The ellipse of a complement is tolerable because *per partes se fructus eius accommodant* is elaborating the statement *non semel maturant*, which anticipates and guides the interpretation of this phrase.

fructus eius: *eius* must refer to the almond tree (*amygdalae* or *amygdali*); there is no singular antecedent for it in context, but it can be inferred from *amygdalas*, which also provides a (plural) subject for *maturant* and *accommodant*. G. has a tendency to use the singular and plural loosely when referring to the tree *qua* species: here, the point of *eius* is to clarify that it is not just trees planted along the crosspath whose fruits mature piecemeal, but the almond *qua* species. For other instances of looseness in number, see 2.4.1n. *persicorum arborem . . . laetissimas*.

nec possint ita: *ita* is continuative (*OLD* s.v. 6a): 'after that has been done [*ita*, i.e. planting them along the crosspath], nor could, etc.'

uel quia crudum uidetur: There is an ellipse of *si aliquid relinquatur*, vel sim., in the parenthesis that must be supplied from *non aliquid . . . relinquatur*. The point is that if the thieves do not strip everything from the almond trees, it will only be because they found certain fruit too unripe to merit picking.

crudum: *crudus* is used in two ways by the agronomists: (1) 'unripe' or 'immature' (*TLL* IV 1235.65–72 (Hoppe)); (2) 'unploughed' or 'unworked' (*TLL* IV 1235.13–15; cf. Pallad. 1.6.8 *qui arando crudum solum inter sulcos relinquit*, etc.). It does not appear in the second sense in *Pom.*

domino: The reference is pointed, since it was just the *dominus* (the owner of the slave-worked farm) who stood to lose if profits were blunted by theft. Moreover, because all of the agronomists take it for granted that they address themselves to *domini*, the appearance of the word here is a means of indirectly addressing the reader. *dominus*, or a similar title with which the reader would be identified, such as *paterfamilias*, is common as an obliquely didactic form of address in the prose of Cato, Varro and Columella,[84] but it is not a favoured device in *Pom.* and the use here may be influenced by his report of Diophanes' words.

The *dominus* himself was at the top of the rigid hierarchy of labour that structured the intensive farming operations described by the agronomists: although he might often be absent from the estate, it was he who had ultimate control over all of its operations and workers (cf. succinctly Spanier (2010) 100, White (1973) 453–6; more fully White (1970) 332–83), and it was he whose capital funded the enterprise. The *dominus* could also be invoked in the

[84] Many examples could be adduced, but for a representative instance, cf. Col. 3.1.6, quoted at 2.3.1n. It is worth emphasizing that this usage of *dominus* among the agronomists very much preserves the 'basic lexical meaning of *dominus* [as] owner' (Dickey (2002) 78), specifically as owner of the agricultural enterprise; quite different is the development of *dominus* as a means of direct address, for which see Dickey (2002) 77–109; also Sherwin-White (1966) 557–8, Adams (1995*b*) 118–19, Coleman (2012) 194–9.

guise of the *paterfamilias*, although the term *paterfamilias* is used more sparingly among the agronomists: in Cato the proportion is especially lopsided, with 55 appearances of *dominus* to only 4 of *paterfamilias* (Spanier (2010) 120; cf. Richter (1978) 61–4). Although the connotative spectra of the words are different, their practical overlap for the purposes of farm ownership is great enough that they can at times be used interchangeably.

3.5–7 We come now to questions of maintenance and improvement. Broadly speaking, these three sections are unified by their focus on the care of the almond tree with an eye to its productivity. But despite their thematic continuity, the reader may at first have the impression that G.'s movement within the sections is desultory: 3.5, for example, turns abruptly from advice on pruning to a warning against allowing animals to approach the trees, and then to an explanation of the dangers that digging (*fossio*) presents. We also find scattered across multiple sections material that one might have expected G. to group together: thus, nearly identical warnings about the dangers of the animal bite are found in 3.5.2 and 3.7.5, and observations about the effect of the climate on the almond are spread across 3.5.3 (pernicious winds/weather), 3.6.4 (frost) and 3.7.4 (sun).

As one inspects the content of the sections more closely, however, these points of initial confusion give way to an appreciation of a sensible organization of material whose structure remains rather implicit than explicit (cf. Introduction §III(a)). The precepts in each of 3.5, 3.6 and 3.7 are in fact loosely grouped according to different aspects or stages of the almond's growth and production and inform the reader of the concerns appropriate to each of those aspects or stages. Consider the following outline:

3.5 Tending the almond during growth to maturity and sustaining the mature tree, containing advice on:

- Pruning the tree (3.5.1);
- Protecting the tree from animals while it grows (3.5.2);
- Avoiding the digging that is often beneficial to trees, but in the case of the almond injurious, especially in conjunction with the dangers offered by pernicious winds (3.5.3);

3.6 Improving the productivity of the mature tree (cf. 3.6.1 *Plurimum generant*), containing advice on:

- Making an unproductive almond productive (*ferax*) (3.6.1–3);
- Mitigating the annual dangers to productivity posed by cold weather (3.6.4);

3.7 Improving the quality of the mature tree's produce, containing advice on:

- Producing soft nuts by pouring warm water on the ablaqueated roots (3.7.1);
- Producing sweet nuts by draining the tree of its bitter liquid and inserting into the wood a honey-coated wedge (3.7.2);

- Producing sweet nuts by offering a potent manure mixture to the tree's roots (3.7.3);
- Avoiding bitterness that comes to the nuts from direct exposure to sun (3.7.4);
- Avoiding bitterness that comes to the nuts from an animal bite (3.7.5);
- Producing superior nuts by culling weaker ones (3.7.6).

Although G. does not give section headings or other signposts like those I have provided, it is reasonable to assume that his audience would have been sufficiently acquainted with the cultivation of fruit trees to recognize the organizing principles of his instructions (cf. Introduction §III(a), §IV). That is to say, an experienced reader would have known enough about (for example) the practice of pruning trees to recognize the context of G.'s remarks on the topic; or, again, the reader may have been pleased to discover that all of the information relevant to enhancing the produce of his trees was collected in the same section (3.7).

Positing some kind of thematic arrangement like that presented above helps to resolve the organizational oddities we have noted. For example, the scattered presentation of G.'s remarks on the climate can be explained by the observation that in each case a different aspect of the climate's influence on the trees is observed: pernicious winds are mentioned at 3.5.3 apropos of the maintenance practices in that section (pruning, digging, etc.); frost appears at 3.6.4 because of the risks it presents to the *feracitas* of the tree (cf. 3.6.2 *ferax*); and the dangers of the sun are mentioned at 3.7.4 specifically because excessive exposure causes the fruit to turn bitter, an effect which it is the section's overriding goal to mitigate (cf. 3.7.2 *fiunt dulcia ex amaris*). So too the warnings against animal bites at 3.5.2 and 3.7.5 have somewhat different force in each context: the risk in both cases is that the animal's bite will make the fruit bitter, but at 3.5.2 the advice is presented as one general caution next to another (the prohibition against digging), whereas at 3.7.5 the cause of the bitterness is emphasized (*cum inopinata in medullas saliua consederit*), mirroring G.'s aetiological interest in the harmful effects of the sun (*quum crebrius exustae dulcior sucus hauriatur*).

3.5 This section contains three recommendations (see previous n.), plus an added caution about climatic hazards for the almond. By and large, the recommendations in this section are not original (see *passim* below), but G. is the first author to gather the previously scattered advice. Palladius follows him, only omitting the mention of hazardous winds and adding a small specification regarding the time of pruning: 2.15.9 *putanda est Nouembri mense, ut superflua et arida et densa tollamus. seruandae sunt a pecore, quia si rodantur amarescunt. circumfodi non debent quotiens florent, quia inde flos eius excutitur.*

3.5.1 Leuissima et minime sumptuosa cultura est earum, putationis annuae scilicet, qua necesse sit ramos superuacuos uel enormiter natos uel aliqua necessitate praefractos falce purgare: These instructions, aimed at shaping the tree and reducing its excess wood, are close to modern recommendations for pruning the almond. Wickson (1891) 152 states that in California pruning was 'almost exclusively intended as a *shaping* process',[85] which is still the case today: 'only limited pruning is applied to [the] almond, usually during the dormant season . . . Usually only crowded, crossing or extremely flat branches are removed, as well as low-hanging branches which obstruct orchard equipment' (*EFN* 712). Modern agronomists also emphasize the importance of the first pruning after transplanting (Wickson (1891) 160–1, 514, *EFN* 712), of which G. makes no special mention. On pruning the almond, see further Wickson (1891) 152–78, *AOM* 42–5, *EFN* 712.

Leuissima et minime sumptuosa: *lĕuis* means that the task will be easy to perform (*OLD* s.v. 9a) and *minime sumptuosus* that it will not require great outlay or expense (*OLD* s.v. *sumptuosus* 1a; for *minime* as a negator, see Pinkster 699–700). For *sumptuosus* in this sense, compare Cato's exhortation to the prospective farm buyer: *Agr.* 1.1.5 *uideto quam minimi instrumenti sumptuosusque ager ne siet* (Dalby 'wasteful').

putationis annuae scilicet: *scilicet* is epexegetic (a late development of the word; see esp. Schrickx (2011) 179–81, also *OLD* s.v. 5b, L&S s.v. II), rendered in the English translation by the colon after 'costly'. *putationis annuae* is a defining genitive after *cultura* (*NLS* §72.5, Pinkster 1023–5).

putationis: Despite the fortunes of its parent verb *puto* and its sibling compound nouns *disputatio*, *computatio*, etc., *putatio* remains almost entirely literal and technical until the time of the Christian writers: for a synopsis, see *TLL* x 2.2744.65–2475.43 (Krebs).

ramos superuacuos uel enormiter natos uel aliqua necessitate praefractos: The list of parts for pruning is full but not redundant, identifying limbs that (1) are superfluous or unproductive, (2) are deformed (lit. grown 'irregularly', *enormiter*; see *OLD* s.v., *TLL* v 2.606.28–38 (Kapp-Meyer))

[85] Wickson (1891) explains this fact by adding, '[o]ur fruit trees are naturally so prone to bear fruit that pruning to produce fruitfulness is seldom thought of, and still more rarely practiced, while pruning to reduce bearing wood, and thus decrease the burden of the tree, is quite widely done, to take the place, in part, of thinning out the fruit'. The opening words of Wickson (1891) 514 on pruning the almond, 'The pruning of the almond is very simple', could virtually be a translation of G.!

or (3) have been marred by some external difficulty (cf. *aliqua necessitate*). The emphasis is clearly on producing a shapely and, so to speak, well-groomed tree, as Pliny might say: cf. *HN* 17.248 *plurimae autem* (sc. *arbores*) *amputari sibi uolunt onerosa ac superuacua, sicut nos ungues et capillum.*[86] The reason for this is that dead or superfluous growth 'hinders growth and nourishment' (Theophr. *Hist. pl.* 2.7.2 βελτίω γὰρ τῶν αὔων ἀφαιρουμένων ὥσπερ ἀλλοτρίων ἃ καὶ τὰς αὐξήσεις καὶ τὰς τροφὰς ἐμποδίζει) – hence all trees require at least some pruning. Theophrastus observes, however, that the amount of pruning, as for irrigation and manuring, varies depending on the species (*Hist. pl.* 2.7.1 διαφέρουσι δὲ τῷ μᾶλλον καὶ ἧττον): for the almond, he prefers that it be avoided when the tree is bearing fruit, except for withered wood (see 3.5.3n.).

ramos superuacuos . . . purgare: Cf. *Arb.* 16.3 *superuacuos ramos amputabis*; Pallad. 11.5.1 *ut amputentur radices superuacuae.*

falce purgare: The word *falx* refers basically to a 'curved blade, equipped with a single cutting edge' (White (1967) 72). To this class of implements belong instruments ranging from mowing scythes and reaping sickles to the vine-dresser's knife (White (1967) 72–6). White (74) counts at least 12 epithets that the Romans gave to the *falx*, representing different kinds of blades (although some of the names refer to the same type). The tool G. has in mind here is the *falx arboraria* of Cato and Varro, later called the *falx putatoria* by Palladius (Le Gall (1959) 57–8, White (1967) 75, 86), the 'billhook'. The billhook was (is) used for lopping and pruning trees: for discussion of its form and function, and drawings, see White (1967) 85–8.

3.5.2 Exigunt sane propter incursus pecorum custodiam diligentem, quorum dente contacta in amarum saporem ex dulcibus transeunt: The prohibition against animal bites may owe something to the more general claim that the almond once cut back grows bitter: cf. Theophr. *Caus. pl.* 5.17.5 χείρω δὲ (sc. αἱ κολούσεις) ποιοῦσι πλείω· καὶ γὰρ ἡ ἀμυγδαλῆ πικρὰ γίνεται, καὶ ἡ ῥόα σκληροτέρα καὶ ἕτερ' ἄττα μεταβάλλει (cf. also *Caus. pl.* 2.15.1). The danger is augmented, however, when the plant is cropped by animals: *Caus. pl.* 5.17.6 χαλεπαὶ δὲ καὶ αἱ ἐπιβοσκήσεις ὅτι συνεπικάουσιν ἅμα τῇ τομῇ καὶ ἀφαιρέσει· διὸ καὶ ὁ πόνος πλείων. The αἰτία for the corruption that G. offers at 3.7.5 relates to the saliva: *ex morsu pecorum omnia haec arbori amaritosa contingere, cum inopinata in medullas saliua consederit.* The general belief that animal bites are especially injurious to plants is widespread among the agronomists (cf. Verg. *G.* 2.376–9). Varro (*Rust.* 1.2.18–20) thus goes to some

[86] For Pliny's anatomizing personification of the tree, cf. also *HN* 16.181 *atque in totum corpori arborum, ut reliquorum animalium, cutis, sanguis, caro, nerui, uenae, ossa, medulla. pro cute cortex.*

length to explain that the ritual enmity between the goat and Bacchus and Minerva results from the fact that the goat's grazing destroys the vine and olive (cf. *Rust* 1.2.18 *quaedam enim pecudes culturae sunt inimicae ac ueneno, ut istae, quas dixisti, caprae. eae enim omnia nouella sata carpendo corrumpunt, non minimum uites atque oleas*). Pliny is the closest witness to G.'s specific claim that the animal bite makes the almond bitter (*amarus*); after reporting Varro's remarks on the goat, he adds: *HN* 17.237 *quaedam* (sc. *arbores*) *hac iniuria* (sc. *morsu pecorum*) *moriuntur, aliqua deteriora tantum fiunt, ut amygdalae – ex dulcibus enim transfigurantur in amaras.*[87]

Superstitious beliefs about venomous bites aside, grazing animals could certainly cause plenty of damage to plants cultivated at any scale. The dangers posed by intruding animals were a source of concern for the agronomists, who recommend the creation of a fence or protective hedge as a preliminary to growing fruits and vegetables. This barrier would prevent both animals and humans from damaging the plants: Col. 5.10.1 *modum pomarii, priusquam semina seras, circumuenire maceriis uel saepe uel fossa praecipio nec solum pecori sed et homini transitum negare, quoniam si saepius cacumina manu detracta aut a pecoribus praerosa sunt, in perpetuum seminaria incrementum capere nequeunt* (cf. Cato, *Agr.* 46.1, Verg. G. 2.371–9, *Arb.* 18.1, *Geop.* 9.6.1). Varro has Scrofa speak of four types of wall for the garden (*Rust.* 1.14.1 *unum naturale, alterum agreste, tertium militare, quartum fabrile*), besides the option of planting rows of pine, cypress and elm along property boundaries (see also Pallad. 1.34.4 on varieties of *munitiones*). The agronomists are especially taken by the idea of the 'living hedge' (*uiua saepes*) of brambles and thorns – Columella (11.3.7) says *haec quidem claudendi horti ratio maxime est antiquis probata* – , an 'inexpensive and also indestructible boundary . . . whose roots are alive and cannot be destroyed by fire' (Bergmann (2014) 263): see esp. Col. 11.3.3–7; also Varro, *Rust.* 1.14.1, Col. 10.27–8, Pallad. 1.34.4–6. White (1975) 25–6 likewise observes the benefits of the living hedge, remarking that 'in the absence of any references to creosote or pitch to arrest decay . . . the dead fence would only be preferred where it was required to be moveable (enclosures for stock), or where a solid barrier was needed'. The hedge would typically have been planted out from 'vigorous shrubs such as butcher's broom (*ruscus aculeatus*) or from the numerous types of thorn-bush (*spina*), such as Christ-thorn (*paliurus australis*), wild rose (*rosa semperuirens*), and brambles (*rubi*)' (White (1975) 26). For more on the need for fences, and their various forms, see Frayn (1974) 16, White (1975) 24–8, Bergmann (2014) 260–72. There are many depictions of walls and fences,

[87] Pliny's statement should of course also be compared with Theophr. *Caus. pl.* 5.17.5 (quoted above), to which add Theophrastus' opening words αἱ δὲ κολούσεις φθείρουσιν ὅλως ὀλίγα (= Plin. 17.237 *quaedam* (sc. *arbores*) *hac iniuria* (sc. *morsu pecorum*) *moriuntur*).

most often wicker (Mugione et al. (2012) 214), in ancient garden paintings (see Plate 2): see Bergmann (2014), Blanc (2015), Bergmann (2018) *passim*.

sane: See 1.1.2n. *Sane*.

in amarum saporem ex dulcibus transeunt: For the locution, cf. Plin. *HN* 25.103 *initio gustus dulcis mox in amaritudinem transit*. For *ex dulcibus*, cf. also 3.7.2 *Fiunt dulcia ex amaris* (after Plin. *HN* 17.252).

3.5.3 Fossio contraria illis habetur, nam florem statim amittunt: Cf. Plin. *HN* 17.247 *amygdalae, si colantur fossione, florem amittunt*. The claim may go all the way back to Theophrastus, who maintains that, uniquely among plants, almonds ought not be watered, manured, pruned or otherwise cultivated when they bear fruit in order to avoid promoting the riotous growth (cf. ὑπερισχύσασα) that will cause them to drop their fruit prematurely: *Caus. pl.* 3.18.2 τὴν ἀμυγδαλῆν, ὅταν ἄρχηται καρποφορεῖν, οὔτε ὑδρεύουσιν οὔτε κοπρίζουσιν οὔτε διακαθαίρουσιν, εἰ μὴ τὰ αὖα μόνον, οὐδ' ἄλλο τῶν τοιούτων οὐδὲν δρῶσιν· πάντα γὰρ ἀφαιροῦσιν ὅπως μὴ ὑπερισχύσασα ἀποβάλλῃ τοὺς καρποὺς ἐν τῷ ἄνθει, δι' ὃ καὶ χώραν οὐκ ἀγαθὴν ζητοῦσιν.[88] If Theophrastus' claims do in fact form the basis for the later prohibitions against digging around the tree – an attractive inference that would explain a terse and otherwise mysterious recommendation –, then his advice was altered at some point in the transmission: Pliny and G. say only that digging will cause the tree to drop its flowers (*florem amittunt*), G. adding that this will happen immediately (*statim*), whereas Theophrastus had feared that the tree's excessive growth would cause it to lose its fruit when in flower (ἀποβάλλῃ τοὺς καρποὺς ἐν τῷ ἄνθει). (The alleged injuries would of course amount to similar losses in productivity, since without the flowers the trees could not be pollinated.)

quod malum . . . uel a nobis: *quod malum* is *amissio florum*, not the digging itself: hence *uel*, ('*at least* not by us'; see *OLD* s.v. *uel* 6), which anticipates the additional sources of danger not within our control that will be introduced momentarily.

cum et alias periclitentur in flore, nubilo prius marcidae deinde Austris insequentibus concrematae: The southwind *Auster* (= Νότος/*Notus*) had a variety of aspects: see concisely *RE* (1) 33.1116.55–1117.2 s.v. 'Notos' (Capelle). The detail *concrematae* suggests that the main threat posed here by *Auster* is heat or aridity (cf. Mazzini 113 n. 33: '[l]'Austro è un vento caldo'). The winds blowing from the south may indeed be warm but, *concrematae* notwithstanding (see on the word below), it is more likely that what

[88] Poor soil is recommended for the almond on the same grounds also at *Caus. pl.* 2.16.8, 3.6.8. For Theophrastus' reasoning, see Michelini (1978).

G. has in mind here is the wet and stormy weather that they bring and for which they were above all known: see *RE* (1) 33.1117.2–66, *TLL* II 1552.74–8, 1553.38–44, 53–8, 60–5, 70–1, *passim* (Ihm). That this is the relevant aspect is perhaps already suggested by *nubilo*, but confirmed by two passages in Pliny that are the source (or witness to the source) for G.'s caution: cf. *HN* 16.109 *pirus et amygdala, etiam si non pluat, sed fiat austrinum caelum aut nubilum, amittunt florem et primos fructus, si, cum defloruere, tales dies fuerint*; 17.11 *nam si, cum defloruere, protinus sequantur imbres, in totum poma depereunt, adeo ut amygdalae et piri, etiam si omnino nubilum fuit austrinusue flatus, amittant fetus*. G.'s *alias* anticipates *prius . . . deinde*, which function as an elaborated disjunction following Pliny's *aut* (*HN* 16.109) and *-ue* (17.11). Pliny's passage in turn depends (as André observes) on Theophr. *Caus. pl.* 2.2.1–3, whose subject is the effect of rain on crops. Theophrastus precisely emphasizes that rain is most injurious when the tree is in flower (cf. G.'s *periclitentur in flore*): *Caus. pl.* 2.2.2 χείριστα δὲ καὶ παρακαίροτατα (sc. τὰ ὕδατα) τὰ περὶ τὰς ἀνθήσεις. (At 2.2.3 Theophrastus states that southerly rains are worse than northerly rains, as a general rule.) Pliny and Theophrastus are important here for another reason. Their emphasis on the fact that *nubilum* and *Auster* are feared because they can strip the plant of its flower and fruit explains G.'s decision to append this information to the dangers of digging: the presentation is motivated by the fact that both weather and digging pose a similar danger, i.e. causing the almond to lose its flower (cf. *florem amittere*).

But a question remains: if *Auster* refers especially to cloudy, wet or stormy weather, why is it that G. uses the verb *concremo* here to describe its effects? Of course, G. may be thinking of a wind that is also hot, but even if the idea of warmth were wholly absent the usage would not be so strange as it might first appear: the agronomists very often use verbs that properly mean 'burn' to describe the effect on plants of things that are neither hot nor fiery. Present-day English 'burn' is often used in the same sense, especially in agricultural language, and the definition offered by the *OED* (s.v. 13d) is entirely germane for Latin words of the same semantic field: 'To produce (on anything) an effect resembling that of burning; e.g. (of the sun) to wither, dry up (vegetation), to parch, dry up (the ground) . . . Sometimes said of cold, and of certain manures and crops, to express their effect on the vegetation or on the soil'. Compare especially in this regard 3.6.4 *metus est ne in flore deprensas [aut] pruina comburat*, where the frost is said to 'burn up' (*comburat*) the trees. Although G.'s use of *concremo* is apparently unique in this application, it is not difficult to understand;[89] *uro* and its compounds are often applied in these slightly

[89] *TLL* IV 93.26 (Lommatzsch) mistakenly places it under the lemmata 'igni delere' and 'in medicamentis', where the reference is also erroneously given as *Pom.* 2.5 instead of *Pom.* 3.5.

extended senses: see *OLD* s.v. *uro* 4c, 9a; for *aduro*, *TLL* I 898.59–85 (Klotz), *OLD* s.v. 6; for *comburo*, *OLD* s.v. 4; for *exuro*, *TLL* V 2.2127.14–23 (Schmeck).

nubilo . . . marcidae: The verbs *marceo* and *marcesco* are properly used of the wilting of plants (*TLL* VIII 372.35–44 (Bachmann), VIII 374.10–33 (*id.*)), but *marcidus* only rarely so: cf. Ov. *Met.* 10.191–2 *lilia . . . / marcida demittant subito caput*. Bachmann in his *TLL* article (VIII 375.66–70) places *marcidus* here under the lemma 'i. q. putidus' (next to Vitr. *De arch.* 2.8.20, of rotting wicker baskets, and Plin. *HN* 23.7, of purulent wounds), but it is preferable to see *marcidus* as retaining its basic meaning: the trees are 'wilting' or 'drooping' under the cloudy weather (as at Ov. *Met.*, above). The notion of decay is accessory, as it is also when G. says at 2.12.4 that the stored peaches 'wither' (*flaccescunt*).

3.6 This section contains recommendations for improving the almond's productivity (*feracitas*) and protecting it from adverse climatic conditions. Although the advice does not originate with G. in most cases (see *passim* below), it continues to be valued highly by the later agronomists, who follow it closely: cf. Pallad. 2.15.9–10 *in uetustate plus affert. si ferax non est, taedae cuneum terebrata radice mergamus uel silicem sic inseramus ut libro tegente claudatur. locis frigidis, ubi metus est de pruina, Martialis dicit hoc remedio subueniri: antequam floreant, radices nudantur et albi lapides minutissimi mixti harenis congeruntur et, ubi iam tempus uidebitur ut debeant germinare, effossi iterum lapides summouentur*; *Geop.* 10.61 εἰ δὲ ἐπιμένει φύλλα μὲν φέρουσα, καρπὸν δὲ μή, τρυπήσας τοῦ πρέμνου τὸ μέρος τὸ περὶ τὴν γῆν, καὶ δαδὸς λιπαρᾶς ἐπίουρον εἰς τὸ τρύπημα κατακρούσας, οὖρον ἀνθρώπινον ἐπίχεε, καὶ οὕτως προσχώννυε.

At the beginning of this section most editors emend to *generant*, but Mazzini retains the subjunctive *generent* and posits a lacuna in order to explain it. To do so is not unreasonable, whether we emend to *generant* or not, if we consider that Palladius (2.15.9) has the more verbose *in uetustate plus affert* (cf. also Plin. 16.117 *amygdala enim et pirus* <u>*in senecta*</u> *fertilissimae*). If we suppose that his *in uetustate* depends on something that was written by G., then with *generant* as few as one or two words could be missing, e.g. *in senecta/uetustate* (Plin./Pallad.), or, if we keep the subjunctive *generent* (**N**) with Mazzini, we might posit a more extensive loss. Condorelli xxxii, however, calls attention to the gap in **N** that divides the paragraph that ends with *concrematae* from the paragraph beginning with *Plurimum generant*, and, comparing similar empty spaces that divide other sections, argues that they were intended for rubrics. As regards the present case, this implies that there is no lacuna before *Plurimum* (thus Condorelli xxxii n. 11), in spite of the plausibility of the suggestion, and hence nothing to explain the subjunctive. Consequently, we must prefer *generant* and otherwise leave the text undisturbed.

3.6.1 Plurimum generant: For *genero*, see 4.2.4n. *uirtute generandi*.

3.6.2–3 Si ferax non erit, terebrari in radicibus debet et taxillus e taeda in foramen aptari; alii in locum taedae cuneum qui esset ex robore. Quidam radicibus perforatis silicem adiciunt et ita arboribus librum patiuntur inolescere: The recommendations given here have many parallels in the agronomists. This class of remedies appears among the Roman agronomists first in Columella but can be traced back in substance to at least Theophrastus. The basic technique is of placing or inserting an object into or around the tree in order to affect its growth, productivity or nature. Elaborations of the method show a 'complicated mixture of logics and outward forms, combining regular farming techniques with overt magic' (Ager (2010) 187). In order to set G.'s recommendations in context, I outline below three of the principal variations on the technique that can be identified in the agronomists. These groupings aid in organizing the messy and diverse evidence but are imperfect and are not intended to show the logical or chronological development of the techniques in antiquity.

We start from the simplest variation, in which stones were placed below the roots of a tree in order to 'anchor' or 'restrain' it (cf. Ager (2010) 185–6). This is of the type recommended by Columella for the pomegranate: 5.10.16 = *Arb.* 23.2 *mala Punica ne in arbore hient* [*Arb.*: *ne rumpantur in arbore*], *remedio erunt* [*Arb. placuit*] *lapides tres, si cum seres arborem ad radicem ipsam collocaueris.* This recommendation alone of the three kinds can be tentatively analysed into both natural and supernatural elements. From the standpoint of magic, the stones may play an anchoring function, preventing the fruit from dropping before they reach maturity (see also below). On a natural basis, the stones could be intended simply to check the growth of the tree by restricting its roots; we have seen already (3.5.3n.) that it was believed that excessive flourishing of a plant could cause it to drop its fruit prematurely.[90] Note, however, that Columella adds that squill (a lily-like plant) may be installed around the roots of the tree, if it is not possible to place stones underneath: 5.10.16 = *Arb.* 23.2 *at si iam arborem sitam* [*Arb.*: *satam*] *habueris, scillam secundum radicem arboris serito.* The squill could be intended to block the growth of the roots in the same way as the stones that were situated before planting, but its influence is more likely to be magical, for the *Geoponica* in repeating the advice attributes the effect to a natural 'antipathy' to pomegranates that the squill possesses: 10.30 ἐν τῷ φυτεύειν προέμβαλλε εἰς τὸν βόθρον κόχλακας. ἐὰν δὲ

[90] In this regard, cf. also *Geop.* 9.10.7 τὰς δὲ ὑλομανούσας τῶν ἐλαιῶν, τουτέστι τὰς πολλὰ μὲν φύλλα φερούσας, ὀλίγιστον δὲ καρπόν, θεραπεύσεις οὕτως. πάσσαλον εἰς τὰς ῥίζας ἐμπήξας ἐλαίας ἀγρίας, ἢ πίτυος, ἢ δρυός, ἢ λίθον.

ὦσι προπεφυτευμέναι, σκίλλας παραφύτευσον. αὗται γὰρ ἀντιπαθοῦσαι οὐκ ἐῶσιν αὐτὰς χαίνειν (i.e. the fruit to split open).[91]

The second variation on the technique is the insertion of a stone into the trunk or roots of a tree in order to render it fruitful (Ager (2010) 186). Such is Columella's advice for the almond: 5.10.20 = *Arb.* 25.1 *amygdala si parum ferax erit, forata* [*Arb.*: *perforata*] *arbore lapidem adigito et ita librum arboris inolescere sinito.* Linguistic echoes suggest that the second of G.'s recommendations (3.6.3 *Quidam radicibus perforatis silicem adiciunt, et ita arboribus librum patiuntur inolescere*), which is also of this type, may depend at some remove on the passage in Columella, although G. has added the specification that the hole ought to be made in the roots – perhaps in order to harmonize the advice with his first recommendation (cf. *terebrari in radicibus*), but perhaps also synthesizing Pliny, Theophrastus or another source descended from them.[92] It is hard to see how inserting a stone into the tree could be explained on the basis of ordinary farming technique (even if we cannot preclude the possibility), and it seems likely that the stone's primary function is now as a magical token, both preventing the loss of and, in some fashion, encouraging growth of fruit (hence described as a remedy for the tree that is *parum ferax*). The anchoring role of the stone in this variation of the technique is particularly clear at Pallad. 3.25.15 (on the apple): *si caduca sunt poma, fissae radici lapis iniectus poma retinebit*; and the fertility aspect at Pallad. 12.9 (of the vine): *uites quae fructu carent fronde luxuriant et pauperiem fetuum compensant ubertate foliorum . . . quidam lapides inserunt inter flexuosa radicum.*[93]

The third variation on the technique (Ager (2010) 186–7) also concerns the insertion of an object into the tree, but the objects now show an increasing variety of forms (wedges, pegs, nails, etc.) and materials (different species of wood and metals) that are designed to obtain a specific magical effect. To this

[91] From an early time (that of the Hippocratic writers), squill was considered a baneful plant: see von Staden (1990) 18–19.

[92] Pliny is quite close to G.'s advice: 17.253 *pomiferis quae germinant nec ferunt fructum* (= G.'s *Si ferax non erit*) *fissa radice inditur lapis fertilesque fiunt, hoc idem in amygdalis e robore cuneo adacto, in piris sorbisque e taeda ac cinere et terra cooperto.* Theophrastus does not make specific reference to the almond but is a likely witness to the common source: *Hist. pl.* 2.7.6 ἐὰν δέ τι μὴ φέρῃ καρπὸν ἀλλ' εἰς βλάστησιν τρέπηται, σχίζουσι τοῦ στελέχους τὸ κατὰ γῆν καὶ λίθον ἐντιθέασιν ὅπως ἂν ῥαγῇ, καί φασι φέρειν; cf. also Nic. Dam. *De plantis* 821a12–14.

[93] Palladius' recommendation at 12.10 (copied by *Geop.* 5.35) to wedge a stone into a fissure in the vine's roots probably does not belong to the class of remedies under discussion, since the purpose there is mechanical, to keep the roots pried open so that manure and urine can penetrate down to them (*ut ad radices instillatio ipsa descendat*).

class belongs G.'s recommendation to insert a die of pine (3.6.2 *taxillus e taeda*) or a wedge of oak (3.6.2 *cuneum qui esset ex robore*) into the almond's roots. Other places in the agronomists where a specific wood is named include Col. 5.10.17 = *Arb.* 24 (pine into pear), Pallad. 2.15.3 (pine into sorb apple), 3.25.4 (pine or oak into pear), 3.25.30 (terebinth and mastic into mulberry), 11.12.2 (willow into date palm), 12.7.5 (terebinth, mastic or willow into peach), 12.7.15 (wild olive into plum), *Geop.* 9.10.7 (wild olive, pine or oak into olive), 10.16 (willow or dogwood into peach), 10.64.5 (elm into walnut). The wood so inserted could impart the qualities of its parent tree to the barren plant: for a clear example, see Col. 5.9.16, where shoots from a fruitful olive are said to reinvigorate a sterile one (so too *Geop.* 9.8, 9.10.4). It is evidently also to this end that G. recommends inserting a honey-covered wedge of wood into the tree to make the fruit sweet: 3.7.2 *post media trunci parte terebrata cuneum qui esset ex arboris quidem radicibus simplici delibutum melle fixerunt.* So far as inorganic materials go, driving metal nails into a tree could effect a binding spell (on which see Faraone (1991), Jordan (1985) 151 with Versnel (1991) 61, Collins (2008) 64–6), whereby harmful properties were restrained. Consider thus Pliny's report (*HN* 16.51) that a bronze nail put into yew takes away the tree's fearsome ability to kill those who sleep under it (*repertum innoxiam fieri si in ipsam arborem clauus aereus adigatur*) or, in the same vein, how an iron peg driven into an almond bearing leaves and wood at the expense of fruit will 'punish' it, curbing its ὕβρις: Theophr. *Hist. pl.* 2.7.6 ἀμυγδαλῇ δὲ καὶ πάτταλον ἐγκόψαντες σιδηροῦν, ὅταν τετράνωσιν, ἄλλον ἀντεμβάλλουσι δρύινον καὶ τῇ γῇ κρύπτουσιν. ὃ καὶ καλοῦσί τινες κολάζειν ὡς ὑβρίζον τὸ δένδρον (cf. *Caus. pl.* 1.17.9).[94] For other metal objects, see also *Geop.* 5.36.3, 10.67.2, Ager (2010) 187.

3.6.2 terebrari . . . foramen: *foramen* is used by the agronomists from Cato to Palladius in a semi-technical sense of a hole made in a tree by *terebratio*, 'boring' (see examples *passim* at *TLL* VI 1.1032.50–79 (Bauer)). For the collocation of the terms, cf. Col. 9.8.9 *terebratur ab latere talea, per quod foramen*, etc. The verb *terebro*, although also used commonly by the agronomists from Cato onwards, is not itself technical, but the derived nouns *terebratio* ('boring') and *terebra* (a 'drill' or instrument for boring holes) are: see *OLD* s.vv.

taxillus: The rare noun, a diminutive from *talus*, is used for precision here, as at Vitr. *De arch.* 10.8.2 (in a hydraulic organ): *taxilli alti circiter digitorum ternum*

[94] For the use of ὑβρίζω in this passage, see Michelini (1978). The tradition of 'punishing' the almond in particular does not appear again after Theophrastus, unless we see its influence lurking behind the various recommendations in the Roman agronomists to insert stones and wood into the trees. Girdling, root pruning and driving nails into the trunk are still practised today in order to shock trees into producing more and larger fruit.

suppositi. It should not therefore be attributed to the preference for diminutives that is a feature of colloquial and later Latin: on that phenomenon, see Väänänen (1981) 80, Herman (2000) 99–100; cf. also Adams (1976) 105, (2013) 788–9 on diminutives with anatomical terms. Maire (1999) 434, (2002*a*) xxxviii observes the frequency of diminutives in *Med*., some of which are to be attributed to the influence of medical Latin.

taeda: 'Pine', although another meaning of *taeda*, 'torch(wood)', seems to be reflected in the *Geoponica*: cf. 10.61 δαδὸς λιπαρᾶς ἐπίουρον.

alii in locum taedae cuneum qui esset ex robore: For *alii . . . cuneum* the MS reading is *illi . . . cuneum*, which does not make sense. The solution of early editors (Mai, Scotti, Antonelli) was to emend to *uel . . . cuneus* ('or, in place of pine, a wedge of oak'). Mazzini follows them in adopting *cuneus* for *cuneum*, but retains *illi*, punctuating with a semi-colon after it (improbably, since word order and rhythm suggest that the colon end is after *forāmĕn āptāri*) so that it will be taken with the preceding *in foramen*. But the way to the true reading lies elsewhere: Mai, despite printing **N**'s *illi* in his edition of 1846, remarks that he recalled reading *ali* in his collation of the palimpsest (cf. Condorelli xl); when Condorelli inspected the remains of the manuscript, he thought he could see *uli*, which (it seems) convinced him that Mai's recollection of *ali* was correct. Mai's reading is possible: the asyndeton is adversative (for which see 2.4.6n.); supply an active verb from *aptari* for *alii* (corrected from *ali* on the supposition of haplography before the following *-i*[95]).

in locum taedae: We might expect *in loco* here (*TLL* VII 2.1586.4–33 (Kuhlmann), *OLD* s.v. *locus* 18c), but variation between the ablative and accusative after prepositions is very common in later Latin (Herman (2000) 53–4, Haverling (2006)), especially after *in*: see *TLL* VII 1.794.80–799.11 (Bulhart-Hofmann), Svennung (1935) 382–3, H–Sz §156.b.ζ, Adams (2016) 327, 439 and more generally Adam's index s.v. '*in*, + ablative for accusative and vice versa'. Confusion with *in* had in fact already begun in the period of early Latin and is not to be automatically considered Vulgar. There are other places in *Pom*. with an accusative after *in* where we might have expected the ablative: cf. 3.1.4n. *in posteriores dissentiunt*; 3.7.5 *in medullas . . . consederit*; 4.7.1 *in pastinatos*; possibly 2.12.4n. *in breue tempus*. For such confusion in *Med*., see Maire (2002*a*) xli.

cuneum qui esset ex robore: The imperfect (*esset*) is repeated also at 3.7.2 *qui esset ex arboris quidem radicibus*. In neither case is the reason for its use pellucid, but here it might be rhythmically motivated (dicretic *ēssĕt ĕx rōbŏre*) and at 3.7.2 it could depend on the tense of *fixerunt*.

[95] Condorelli xl retains the form *ali*, for which see *TLL* I 1622.54–7 (Hey), Grandgent (1907) 95–6.

3.6.3 radicibus perforatis silicem adiciunt: *radicibus perforatis* is probably an ablative absolute (as translated), not a dative governed by *adicio*, for when used in the literal sense of 'add or thrust into', as here, the verb always takes *in* + accusative or *in* + ablative (*OLD* s.v. 4a, *TLL* I 666.38–52 (Weynand)). The syntax would not however be difficult on the alternative construal, for *adicio* has other concrete applications governing the dative, as when it means to 'mix in' or 'add to' of ingredients: see *OLD* s.v. 5, and cf. Scrib. Larg. 160 *aquae purae feruenti salem adicito.*

et ita: For the continuative force of *ita*, see 3.4.3n. *nec possint ita.*

ita arboribus librum patiuntur inolescere: The clause is equivalent to Col. 5.10.20 = *Arb.* 25.1 *ita librum arboris inolescere sinito.* Note the substitution of the imperatival indicative *patiuntur* for Columella's future imperative *sinito* (for the meaning of *patior*, see 2.2.2n. *recentia* . . . *mergi patiuntur*) and the dative *arboribus* in lieu of the genitive *arboris*. The latter alteration requires some comment. Both context and the Columellan source show that G.'s *arboribus* contains the idea of possession, 'the trees' bark'. Despite this fact, convention dictates that the dative is not to be called possessive (*datiuus possessiuus*) in the strict sense, which on most accounts is held to require the verb *esse*: so Löfstedt (1963) 64–5, Adams (2013) 288, Pinkster 107–10; but for a broader definition, see Ernout and Thomas (1953) 73, discussed by Adams (2013) 288–9. *arboribus* falls, rather, into the heterogeneous category known since Havers (1911) as the 'sympathetic dative' (*datiuus sympatheticus*; see also Bennett (1910–14) II 134–45). This dative is often said to be of the type <*mi*> *misero* . . . *malae dolent* (Pl. *Am.* 408) or *quo pacto ei pugno praefregisti bracchium* (*Mil.* 26), although, as Adams (1995*a*) 449–50 points out, there was from the first a tendency to lump together under its name examples of quite different types: Havers (1911) 3–4 himself uses it as a label for six categories of the dative, of which the kind expressed by the examples above is the most common. Although he does not attempt to reconcile every alleged example of the sympathetic dative, Pinkster 919–20 helpfully cuts through much of the confusion by identifying the essential characteristic of such datives as the relationship of 'inalienability' in which the constituent dative stands to another noun in the sentence.[96] This is clearly the case with G.'s *arboribus librum patiuntur olescere*, where the dative *arboribus* stands

[96] Nevertheless, 'it is often difficult to decide whether a dative constituent should be regarded as a dativus sympatheticus (that is, a satellite) or as an argument of a complex expression' (Pinkster 922). Another definition for the sympathetic dative is offered (cautiously) by Adams (2013) 289–90: 'some may find it preferable to classify them with Ernout and Thomas simply as one type (adnominal) of possessive dative, with a special tendency to occur with anatomical terms'.

in just such an implicative, i.e. inalienable, relationship to *librum*; although the sympathetic dative has most often been connected with human (or animal) body parts, it occurs not infrequently in reference to plants or trees: see Pinkster 920 s.v. 'inanimate dative constituents'; also Pinkster (2005) 242–3, Reinhardt (2010) 218–19.[97]

Several of the orthodox opinions regarding the sympathetic dative have by now been discarded or replaced with more modest conclusions. The sympathetic dative was previously alleged (see Pinkster 921–2) to be semantically equivalent to a possessive genitive, except in that it (1) is more 'lively' or indicates a special interest (Havers (1911) 2, *NLS* §63, Chantraine (1963) 71) and (2) is characteristic of Vulgar or colloquial language (Svennung (1935) 221, *Syntactica* I 232–5, Önnerfors (1963) 201, H–Sz §67.b). On the first point, Pinkster 921–2 has refuted the claim of semantic equivalence with the genitive, and it is further doubtful whether the notion of 'special interest' is precise enough to be useful (cf. Adams (2013) 289). As for the second, it must be rejected as an over-generalization: see Adams (1995*a*) 449–60, (2013) 290, Pinkster 922, Pruvost-Versteeg (2015). Therefore it cannot, we should add here, be adduced as evidence of Vulgar influence on G.'s Latin.[98]

3.6.4 In locis frigidis, in quibus metus est ne in flore deprensas [aut] pruina comburat, hoc modo tradunt amygdalis subueniri: antequam floreant nudare radices iubent et albi generis lapides quam minutissimos cum harenae cumulo aggregant: The almond is 'extremely subject to frost injury where moderately late spring frosts prevail' because of its early bloom (Duke (1989) 250–1). The mixture of granular stones and sand is evidently meant to serve as a kind of insulating blanket to prevent the penetration of frost and cold to the roots: for the vine, cf. Col. 3.15.3–4 (reporting Mago), *Arb.* 4.4–5. The purpose of laying bare the roots is thus quite different from that of Theophrastus' recommendation: *Caus. pl.* 3.18.2 ἐάν . . . ἀκαρπῇ, τάς τε ῥίζας γυμνώσαντες παραδιδόασιν τῷ χειμῶνι καὶ τὰς ἄλλας κολάσεις προσφέρουσι τὰς εἰρημένας (cf. *Geop.* 10.61 ἄκαρπον δὲ (sc. ἀμυγδαλῆν) καρποφορῆσαι ποιήσεις, τὰς ῥίζας χειμῶνος γυμνώσας). Those treatments are intended to compel the tree to bear fruit through curbing or 'punishing' it: cf. Theophr. *Hist. pl.* 2.7.6, quoted at 3.6.2–3n.

[97] Examples: Lucr. 1.251–2 *at nitidae surgunt fruges ramique uirescunt / arboribus, crescent ipsae fetuque grauantur*; Plin. *HN* 12.40 *nulli arborum folia ibi decidunt.*

[98] For a sensible take on the possible 'colloquial ring' of the sympathetic dative, see Reinhardt (2010) 219 (with Petersmann (1977) 84 on the evidence of Petronius).

in flore deprensas: 'Caught' in flower, a vivid metaphor not found elsewhere in the agronomists; for *deprehendo* with an abstract agent, see *TLL* v 1.608.12–35 (Bögel), *OLD* s.v. 5a.

[aut] pruina: Mazzini 144–5 defends **N**'s *aut pruina* on the ground that *aut* is sometimes used for *et* in later Latin (H–Sz §269.c.β); although this is true, *et* seems no more wanted than *aut* (*pace* Mazzini 145, who states that '"*et*" nel senso di "anche" si adatta bene al contesto'). We can either suppose that one of the disjuncts belonging to *aut* has fallen from the text (e.g. <*nix*> *aut pruina*, Scotti,[99] Condorelli, Winterbottom; Mazzini 144–5 *contra*), or simply delete *aut* (Mai). That Palladius has only the frost (*ubi metus est de pruina*) argues for deletion, although it is admittedly hard to see how *aut* could have made its way into the text. Condorelli attempts to find an *explanans* for the corruption when he moves Scotti's *nix* after *ne*.

pruina comburat: For the application of *comburo* to the effects of frost, see 3.5.3n. *cum . . . concrematae.*

hoc modo tradunt amygdalis subueniri: For the metaphor and the force of the *sub-* prefix in *subueniri*, see 2.3.5n. *subsidium teporis et umoris.*

antequam floreant: For the subjunctive, see 2.3.4n. *priusquam . . . committantur.*

nudare radices: *nudo* is used in two senses by the agronomists: (1) of stripping the bark, leaves or other outer part of a plant (e.g. Col. 4.28.1, 5.11.10 = *Arb.* 26.8; cf. *nudus* at Pallad. 4.1.2); (2) of laying bare the roots of a plant by excavating the soil around them (e.g. Col. 4.14.2, 5.5.19; cf. *nudus* at Pallad. 2.6). Joined with *radices* here, the word describes the procedure of *ablaqueatio* (2.8.1n.), which it stylistically varies.

albi generis lapides: The genitive is of description or quality. It is not clear what sort of stone G. has in mind, but at the very least the stipulation that the *lapides* be white would ensure that the cultivator not mistakenly substitute a soil mixture in place of the stones.

lapides . . . cum harenae cumulo aggregant: *aggrego* has a late meaning 'heap up', first in the *Vetus Latina* then not again (except for the present instance) until Augustine: see *TLL* i 1322.36–62 (Vollmer). While the meaning of *aggrego* is therefore unobjectionable, perhaps it would be better to read *adaggerant* (or the simplex *aggerant*). *adaggero*, a purely technical term first found in Cato (*Agr.* 94.1), has in Columella and Pliny just the meaning we are looking

[99] For Scotti's supplement *nix*, compare the Ciceronian (*Cat.* 2.23, *Nat. D.* 1.24, 2.26) collocation *nix pruinaque*.

for, 'heap onto (roots, etc.)': cf. Col. 5.11.8 *deinde terram circa arborem adaggerato usque ad ipsum insitum*; Plin. *HN* 17.263 *huius praecipue uere terram adaggerari radicibus.*

cumque iam tempus uidebitur ut debeant germinare . . . praecipiunt: For the tense of *uidebitur*, see Introduction §v(b). When *tempus* means 'the proper or due time' (see 1.1.7n. *tempus*), it can take an epexegetic infinitive, an accusative-and-infinitive phrase, or, as here, an explanatory *ut*-clause (i.e. noun clause).

debeant germinare: Late uses of *debeo* where the word has been bleached of its original meaning (for the general phenomenon, see 2.1.3n. *Graecis . . . siquidem*) might lead us to suspect weakening here and in two other places in *Pom.*: cf. 4.3.4 *si quae forsan integrae debent credi*; 4.4.4 *si tamen fas est ut admittere in animum debeamus.* It is not unreasonable to see these three instances of *debeo* as weakened to some degree, but it must also be admitted that the idea of necessity or obligation carried by the word is not out of place in any of them: at 3.6.4 (here) stress is laid on the time (*tempus*) at which the plants 'ought' (sc. naturally) to undergo germination; at 4.3.4 it describes implicitly the properties or appearance of the chestnuts that 'ought' to produce a belief about their soundness; and at 4.4.4 there is adopted the perspective of 'we [farmers]' who 'ought' (or ought not) to adopt a course of action.

germinare: The verb is rare before the Imperial period, but then a common technical term in the agronomists and, later, generally widespread in both literal and metaphorical applications: see *TLL* VI 2.1926.13–1927.10 (Schuster). Cf. also 2.3.2n. *germinis.*

quae fuerant infusa: The neuter plural stands in for the mixture *albi generis lapides . . . cum cumulo harenae* (for the gender see H–Sz §232.D.d.β).

3.7 As for 3.6 (n.), most of this section is reproduced almost verbatim by Palladius: 2.15.11 *teneras nuces amygdalus creabit, ut dicit* (sc. *Martialis*), *si ante florem radicibus ablaqueatis per dies aliquot calida aqua ingeratur. ex amaris dulces fiunt, si circumfosso stipite tribus digitis a radice fiat cauerna per quam noxium desudet umorem, uel medius truncus terebretur et cuneus ligni melle oblitus imprimatur uel si circa radices suillum stercus affundas.* Much of the advice also recurs in the *Geoponica*, e.g. that on producing soft almonds by the application of warm water: 10.59.3 ἁπαλὴν δὲ καὶ γλυκεῖαν ἐκ σκληρᾶς καὶ πικρᾶς ποιήσεις, εἰ τὰ περὶ τὸ στέλεχος ἕως τῶν ἐπιπολῆς ῥιζῶν γυμνώσειας, καὶ ἀρδεύσειας συνεχῶς ὕδατι θερμῷ, πρὸ τῆς ἀνθοβολήσεως.

3.7.1 utile erit . . . aqua calida . . . irrigata: It looks at first as though *utile erit* is impersonal and will govern a clause (compare the remarks in the Introduction §v(d) on 3.7.3 *utile est si effundatur*), but at the end of the sentence we see that it is a predicate of the clause-like participial phrase *aqua calida per aliquos dies irrigata*. It is not necessary to emend with Mazzini to *utilis*; the predicate adjective is regularly made into a substantive neuter singular: see H–Sz §238.a. Cf. also 4.2.3 *nux reseruata difficile sit contra putredinem uindicare* (n.).

ablaqueatis radicibus: See 2.8.1n. *Ablaqueare*.

3.7.2 Fiunt dulcia ex amaris si circumfosso stipite tribus a radice digitis latior cauerna ponitur qua pituita decurrat: The linguistic parallels make it almost certain that Pliny is a close witness to G.'s source, if not the source himself: *HN* 17.252 *amygdalae ex amaris dulces fiunt si circumfosso stipite et ab ima parte circumforato defluens pituita abstergeatur*. The problem, however, is that *pruina*, the easy and universally accepted fix for *pruna* (**N**), renders G.'s passage unintelligible. Whereas the rationale of Pliny's advice to bleed the tree of its noxious sap (*pituita*) in order to make it sweet is pellucid, there is no explanation in G.'s passage where the frost (*pruina*) would come from or why its flowing down into a ditch near the roots (if *decurrat* could even be used to describe the motion of frost) should have any benefit for the almond. The basic difficulty is that, no matter how far one is willing to stretch the meaning of the word, *pruina* cannot mean *pituita*. Indeed, Pliny's nearly singular application of *pituita* to a plant (*TLL* x 1.2227.38–41 (Rzepiela), *OLD* s.v. 3) is an attempt at a precise translation of δάκρυον/ἐπιρροή of Theophrastus, who is the ultimate source of this advice for the almond: *Hist. pl.* 2.2.11 ἀμυγδαλῆ δὲ (sc. τῇ θεραπείᾳ μεταβάλλει) ὅταν πάτταλόν τις παίῃ [*coni. Amigues, codd.* αἴῃ *uel* ἐνθῇ], καὶ τὸ δάκρυον ἀφαιρῇ τὸ ἐπιρρέον πλείω χρόνον καὶ τὴν ἄλλην ἀποδιδῷ θεραπείαν; *Hist. pl.* 2.7.7 ἀμυγδαλῆν δὲ καὶ ἐκ πικρᾶς γίγνεσθαι γλυκεῖαν, ἐάν τις περιορύξας τὸ στέλεχος καὶ τιτράνας ὅσον τε παλαιστιαῖον πανταχόθεν, τὸ ἀπορρέον δάκρυον ἐπὶ ταὐτὸ ἐᾷ καταρρεῖν. τοῦτο μὲν οὖν ἂν εἴη πρός τε τὸ φέρειν ἅμα καὶ πρὸς τὸ εὐκαρπεῖν; *Caus. pl.* 1.17.9 τῆς δὲ ἀμυγδαλῆς ἐάν τις ἐγκόψας τὸν πάτταλον ἀποκαθάρῃ τὴν ἐπιρροὴν τῆς ὑγρηότητος ἐπὶ ἔτη δύο ἢ τρία, καὶ γλυκεῖαν ἐκ πικρᾶς γίγνεσθαί φασιν.[100]

Even if *pituita* is what G. should have written, what is to stop us from supposing that he encountered Pliny's unique application of *pituita* and, through ignorance or carelessness, wrote *pruina* for it? There are two strong arguments against this hypothesis: (1) the evidence of *Med.* shows that G. knows the word *pituita* (5x: 13.13, 23.6, 29.4, 32.4, 38.4); and (2) Palladius' version is not garbled with *pruina* or another nonsense word: 2.15.11 *ex amaris dulces fiunt si circumfosso*

[100] Cf. Nic. Dam. *De plantis* 821a39–40. Perhaps also germane are Vitruvius' detailed instructions to bleed trees intended for lumber of their *uitiosus liquor*: cf. *De arch.* 2.9.3–4 with Callebat et al.

stipite tribus digitis a radice fiat cauerna per quam noxium desudet umorem.[101] As regards (2), consider that Palladius' sentence otherwise closely follows G.'s in vocabulary and syntax and indeed stands in a passage that he has explicitly attributed to him (*ut dicit* (sc. *Martialis*); for context see Pallad. 2.15.9–11, quoted at 3.6n. and 3.7n.). Should we believe that Palladius read *pruina* in G. and was inspired to write *noxius humor*? Much more plausible is the hypothesis that, encountering the recondite *pituita*, he wrote an accurate paraphrase. (There is no evidence that Palladius had his Pliny or Theophrastus to hand against which to check G. here.)

Scotti claims that the various parallels quoted above actually show that *pruina* is a synonym for *lacrima* (i.e. δάκρυον/ἐπιρροή), *gummi*, *pituita*, *noxius*, *umor*, etc., but this argument begs the question: we cannot begin from the premise that *pruina* was in the text and, bringing together the parallel passages, accommodate its meaning to them, but must rather examine those parallels to ask what, on their evidence, we can conclude about the original form of the text. Wick (*TLL* x 2.2382.60–2) remains sceptical of *pruina* here, and it is better to emend than impose a fantastical meaning on the word. The rare *pituita* is quite plausibly corrupted to *pru(i)na*, especially if the scribe had in mind the appearance of *pruina* at 3.6.4.

dulcia ex amaris: The neuter *dulcia* is used for Pliny's *dulces* (*HN* 17.252, quoted previous n.) because G. picks out the nuts (*amygdalum*) rather than the tree (*amygdala*).

tribus a radice digitis: The ablative is one of extent (or measure) denoting relative position (thus *a radice*): see Pinkster 802–3, and cf. Caes. *BGall.* 1.43.2 *legionem Caesar quam equis deuexerat passibus ducentis ab eo tumulo constituit* (an example from Pinkster and *NLS* §82.ii). It does not depend on *latior*.

cauerna: The word is used commonly and in a variety of contexts by Pliny but is found nowhere else in the agronomists with the exceptions of Pallad. 2.15.11 (reproducing this passage) and 4.9.4 (of the holes made by moles).

qua pituita decurrat: For *decurro* of liquids (very common), see *TLL* v 1.227.67–288.50 (Rubenbauer). *qua* picks up *cauerna* and designates the route or path: see *OLD* s.v. *qua* 3, Pinkster 829–33.

post media trunci parte terebrata cuneum qui esset ex arboris quidem radicibus simplici delibutum melle fixerunt: For the substance of this magical advice, see the second category of remedies discussed

[101] Cf. *Geop.* 10.59.1 γλυκείας δὲ ἐκ πικρῶν ποίησεις, εἰ ἐκτρυπήσας τὸ πρέμνον ἐπὶ παλαιστὴν ἐν τετραγώνῳ, τὸ δάκρυον ἐκδέξῃ κατ' ἐνιαυτόν, ἕως ἂν γλυκανθῇ. This passage too may depend proximately or remotely on *Pom.*

at 3.6.2–3n.; for the attempt to affect the sweetness of the produce with honey, compare also the use of *aqua mulsa* at 3.1.1n.

cuneum qui esset ex arboris quidem radicibus: Cf. 3.6.2 *cuneum qui esset ex robore.* The position of *quidem* is unusual. Provided that the text transmitted by **N** is correct (Reeve's *eiusdem* is attractive), the easiest way to understand the clause is to attach the word closely to the preceding *arboris* and suppose that it has an emphasizing or focalizing function (cf. H–Sz §259, *OLD* s.v. 1a, Adams (1994) 3–5, Spevak (2010) 20).[102] G. would then be stressing that the wedge that is inserted into the trunk must come from the roots of the same (concretely imagined) tree that is the subject of discussion in the main clause (sc. and not simply of any tree, even of the same species).

delibutum: The adjective *delibutus* is formed as though from *delibuo* (<*libo*); that verb does not appear in any finite form, however, until Tertullian (*Coron.* 12): see *TLL* v 1.442.35–7 (Lommatzsch).

3.7.3 Nec minus utile est si effundatur stercus suillum, humano lotio resolutum et radicibus superfusum: The tradition of applying swine manure and urine to the roots of the almond, pomegranate and other trees in order to sweeten their produce was old, and variations of the technique are repeatedly mentioned by Theophrastus as a good example of the way that cultivation can change the nature (φύσις) of a plant for the better: cf. *Hist. pl.* 2.2.11, *Caus. pl.* 2.14.2, 3.9.3, 3.17.5–6. Cato (*Agr.* 7.3), Columella (5.10.15 = *Arb.* 23.1) and Pliny (*HN* 17.259) all mention the procedure for the pomegranate, but G. is the first after Theophrastus to mention it in connection with the almond; it is subsequently picked up by Palladius (2.15.11) and the *Geoponica* (3.3.4, 10.59.2).

[102] The references given here should be supplemented by the results of several important investigations of *quidem* beginning from the monograph of Solodow (1978). Kroon (2004), (2005), (2009) 151–6, (2011) 185–6 argues that *quidem* is embedded in autonomous units and plays a discourse-structuring role by signalling that a unit is 'informationally not complete and needs, for its proper interpretation, another piece of information with which it forms a conceptual whole' ((2011) 185). This pragmatic view largely agrees with the interpretation of *arboris quidem* offered here. Danckaert (2014), (2015) both builds on and revises Kroon's thesis: in (2014), he explores its syntactic and semantic properties and provides further evidence that it is a focus particle (but he argues that it does not focalize a constituent or word, as suggested here), while (2015) is concerned with the pragmatics of *quidem* and argues (*pace* Kroon) that *quidem* does not so much structure discourse as ground it by indicating the 'information status of a particular message' (137).

In general, the recommendation to expose the roots and apply potent manures or special fertilizers as a way of affecting the produce or remedying barrenness is common in all of the agronomists from Theophrastus on, but becomes especially so in the later authors (Palladius, *Geoponica*). Particularly favoured for this treatment was *amurca* (cf. 1.1.10n. *cum amurca*): cf. Cato, *Agr.* 93, Col. 2.14.3, 5.9.14, 5.10.10 = *Arb.* 21.2, Plin. *HN* 17.256, 17.259, Pallad. 3.25.23, 4.10.20, 4.10.30, 11.8.3, 12.7.14, *Geop.* 3.15.6, 9.10.1, 9.10.8, 10.48.4. Also recommended are tanner's refuse, i.e. water from the soaks and some solid matter (e.g. Theophr. *Caus. pl.* 3.9.3, Plin. *HN* 17.258), urine (aged or otherwise, e.g. Theophr. *Caus. pl.* 3.9.3, Cato, *Agr.* 7.3, Col. 5.10.15 = *Arb.* 23.1, Plin. 17.259, Pallad. 3.8, 4.7.4, 4.10.3, 4.10.20, 12.10, *Geop.* 5.35.1, 10.34, 10.84.2), old wine or wine lees (e.g. Plin. 17.259, Pallad. 3.25.5, 3.25.30, *Geop.* 3.13.3, 10.23.6), ash (e.g. Plin. 17.261, Pallad. 3.25.4, 4.10.4, *Geop.* 3.13.3), salt (e.g. Theophr. *Caus. pl.* 3.17.2–4, Plin. 17.261), various kinds of ordure (human, swine, etc.) and compounds formed from these and many other ingredients (lizards, beans, etc.). Although in general the point of these remedies was to offer additional nourishment to the tree, some of the fertilizers had unique effects in relation to plants. Theophrastus speaks of the strongest manures as possessing a kind of 'bite' or 'stimulation' (*Caus. pl.* 3.17.5 ἀνάδηξις) that can a change a tree's produce by altering the nature of its roots (3.17.6 τῷ τὰς ῥίζας ἀλλοιοῦσθαι).[103] This shock brings a barren tree to produce fruit or makes one that is bitter turn sweet by making the roots more 'attractive' (3.17.5 ἑλκτικωτέρας) of nutrients.

si effundatur stercus suillum, humano lotio resolutum et radicibus superfusum: If *superfusum* is not to be redundant with *effundatur*, a pause is required after *suillum*: the participial phrases *humano lotio resolutum* and *radicibus superfusum* (note the accentual parallelism, *cursus uelox*) are thus specifying, telling us how to prepare and where to put the manure. The collocation *radicibus superfusum* has a Columellan precedent: 4.8.3 (sc. *urinam*) *radicibus superfundes*.

stercus suillum: Swine manure was considered potent, and polarized the agronomists. Theophrastus (*Hist. pl.* 2.7.4, reporting Chartodras) assigns it a high value, but Columella considers it the worst (2.14.4 *deterrimum ex omnibus suillum habetur*): see further Appendix II. Beal (1894) 10 remarks that pig manure is 'generally rich although containing a high percentage of water'.

humano lotio: *lotium* is a 'rather vulgar word' (Adams (1992*b*) 503). Adams (1982) 247 summarizes the distribution of *lotium* against *urina*, the polite

[103] Pliny paraphrases the idea, translating Theophrastus' ἀνάδηξις with *morsus*: *HN* 17.261 *quaedam* (sc. *arbores*) *poena emendantur aut morsu excitantur.*

equivalent, thus: '[i]n Cato *lotium* is the standard word, but by the end of the Republic and the early Empire (Varro, Cicero, and most notably Celsus) *urina* was clearly the polite term'. Although *lotium* is absent from Varro, and only scantily represented against *urina* in Columella and Palladius (Adams (1982) 247), it may not have had a vulgar or offensive tone in agricultural Latin. On *lotium* and *urina*, see further Adams (1982) 246–9. Although neither *urina* nor *lotium* appears again in *Pom.*, *urina* is the regular term in *Med.* (*CG* s.v.).

3.7.4 fraglanti soli: There was some variation in the spellings *flagr-* and *fragl-* in antiquity (*TLL* vi 1.846.26–30 (Bacherler), Paratore (1948) 26–9, Zimmerman (2012) xxxii), for which the grammarians claim a learned distinction: cf. Non. 438.16–18 *flagrare et <fraglare> ita discernuntur, quod <flagrare> <est> ignescere, incendi et ardere, fraglare uero olere. et in plurimis uenitur ista discretio*; Serv. *ad* Verg. *Aen.* 1.436 *quotiens incendium significatur, quod flatu alitur, per l dicimus, quotiens odor, qui fracta specie maior est, per r dicimus.* Were G. to observe this distinction, we should expect here *flagranti*, not *fraglanti*; but we need not suppose that the situation was so clear-cut as Nonius and Servius make it out to be (note anyway Nonius' *in plurimis*): see the following n. For Scotti's conjecture *flagrantiae solis* (a rhythmically attractive cretic-trochee *flagrāntĭae sōlis*), cf. Apul. *Met.* 4.31 *ille serico tegmine flagrantiae solis obsistit inimici.*

quum: Cf. Quint. *Inst.* 1.7.5 *illa quoque seruata est a multis differentia, ut . . . cum, si tempus significaret, per qu, si comitem, per c ac duas sequentis* (sc. *litteras*) *scriberetur.* The claim is repeated by many of the grammarians (*TLL* iv 1339.65–81 (Hey)), and Velius Longus and some others say that the distinction was preserved by the *antiqui* in particular: cf. *De orth.* Keil 7.70.15–17 *q quoque littera facit differentiam uocum ab antiquis maxime obseruatam. <nam cum> quotiens pro aduerbio temporis scribebant, q littera utebantur; <quotiens> pro praepositione, c ponebant*; see also Mar. Victorin. 4.30–4 with Mariotti. Mazzini 59 curiously observes that the juxtaposition of the (supposedly learned) *quum* with the (supposedly Vulgar) ἅπαξ *amaritosus* and substitution of the accusative for ablative (*in medullas . . . consederit*) confirms the 'mal riuscite ambizioni di Gargilio'. The judgement is unnecessarily harsh: neither *amaritosus* nor the use of the accusative after *consedo* + *in* is necessarily Vulgar (see respectively Introduction §v(d) and 3.6.2n. *in locum taedae*) and *quum* is not necessarily a learned archaism. As concerns *quum*, it is of course true that the semantic distinctions that the grammarians attached to orthographic differences need not have had any basis in reality in order to influence the style of their contemporaries; and again, it is true that, if a particular spelling was held to be archaic, or to preserve a learned distinction, we may be able to infer something about an author's literary intentions on the basis of his usage of it, regardless of its historical authenticity. But the fact remains that the situation in late antiquity was probably a great deal more complicated than the grammarians allow, and, when we further consider that the variation in forms is very common in manuscripts and may as easily be attributed to scribal normalization (in either direction) as to what the author wrote, we should not attach much importance to the isolated occurrence of the phenomenon: G. (or his copyist) hardly makes a habit of writing the conjunction *cum* with *qu*, doing so only here.

dulcior sucus hauriatur: *dulcior* specifies that portion of the sap which is responsible for the sweetness of the fruit (the standard for the comparison is implicit; cf. 1.1.15n. *Diligentioribus*), and hence, once consumed (*hauriatur*), leaves it bitter. For *haurio* in the sense of 'consume' or 'devour', often applied to fire, see *OLD* 7a, *TLL* VI 3.2570.83–2571.41 (Brink).

3.7.5 Illico [praesto] uidetur ex morsu pecorum omnia haec arbori amaritosa contingere: A difficult sentence, even though the general meaning is obvious. Winterbottom 143 n. 1 wonders whether *illico praesto* (*pres-* **N**) hides a proper name; a more likely assumption (Reeve) is that one word is a gloss on the other. Deleting *praesto* is on balance preferable, for the word is not found elsewhere in G. but *il(l)ico* appears twice more in the requisite sense (3.8.2, *Med.* 3.12). If the rest of the text is left undisturbed, *uidetur* must be impersonal (governing accusative + infinitive) and *omnia haec . . . amaritosa* (the subject of *contingere*) taken substantively as an involved kind of *uariatio* for *amaritudinem* of the previous sentence, to which *haec* now points back. For the general structure of the sentence, cf. 4.6.4 *Iulius Atticus opacis et septentrionem spectantibus cliuis <u>laetissimum fieri castaneae pollicetur</u>*. Note in both sentences the accusative + infinitive (*uidetur . . . contingere / fieri . . . pollicetur*) and the use of a neuter substantive adjective (*omnia haec . . . amaritosa / laetissimum*) paired with a verb meaning 'happen' or 'occur' (*contingere / fieri*) to describe an effect on the plant, which is put into the dative as the patient (*arbori / castaneae*). For *omnia haec . . . amaritosa*, cf. also *Med.* 46.1 *pruni recens pomum uentrem resoluit, stomacho contrarium iudicatur sed <u>omne hoc incommodum</u> celeri digestione finitur.* It is unnecessary to alter the text further here, although a number of changes may be considered as producing an easier reading: perhaps *uidentur* for *uidetur*; *huic* for *haec* (cf. 2.4.5 *huic arbori*); and *calamitosa* ('destructive') for the ἅπαξ *amaritosa* (for which see Introduction §v(d)). Germane to the last suggestion (Reeve's) is perhaps the 'old rustic use' of *calamitas* 'specifically denoting disaster befalling crops, whether from disease or the weather' (Adams (2016) 76), mentioned by Serv. *ad* Verg. *Aen.* 1.151 and evidenced in Cato (cf. e.g. *Agr.* 1.2, 35.1, 141.2).

uidetur . . . cum . . . consederit: For the tense of *consederit*, see 1.1.7n. *ubi tempus erit . . . pollicentur.*

ex morsu pecorum: For the danger thought to attend the animal bite, see 3.5.2n.

3.7.6 uiridibus auferas: *uiridibus* refers to unripe nuts (*OLD* s.v. 3a). For *aufero* of thinning or removing plant parts, cf. Pallad. 3.24.4 (of onions) *si capita uoluerimus his esse maiora, folia omnia debemus auferre.*

tenuiora: Like 3.7.4 *dulcior*, *tenuiora* (sc. *amygdala*) assumes an implicit standard in context, specifying in particular those unripe nuts that are the 'weaker [of the lot]'.

quaeque liqueris: Although *linquo* is by no means unknown to prose, the ordinary word is *relinquo* (cf. *TLL* VII 2.1460.36–9 (Balzert)) and it would not have been difficult for *re-* to fall from the text; perhaps it is even concealed in *-que* (Reeve). It should give us pause, however, that the verb *linquo* appears at *Med.* 16.1 (where it is a secure conjecture of Rose for the transmitted *liquitur*): *quibus anima linquitur odore succurrit* (sc. *cucumis*). See also on this usage *TLL* VII 2.1461.50–4 (Balzert). Maire (2002*a*) xxxvi n. 139 discusses *linquitur* in the context of a number of poeticisms in *Med.*, pointing out that G. in adapting Plin. *HN* 20.12 *defectum animi* 'choisit une expression poétique et n'utilise pas comme d'ordinaire un verbe composé'. It is true that *linquo* in the present instance is less poetical than at *Med.* 16.1, where it describes the soul or life temporarily 'departing' the body, but there is no doubt at least that the word is in G.'s vocabulary.

3.8 The truncated eighth section covers the harvesting (see Plates 5, 6) and preservation of almonds. G. reports that they begin to grow ripe in early April and adds that they give a sure sign of maturity when they slough their hull (3.4.3n.). Celsus is brought in (3.8.1 *ut putat*) as an additional, authoritative witness to the time of their ripening. If the hull (3.8.2 *cortex*) is not easily removed, then further dehydration, achieved by sinking the nuts in moisture-absorbing chaff, will allow it to be detached.

The missing portion of this section (on preserving) may be cautiously supplemented from Palladius, who sticks closely to G. here as elsewhere in *amygd.*: 2.15.12 *amygdala ad legendum maturitatem fatentur, cum fuerint spoliata corticibus. haec sine cura hominis seruantur in longum. si difficulter corium dimittent, paleis obruta continuo relaxabunt. item decoriata si aqua marina lauemus aut salsa, et candida fiunt et plurimum durant* (cf. also *Geop.* 10.58). Palladius' claim that the almond is easy to preserve (*haec sine cura hominis seruantur in longum*), not found in *Pom.*, is true: as for other nuts (see Plate 13) there is no need for the elaborate techniques of drying, pickling and conserving that the Romans recommended for grapes, olives and fleshy fruits (see 1.1n., 2.12n.). In fact, the simple salting that G. mentions here is the first advice specific to preserving the almond found among the agronomists.

3.8.1 A Kal. Aprilis, ut putat Celsus, incipiunt esse matura: The timeframe claimed by Celsus is peculiar. Almonds grown in California (and similar climates, e.g. India) are not fully developed until the beginning of May, and not ready for picking for another two months after that (*AOM* 48, *EFN* 712), and it is unlikely that the timeline would have been much different in Italy; so Palladius (8.3.2) tellingly says *nunc* (sc. *mense Iulio*) *locis temperatis amygdala matura sunt ad legendum*.

incipiunt esse matura: For expressions denoting ripeness, see 3.4.3n. *maturant*.

si asseruanda deligi placeat: The thought contained in the conditional is otiose, strictly speaking: *A Kal. Aprilis . . . incipiunt esse matura* and the following *signum maturitatis ostendunt cum* show that the issue at hand is the proper time for harvesting almonds, not the question whether they are to be gathered in order to be stored up or preserved at all. The time of the almond's maturation has no bearing on this second question, and the matter of preserving them is otherwise explicitly deferred until after the almonds have already been picked and freed from their hulls, hence 3.8.3 *Tum corio liberatae* ('only at that time'). The *si*-clause does not in fact introduce a new thought so much as effect a smooth transition between the more general rule that precedes and the specific guidelines that follow. This fact explains the mood of *placeat* (against the indicative main verb *ostendunt*), for the subjunctive is regular when a *si*-clause is parenthetical, i.e. when the 'the statement of the main clause is true without regard to the condition' (Nutting (1923) 159). For parenthetical conditionals, see further Nutting (1923) 159–64, (1925) 91–3.

asseruanda: For *asseruo* of preserving or storing up, an application especially favoured by Pliny, see *TLL* II 872.82–873.32 (Oertel), *OLD* s.v. 4. The verb is used also in this sense at 2.12.1.

cum corticem remiserunt: For *remitto* of the loosening of a plant's bark, cf. Col. 11.2.37 *possunt . . . si iam librum remittunt, inseri oleae* (also 11.2.41).

3.8.2–3 cortex . . . corio: The substitution of *corium* for *cortex* is motivated by *uariatio*. *corium* is less common in the sense of a rind or shell of fruit (before G. only in Scribonius Largus and Pliny; see *TLL* IV 953.84–954.26 (Hoppe)), for which *cortex* is the preferred word. Although both words can be used to refer to animals, plants and a variety of artificial things, *corium* is more frequently used of the skins of animals or humans, *cortex* of vegetative coverings: note thus 15 occurrences of *cortex* in *Med.* against only one of *corium* (*CG* s.vv., *corium* at 41.6 referring to animal skins). On *cortex* see also De Meo (1986) 39–40.

3.8.2 paleis: For the use of chaff as a desiccant, see 1.1.9n.

3.8.3 corio liberatae: This metaphor seems to have been favoured by G. – at least, that is what we might infer from Palladius (2.15.19), following G.'s (now lost) chapter on the walnut: *Martialis expertum se ait uirides nuces tantum liberatas putaminibus suis melle demergi*, etc.

si quis aqua marina*:** For the likely content of the missing portion of the sentence, see 3.8n. Palladius (the *Geoponica* follows) says that the seawater is supposed to bleach the nuts (*candida fiunt*) and cause them to keep for longer (*plurimum durant*). Bleaching may have been desirable from the point of view of the market (cf. Wickson (1891) 515), but it is not clear how a simple salt bath could have this effect (see the following n.).

aqua marina: Seawater had a variety of purposes, especially in the seasoning or preservation of wine or must (e.g. Cato, *Agr.* 112, Col. 12.22, 12.25). Columella (12.25) explains how to boil it down for use in preserving wine, or, if the farm should not be located near the sea, how to make a brine that could substitute for it. (It is different from ordinary brine, for which see 2.12.1n. *muria.*) Seawater was also heated and used as a dip before desiccation to remove the waxy bloom on fruits that prevents moisture loss (Col. 12.46.5 on the pomegranate; cf. Pallad. 4.10.7), although it is not the salt that is effective in this process but the high temperature of the water: see Thurmond (2006) 174, 185.

<IV DE CASTANEIS>: Fr. 12 Condorelli. Ch. IV Mai. Ch. III Scotti.

4.1–7 In his typical mixed style (Introduction §III(a)), G. discusses in considerable detail the facts and methods relevant to planting the chestnut (Plate 15). Subjects treated include planting time (4.1–3), preservation and selection of the seeds (4.2–4), alternative methods of propagation (4.4), soil and climate (4.5–6), and preparation of the nursery (4.7). How thorough G. is may be appreciated from the observation that, by the time that **N** breaks off in 4.7, we have not yet finished learning how to plant the seeds.

There are two considerations germane to the chestnut in particular that divert G. and prolong his initial remarks. The first is the difficulty of preserving the nuts until planting time, since they are prone to rot (4.2.3). The problem leads him to weigh some methods for storing the nuts and testing them for soundness (4.2–4). The second is the fact that the chestnut tree was grown not only for its fruit but also as a source of wood to be used in vine props: Pliny says that it is preferred to all other trees for this purpose (*HN* 17.147 *castanea pedamentis omnibus praefertur*), while Columella (4.33.1) says that only the oak is better (cf. Jashemski (2018*a*) 127). G. acknowledges the use in passing at 4.1.5 (*sationem caeduam*) and addresses it more directly at 4.4.4, where he says that some people distinguish propagation from slips (*plantae*) and nuts as suitable for fruit and timber, respectively. It cannot be known whether G. made more of growing the tree for timber elsewhere in *cast.*,[104] but it should be noted

[104] This is less probable at least on the evidence of Columella, who distinguishes between the uses of the tree so far as to discuss it in separate places from the perspectives of growing for timber (book 4) and for fruit (book 5).

that references to growing the tree for timber are strictly subordinate to the subject of growing the tree for fruit. This is shown clearly by 4.1.5 *quamquam ipse sationem caeduam assentiatur*, where G. reports this planting time in spite of (cf. *quamquam*) the fact that Pliny has a crop for felling in mind. At 4.4.4 too, the salient lesson (if it is to be trusted; cf. *si tamen*) is that slips are better suited for growing fruit, something the cultivator ought to take into consideration.

4.1–3 In 4.1 G. reports planting times from no fewer than eight authorities. They range widely from early November (*uergiliarum . . . occasus*) to the middle of March (*Kal. Martias in Idus easdem*). Though G. does not clearly express his opinion as to which if any of them is the most suitable (and the passage is marred by several textual difficulties that complicate the question), according to G.'s technique of 'intensive' didactic layering we should take note of the number of authorities who agree on a winter planting: see the discussion in the Introduction §III(a). The situation is further complicated in 4.2 and 4.3, where G. introduces competing factions of *Graeci* who espouse plantings in autumn (4.2.1 *ab Arcturi* [*et*] *exortu*, early September) and spring (4.3.1 *ex uerna satione*). Palladius drastically condenses G.'s wide-ranging report on the chestnuts' planting time into a single concise statement: 12.7.17 *serenda est ergo ipsis castaneis, hoc est seminibus suis, mense Nouembri et Decembri, item Februario.*

4.1 There is a discernible logic governing the presentation of planting times. Mago and Celsus both give the chestnut the same planting time as the almond. Columella offers two planting times, winter and spring. Spring is dropped for now (to be discussed further in 4.3), and the thread is picked up by Pliny, who also (4.1.5 *etiam*) recommends winter. So too Iulius Atticus and Curtius Iulius, the latter of whom distinguishes better (4.1.7n. *primam*) and worse (*secundam*) plantings. Iulius Fronticus gives no preferred planting time, and the Quintilii offer early November (the commencement of winter).

4.1.1 ***corripiat: If *corripiat* can mean 'pick' or 'gather' – this is by no means certain, for it would be a very forcible word and is not otherwise found in this application – , then it might answer to a protasis that gives some planting condition, so that the shape of the sentence would be: '[sc. if one were to plant chestnuts at such and such a time,] then one would gather fruit of the sort, etc.'. With the quotation from Virgil's *Eclogues* (4.1.1n. *mea quas Amaryllis amabat*), this crop of chestnuts would be idealized as especially desirable through a comparison with those that Amaryllis cherishes. It is uncertain in general to whom the advice is attributed. Later in this paragraph I print *amygdalorum exemplo* rather than *amygdalo eorum exemplo* of **N**, but if the paradosis is retained, the authority in the lacuna would have been in agreement with the advice of Mago and Celsus.

pastor: See Clausen for the *pastor Corydon* who speaks in *Eclogue* 2.

praecinit: Of *praecino* Breimeier notes 'modo minus definito vis verbi simplicis augetur' (*TLL* x 2.439.22–3), citing this among other instances.

mea quas Amaryllis amabat: Verg. *Ecl.* 2.51–2 *ipse ego cana legam tenera lanugine mala / castaneasque nuces, mea quas Amaryllis amabat.*

4.1.2–3 Mago breuiter, ut Poenus et cui peregrinae eiusmodi arboris minus nota cultura sit . . . Cornelium tamen Celsum, Italicae disciplinae peritissimum, decuit aliquatenus edocere quidquid Magonis ignorantiam fugerat: G.'s statements here are important in that they show him assuming an Italian perspective for his instructions: see Introduction §i(a). The claim that the chestnut is foreign to the Carthaginian Mago is accurate, for the tree is native to Europe, Transcaucasia and Asia Minor: see *RE* (1) 20.2338.62–3 (Stadler), André on Plin. *HN* 15.92, Bounous and Marinoni (2005) 298–9.[105] For Italy in particular, the use of chestnut was indeed ancient: Stadler (*RE* (1) 20.2338.63–6) mentions its employment in Bronze Age dwellings (though cultivation for this purpose would come later, and Meiggs (1982) 240–1 notes its absence in Vitruvius' section on timber).

4.1.2 Mago breuiter, ut Poenus et cui peregrinae eiusmodi arboris minus nota cultura sit: For the combination of an adjective and relative clause with a generic subjunctive, see Oakley on Liv. 6.34.11. Scotti's emendation of *sit* to *ait* is unnecessary: the verb introducing Mago's statement can be inferred from *breuiter*. Compare a similar omission at *Med.* 47.1–2: *mespili arborem Dioscorides contentus agrestem nominasse, hoc tantum asseuerans quod poma eius et folia uirtute non discrepent. sed et ceteri diligentius qui et uocabulum eius aliud addiderunt.*

castaneam in umido sicut Graecam nucem serito: *sicut Graecam nucem* refers to the likeness between almond and chestnut in planting time, as context requires. The detail *in umido*, on the other hand, is to be understood with the chestnut only; it does not apply to the almond, which the agronomists hold to prefer hot and dry soil and to fail in moist conditions (see 3.1.2n. *terram . . . minutissimam*). It is noteworthy that the sole imperative in *Pom.* occurs in this direct quotation of Mago; G. prefers other means of communicating his recommendations (for a synopsis, see Introduction §v(b)).

castaneam: *castanea*, from Gk. κάστανον, -άνεια (Beekes 1 655), can refer to either tree or nut. Isidore proffers a curious etymology: *Orig.* 17.7.25 *castaneam Latini a Graeco appellant uocabulo. hanc enim* καστανίαν *Graeci uocant, propter quod fructus eius gemini in modum testiculorum intra folliculum reconditi sunt, qui dum eiciuntur quasi castrantur.*

[105] Theophrastus (*Hist. pl.* 4.5.4) remarks on the density of chestnut trees in Asia Minor, at Mount Tmolus (present-day Bozdağ) overlooking Sardis and at Mysian Olympus (present-day Uludağ, in Bursa): see Amigues *ad loc.*

Graecam nucem: For this older name for the almond, see 2.10.6n. *amygdalum.*

4.1.3 Cornelium tamen Celsum: Only here in *Pom.* is Celsus given his *nomen*; this fact when taken in conjunction with *tamen* splitting the name is interesting. Although it is in general regular to split *praenomen* and *nomen* by sentence connectives, and no additional emphasis is necessarily thrown thereby onto the *praenomen* or *nomen*, it is much less common to do so with *nomen* and *cognomen*, as here: see Devine and Stephens (2006) 272–5. We may thus suspect that the fronted *Cornelius*, an old and illustrious Roman *gens* (*RE* (1) 7.1249.12–62 (Münzer)), followed by the adversative *tamen* is pointed, for G. goes on to state that precisely because he was *Italicae disciplinae peritissimus* Celsus ought to have taught what escaped Mago. The name *Cornelius* reminds us of that.

Italicae disciplinae peritissimum: For *Italica disciplina* in the sense of 'practice adapted to Italy' or 'to the Italian climate', cf. *Italica sententia* at Plin. *HN* 17.93 *uentosis locis crebriores* (sc. *arbores*) *seri conducit, oleam tamen maximo interuallo, de qua Catonis Italica sententia est in XXV pedibus, plurimum XXX seri. sed hoc uariatur locorum natura.* (Pliny goes on to contrast Mago's advice in the same way that G. does.)

quidquid Magonis ignorantiam fugerat: For the expression, cf. Phaedr. 4.5.48–9 *ita quod multorum fugit imprudentiam, / unius hominis repperit sollertia.* The formulation 'escape/flee one's ignorance' is strictly illogical, but has the same meaning as the more common expressions 'escape one' or 'escape one's awareness, knowledge, sim.': see for examples *TLL* VI 1.1492.83–1493.11 (Rubenbauer).

sed et ipse castaneas amygdalorum exemplo reiectat: In the present instance *reiecto*, which means to 'cast back' or 'throw back', vel sim. (L&S, *OLD*), must be taken like *reicio* (*OLD* s.v. 10), 'to refer *x* to *y*', and construed with the dative (Scotti thought of *remitto*.) If the transmitted text *amygdalo eorum exemplo* is sound, then it is the 'almond' to which the chestnuts are referred and it is on 'their example' that it is done. But who would 'they' be? It is at least possible that the lacuna contained another authority who recommended the almond's planting, thus giving a sense to *eorum*, but it seems unlikely: G. is contrasting sharply the instructions of Mago and Celsus, and it is Mago alone whom he has in mind in the preceding sentence (cf. *quidquid Magonis ignorantiam*). Marx's emendation *amygdalorum exemplo* ('to the example of the almond') is desirable. For *exemplum* as an 'example (for imitation)', see *OLD* s.v. 6.

4.1.4 Columella, cui curae fuit diligentius et de castanea conscribere, etiam ipse indifferenter et uarie sententiam promit; fluxe enim ista scribit: affirmat a Nouembri mense per totam hiemem seri posse, nunciat Kal. Martias in Idus easdem ne quid

intersit, et plantat ut <uix> quando quaeque sine culpa manus ueniat scire possimus: G. is basing himself on two places in Columella: 4.33.2 *seritur ab Nouembri mense per totam hiemem*;[106] 5.10.14 = *Arb.* 22.3 *hoc ante Kalendas Martias facito, uel etiam inter Nonas et Idus Martias. eodem* [*Arb. hoc*] *tempore iuglandem et pineam et castaneam serere oportet.* In the former spot Columella is discussing chestnut plantations meant for vine props, in the latter for fruit; one might have thought that this distinction would be sufficient to avoid G.'s allegation of confusion, but it is true at least that Columella does not explicitly state why he offers different planting times for trees of different purposes.

From the vantage point of the whole series of statements we may say something about the necessity of supplying *uix* (Reeve) or *non* (Scotti) in the final clause. As transmitted, *et plantat ut quando quaeque sine culpa manus ueniat scire possimus* means (depending on whether we supply *satio*, *castanea* or something else with *quae . . . ueniat*): 'and his plantings are such that we are capable of knowing when any [chestnut/sowing] will grow without manual error'. (For the translation of *plantat*, see 4.1.4n. *et plantat ut.*) This reading takes the indirect question *quando quaeque sine culpa manus ueniat* to be asking which of the planting times recommended by Columella is the most suitable for the chestnut (i.e. which is *sine culpa manus*; see also 4.1.4n. *ut . . . possimus*). But the difficulty with the text as it stands is that Columella seems to express his opinion on planting times (*plantat*) in just such a way that we *cannot* in fact know which is suitable, because he offers several possible times for planting but makes no distinction among them: that was the point of G.'s insistent criticism that Columella writes *differenter*, *uarie* and *fluxe* about the chestnut's planting time. We expect therefore that *et plantat*, etc., will offer a negative or at best aporetic conclusion. To fix the text *uix* is perhaps better than *non*, for the former if placed after *ut* offers a more evident *ratio corruptelae* for the omission.

cui curae fuit: Scotti's *fuit* is superior to the transmitted *fit*; not simply because the more usual dative construction is with *sum*, for there are some examples of *fio* + dative that can be adduced (*TLL* VI 1.113.36–49 (Hey)), but because *fit* does not have a sense in this context. In the closest examples that can be quoted (e.g. Cic. *Off.* 1.139 *ampla domus dedecori . . . domino fit*,[107] Chiron

[106] With this planting time compare the *Geoponica*'s general advice for trees: 10.2.1–2 ὁ μετοπωρινὸς καιρὸς εἰς φυτείαν ἐπιτηδειότερός ἐστι, καὶ μάλιστα ἐν τοῖς ἀνύδροις τόποις, δι' ὅλου γὰρ τοῦ χειμῶνος ἀρδεύεται τὰ φυτευόμενα. χρὴ οὖν ἅμα τῷ ὄμβρους γενέσθαι, εὐθὺς μετὰ τὴν δύσιν τῶν πλειάδων φυτεύειν, ἄχρι τῆς χειμερινῆς τροπῆς, τουτέστιν ἀπὸ ζ' ἢ ια' τοῦ Νοεμβρίου μηνός, ἕως κ' τοῦ Δεκεμβρίου μηνός.

[107] Note, however, that neither Atzert in his Teubner nor Winterbottom in the OCT allows *fit* into the text, accepting rather *est* of the tenth-century Harley manuscript **L**.

650 *ne deformitas curae fiat*), *fio* retains its meaning, whereas here it would be pointless.

diligentius: See 1.1.15n. *Diligentioribus*.

conscribere . . . scribit: Iteration of a compound verb by simplex: cf. 2.1.2 *disponi . . . ponenda* (n.), 3.3.5 *describit . . . scribit*.

etiam ipse: Not assimilating Columella to a group (as *etiam ipse* often does; cf. *OLD* s.v. *ipse* 6) who wrote *indifferenter et uarie*, but simply contrastive (cf. 2.1.3n. *poma ipsa matura integra*) after the statement *cui curae fuit diligentius . . . scribere*.

indifferenter et uarie sententiam promit; fluxe enim ista scribit: For *promo* of setting forth an idea, see *OLD* s.v. 6a. *indifferenter* and *uarie* describe the variety of Columella's writing; the addition of *fluxe*, which means 'loosely' or 'carelessly' (*TLL* vi 1.984.73–81 (Bacherler)) and is opposed here to *diligentius*, has explanatory value (*enim*) because it locates the variation in Columella's own inattention. For *indifferenter*, cf. also *Med.* 2.3 (of parsley) *unde mirari necesse est quod a quibusdam imperitis indifferenter adiuncta* (sc. *res*) *aquae calidae febricitantibus detur*, etc. (Maire 'équivalent de *neglegenter*, "négligemment", et de *temere*, "à la légère"').

nunciat Kal. Martias in Idus easdem ne quid intersit: With *nunciat* supply *seri posse*. For the form *Kal. Martias* (bare accusative of duration, but simply indicating 'when'), see *TLL* vii 2.757.74–7 (Buchwald). The preposition *in*, attaching *Idus easdem* ('until, up to', *OLD* s.v. 13b), subsumes the whole period of time between the Kalends and Ides of March. *ne* stands for *ut non* (as also at *Med.* 17.12, 30.36, 42.15; see Svennung (1935) 513, H–Sz §347.γ) and should probably be taken closely with *nunciat* to indicate that Columella presents the March dates in particular in such a way that he gives no criterion (*quid intersit*) for distinguishing among them: thus Col. 5.10.14 = *Arb.* 22.3 *hoc ante Kalendas Martias facito, uel etiam inter Nonas et Idus Martias*. It is also possible that *ne quid intersit* has a wider scope, however, and means that there is rather no distinction between the November and March ranges. Curiously, G.'s *Kal. Martias in Idus easdem* looks more like Pliny than Columella: *HN* 17.136 *castaneae a Kal. Martiis <ad> Idus easdem*.

et plantat ut: The translation 'and his plantings are such that' (*et* perhaps connoting *et ideo* or *itaque*, for which see *TLL* v 2.892.70–6 (Hofmann)) takes *plantat* to be an example of the figure of speech whereby someone is said to do what in fact they write about doing: see concisely Coleman on Stat. *Silu.* 4.2.2, and more fully Lieberg (1982). (The plural 'his plantings' follows the present tense of the verb and refers back to the two planting times Columella mentions.) The construction is ubiquitous in *Pom.*, but in most cases is so natural as not to cause difficulty.

ut <uix> quando quaeque sine culpa manus ueniat scire possimus: A reading with *quandoque quae* is not impossible ('hardly ever know which, etc.'), but Antonelli's suggestion (following Scotti) improves the sense. The referent of *quaeque* is uncertain, but the context requires that we take the word to refer to something that will distinguish among the different planting dates offered by Columella: supplying *satio* or *castanea* would work equally well, although *castanea* is easier in context. *uenio* means simply 'grow' (as commonly: see *OLD* s.v. 5, and cf. 1.1.2n. *melius proueniunt*), and takes its point from the predicative *sine culpa manus*. *OLD* s.v. 3b gives the precise meaning of *culpa* here: 'failure to take the right measures, neglect, or faulty judgement; an error, mistake'. This application in an agricultural context is well illustrated by Col. 4.24.7 *transuersae* (sc. *uites*) *plus umoris et recipiunt et continent. eam culpam maxime uinitor fugito*. The idea of *manus* here is that of the hand as the instrument of skilful or productive intervention in nature, extremely common in reference to agricultural work: see the examples at *TLL* viii 355.61–356.8 (Bulhart).

4.1.5 Etiam Gaius Plinius, quamquam ipse sationem caeduam assentiatur, hiemali tamen primum tribuit: Cf. Plin. *HN* 17.148 (sc. *castanea*) *seri nuce diximus, sed nisi ex maximis non prouenit, nec nisi quinis aceruatim satis – refringi solum debet supra –, ex Nouembri mense in Februarium, quo solutae sponte cadunt ex arbore atque subnascuntur*. G. rightly adverts to the fact that Pliny is speaking of a crop grown for felling here (cf. *HN* 17.147), but he seems to overlook an earlier statement by Pliny that would have given him grounds for levelling the same complaint as he had against Columella: *HN* 17.136 *communis quidem Italiae ratio tempora ad hunc modum distribuit: . . . nucibus iuglandi et pineae et abellanae et Graecae atque castaneae a Kal. Martiis <ad> Idus easdem*. In any case, there are some parts of G.'s sentence that need clarification: (1) the force of the concession *quamquam*; (2) the exact point of *etiam*; (3) the force of *assentiatur*; (4) the claim that Pliny prioritizes (*primum tribuit*) a winter sowing.

(1) Antonelli, col. 1702 is unhappy with *quamquam*: 'Perchê adunque il nostro autore attribuisce ciò soltanto a Plinio *quamquam ipse*? O è errata la scrittura del codice, o è da affermare che egli non pose mente al precetto di Columella'. What Antonelli has in mind is that *quamquam ipse* can be read as implying that it is in contrast to Columella that Pliny is writing about the *satio caedua*. But of course it is not true that Pliny alone writes about the timber planting, for Col. 4.33.2 is explicitly concerned with the *satio caedua*, and Pliny in fact agrees with Columella regarding a winter sowing for it. This prompts Antonelli to wonder whether the text is corrupt, or whether G. has made a mistake with the facts. Antonelli's problem is illusory if *quamquam* is read against the larger discursive context: as suggested at 4.1–7n., the purpose of *quamquam* is effectively to apologize for reporting a statement that pertains not to growing the tree for fruit – which is G.'s subject – but for timber. It is thus not strongly adversative with the preceding sentence (although it is true that *ipse* implies a change of

subject). That does not explain why G. does not also observe that Columella's November planting is a *satio caedua* (see 4.1.4n.), but it eliminates the difficulty that Antonelli observes.

(2) *etiam* merely links Pliny's winter planting time with that of Columella (Col. 4.33.2 *ab Nouembri mense per totam hiemem* / Plin. 17.148 *ex Nouembri mense in Februarium*). It does not bear on the purposes for or methods of growing the tree. (3) *assentiatur* does not mean that Pliny agrees with Columella about a *satio caedua*; that reading would only be coherent if *quamquam* were adversative with the preceding sentence, which it is not. (It is not clear anyway whether *sationem caeduam assentiatur* could mean that.) Antonelli and Mazzini take *assentiatur* as tantamount to 'have in mind' ('miri'), but it should rather be given its original force: the simplified thought is, 'Pliny preferred a winter sowing, but *he would agree* (= *admit*) that this sowing applies to timber trees.' That is, the agreement communicated by *assentior* is not with Columella, but rather carries a wider discursive function. Finally, (4) Pliny does not heap any special praise on a winter sowing (except to imply that other plantings will fail; cf. *nisi*), so the claim that he awarded it 'first place' (*primum*) is probably only rhetorical *uariatio* for the statement that it is what Pliny recommended.

quamquam . . . assentiatur: The semantically unmotivated subjunctive with *quamquam* is met often from the first century CE on (H–Sz §325), but on the interpretation offered in the previous note the subjunctive is potential.

sationem caeduam: Cf. Meiggs (1982) 263: '*Silua caedua* is a technical term for coppicing, which involves the encouraging of multiple stems from the base of a tree and cutting them at regular intervals.' See also *TLL* III 64.1–14 (Hoppe), *OLD* s.v. *caeduus*.

hiemali tamen primum tribuit: *sationi* must be mentally supplied from *sationem caeduam*. For the expression *primum tribuo*, see *OLD* s.v. *tribuo* 5: '(w. neut. adj. or pron.) To place (so much) value (on)'. *primus* here is 'the first' in value, as also at 4.1.7 *sationem . . . <u>primam</u> Decembri, <u>secundam</u> Februario assignat*.

4.1.6 Iulius Atticus . . . serere decreuit: The transmitted verb *indecreuit* appears nowhere else in Latin, and the contribution of *in-* is obscure; Mazzini 57–8 attempts to explain it as a Vulgarism, but I am not confident about introducing a new word into the language and hence emend with Scotti to *decreuit* (with a slight improvement to the rhythm, resolved cretic-trochee *sĕrĕrĕ dēcrēuit*).

4.1.7 Curtius Iustus sationem nucis primam Decembri, secundam Februario assignat: *primam* means the 'best' planting, *secundam* 'second best' (*OLD* s.v. *secundus* 10–11). The structure of the statement takes its cue from but varies 4.1.5 *hiemali . . . primum tribuit*: the present *assignat* is used for perfect *tribuit* and the object *satio* with predicate *primam* instead of the adverbial or internal *primum*.

4.1.8–9 Iulio Frontico, cum et soli de genere et alia ratione culturae nonnulla tractaret, non fuit [culturae alia] norma quando rectissime castaneae sererentur. Quintiliis in satione castaneae uergiliarum non omittendus occasus: Editors have adopted various strategies for emending and punctuating the transmitted text. Condorelli xxix–xxxii, for example, prefers the following reading (emending *Quintiliis* to *Quintilii*): . . . *norma. Quando rectissime castaneas serere rentur Quintilii: in satione castaneae uergiliarum non omittendus occasus*. He explains the *quando*-clause as a kind of 'lemma' (xxxii). That would not, however, be in keeping with G.'s style in *Pom.* and *Med.* and leaves the sentence *Iulio Frontico . . . norma* obscure. The MS reading *serere rentur* should not be defended (it is syllabic dittography; see for other examples 2.4.6n. *contrarium* [*tum*] *crediderunt*).The text can be more easily understood if the *quando*-clause is attached not to the Quintilii, but rather to *norma* of the previous sentence, as a type of *ratio, cur* construction. The *cum*-clause is concessive, establishing that although Iulius Fronticus wrote on some topics concerning cultivation of the chestnut (such as the soil, *soli de genere*), he did not in spite of that provide any guidance on the planting time (*non fuit . . . norma quando*). Winterbottom's deletion of *culturae alia*, whose intrusion can easily be explained by the preceding *alia ratione culturae*, is persuasive. (*culturae alia* can be given a sense, but it only obfuscates G.'s point.)

4.1.8 Iulio Frontico: For possible corruption in the name Fronticus, see Introduction §11.

cum et soli de genere et alia ratione culturae nonnulla tractaret: The MS's *solidegenere* presents a challenge. Scotti, Mai, Antonelli and Mazzini suppose haplography after *solide* and print *solide* <*de*> *genere*. Winterbottom would simply delete *soli*, reading *et de genere*. What, in either case, does *genus* refer to? Antonelli (who does not, however, take the *cum*-clause concessively as I do) translates *solide* <*de*> *genere* impossibly as 'che trattò in generale sulle varie specie di coltivazione, e in particolare su questa ebbe a notar qualche cosa'. Mazzini, also printing Mai's text, takes a different tack: 'sebbene abbia trattato ampiamente della riproduzione ed abbia detto varie cose intorno ad altri modi di coltivazione'. But *genus* does not mean 'riproduzione'. Perhaps a better approach would be to suppose that *genus* refers to the chestnut tree itself: 'although he treated some matters concerning this species (*de genere*) and other aspects of [its] cultivation'. (Presumably on this construal *alia* means 'other [than its planting time]'.) Yet it is very doubtful whether *de genere* should mean that: G. never uses the word so vaguely, and in 13 other occurrences in *Pom.* and *Med.* (*CG* s.v.) *genus* appears only once without an anaphoric pronoun or specifying adjective (*Med.* 40.8 *pro discrimine generum*). We should therefore think that *solide* conceals something that will specify the meaning of *genus*. The easiest solution is simply to divide *solidegenere* into *soli de genere*, with *soli* attaching to *genere* to mean 'soil type' (for the *iunctura*, cf. Col. 5.6.5, 11.arg.2, Plin. *HN* 9.132). Nonetheless, the word order is unusual (should *soli* be transposed, e.g. after *genere*?) and there may be a deeper corruption.

norma: 'Rule', 'standard' or 'practice' (L&S s.v. II, *OLD* s.v. 2), a meaning which derives from the basic sense of 'square', the tool used by carpenters, masons, land-surveyors (Campbell (2000) 501, Chouquer and Favory (2001) 446) and so on to create right angles. The word appears nowhere else in the agronomists, although it is printed by the conjecture of Condorelli at 3.3.4 *nouuam <normam> amygdali suadent* (n.). (The related verb *normo* appears once at Col. 3.13.12, where it means to 'square' or 'make perpendicular'.) If Iulius Fronticus is actually the Sex. Iulius Frontinus who wrote on land-surveying (see Introduction §II), and G. is thinking of his works on that subject, then *norma* could be punning.

4.1.9 Quintiliis in satione castaneae uergiliarum non omittendus occasus: The *uergiliae* are the Pleiades, whose morning setting in early November marks the beginning of winter: cf. Plin. *HN* 18.222 *uergiliarum matutinus occasus* (sc. *incohat*) *hiemem*. The apparent setting at Rome was around the ninth of November: see *DGRA* I 227, André on Plin. *HN* 17.128, 130, Varro, *Rust.* 1.28.2, Col. 11.2.78, 84, *Geop.* 1.9.11. For the timing given by the Quintilii, cf. Pallad. 12.7.17 *eligendae sunt castaneae ad ponendum recentes, grandes, maturae. quas si Nouembri mense ponamus, facilem se praesentia fructus ipsius praestat*; *Geop.* 10.2.1 (on fruit trees in general) χρή . . . εὐθὺς μετὰ τὴν δύσιν τῶν πλειάδων φυτεύειν, etc.

4.2 G. reports a group of Greeks who 'follow nature' (4.2.2 *naturam secuti*) in espousing an autumn planting time for the chestnut. The key observation is that the chestnuts naturally reach maturity at this time, when their hulls split and the nuts are ready to drop at the touch of a hand. Both the autumn planting time and the explanation based on nature are *mutatis mutandis* very similar to the recommendations of 'nearly all of the Greeks' for the peach: cf. 2.1.3 *Graecis omnibus ferme <u>autumnalis satio ratione naturali</u> uidetur electa, siquidem eo tempore <u>cum pomis caducis ossa delapsa</u> in nouas plantas saepe uideamus sine culturae adminiculis sua sponte fruticare*. In the present passage, 4.2.3 seems to give a special reason for an autumn planting time in the case of the chestnut, namely, that this nut is difficult to protect against rot. The idea is therefore that an earlier planting will obviate the difficulties of preserving the nut. It is hard to see, however, how the final statement 4.2.4 *Sana igitur*, etc., can follow logically from this statement; *igitur* is rather resumptive (*OLD* s.v. 5), bringing us back to the moment of planting (*exponi* picking up *serere castaneam*) after the intervening explanation for the autumn sowing.

4.2.1 Sed et quibusdam Graecis placet ab Arcturi [et] exortu serere castaneam: *Sed et* is progressive rather than adversative (but with normal adversative force at 4.1.3, 4.5.2), a tendency for the expression noted in *Med.* (Maire (2002*a*) xl). *Arcturus* is the bright star in the knee of Boötes. The apparent morning rising at Rome was around 21 September: see *DGRA* I 228–9, Verg. *G.* 1.68 with Mynors, Col. 9.14.10.

ab Arcturi [et] exortu: Mazzini defends *et* as meaning 'rather' (cf. *TLL* v 2.893.56–894.3 (Hofmann)) and adduces the examples at *TLL* v 2.898.13–21 to show that the postposition is possible. The argument is specious. Corrective *et* can indeed have the force of *potius* or *immo*, but this does not mean that it shares the full range of syntactic properties of those particles: as the examples show, when *et* means 'rather' it does not demonstrably lose its coordinating function; nor, again, does it lose its coordinating function merely by being postponed. But no connective is wanted after *sed.*

4.2.2 Nec immerito naturam secuti quibuscumque tempus hoc placuit: The litotes expression *nec immerito* (for which see 1.1.7n. *non leui pondere*) first and foremost confirms the soundness of the Greeks' natural basis (*secuti naturam*) for an autumn planting, as shown by the explanatory statement *tunc enim*, etc. It may also be a cautious endorsement of the planting time in its own right.

seminis, id est nucis: G. uses the word *semen* to refer to the nut because he is thinking of its generative capacity: so (on the almond) cf. 3.1.5 *Supra nuces terra sesquipedis ingeritur, et si nec ullus imber fuerit, ter in mense seminibus expositis umoris alimenta sumministrentur*, etc. (where *uariatio* may also be a motivation). In order to avoid confusion in the present instance, G. 'glosses' the word with *nux*. For remarks on the variable terminology applied to plant and fruit parts, see 2.1.1n. *ossa*; for the term *nux* in particular, see 3.1.1n. *nuces.*

iusta maturitas . . . adeo ut: The meaning of *iustus* is in essence the same as at 2.5.9 (*plus iusto*), 2.11.2 (*mole iusta*), 3.4.2 (*iusta . . . interualla*) and 4.7.3 (*mensura non satis iusta*), stressing that something is fitting or appropriate vis-à-vis an implicit standard: see 2.11.2n. *mole iusta.* Here the adjective is somewhat weaker, however, and with *adeo* simply anticipates the *ut*-clause in marking the state of ripeness 'appropriate' for the events described in it to obtain. G. also speaks of the *iusta maturitas* of a fruit at *Med.* 43.15 *tunc cydonea iustae maturitatis seminibus et ligno interiore purgata*, etc.

ut hiante tegumento, si legent, tum manus condat in terra, cum caduca labuntur: *tegumentum* (not applied to a plant part by the other agronomists) refers to the spiny burr or husk of the chestnut that covers the nuts (see Plate 15). As the fruit matures, the burr yawns open and separates from the tree (i.e. abscises; see 3.4.3n.); only when it has fallen to the ground is the fruit fully ripe and ready for harvesting (Bounous and Marinoni (2005) 327; see Plate 5). It is this final abscission marking the ripeness of the nuts that G. has in mind in the rather convoluted *ut*-clause, the point of which is to give an observational index for the state of *iusta maturitas*: when the burr opens (*hiante tegumento*), then the nut is so loosely attached to the tree that, if one reaches to gather it (*si legent*), a touch of the hand (*manus*) is enough to 'bury' (*condat*) it in the soil (i.e. because the momentum of the detached nut will cause it to sink into the leaf litter and earth at the base of the tree). The words *cum caduca labuntur* provide an even more general view of the situation, for *labuntur* paraphrases *manus condat in terra* and the word *caducus* when applied to fruit refers specifically to the state of maturity that precipitates abscission: cf. 2.1.3 *cum pomis caducis ossa delapsa* (with 2.1.3n. *pomis caducis*). (Winterbottom's relatively easy emendation of *ut* to *cum* is desirable for establishing the proper relationship between the clauses and avoiding the awkward double *ut*.) The readiness to fall from the tree vividly described in the *ut*-clause in essence defines the state of *iusta maturitas* that supports the Greeks' 'natural' planting time.

hiante: For the meaning of *hio*, see 3.1.1n. *hiascere coeperunt*.

si legent . . . manus: The subject of *si legent* is vague, not restricted to *quidam Graeci*; perhaps the passive *legentur* should be considered (*-tur* being truncated before *tum*). In any case, the purpose of the detail is to focus the description through the eyes of the cultivator – the experienced reader of the text, perhaps – reaching out a hand (*manus*) to pick (*legent*) the chestnuts that are in season.

condat in terra: The construction of *condo* + *in* + ablative is regular (*TLL* IV 148.35–149.9 (Spelthahn)), but the verb is much more commonly used by the agronomists of preserving: see examples at *TLL ibid.*, *OLD* s.v. 2a, b; cf. also 3.4.1n. *recondant*.

4.2.3 nux reseruata difficile sit contra putredinem uindicare: If the text is correct (Reeve reasonably wonders whether there is a lacuna after *reseruata*), it may be explained in one of two ways: either *contra putredinem uindicare* is epexegetic with the adjective *difficile* (for the neuter predicate following *nux*, cf. 3.7.1n. *utile erit . . . aqua calida . . . irrigata*), or the whole construction is an anacoluthon (= *difficile sit nucem reseruatam contra putredinem uindicare*).

nux reseruata: G. discusses the methods for preserving the chestnut until a later planting time in the following sections 4.3–4.

contra putredinem uindicare: The standard construction with *uindico* when it means 'defend' (*OLD* s.v. 4) is *ab*, *ex* or the bare ablative: so cf. *Med.* 45.2 (sc. *semen citrii*) *stomachum a nausia uindicat*. The adverb *contra* may be paired with it in other meanings, however: cf. Sen. *Con.* 1.1.23 *quam innocenter me contra parricidium uindico* ('avenge'); Quint. *Decl.* 338.17 *filium matri uindico contra eum qui* *asserere coepit nuper* ('claim'); Serv. *ad* Verg. *Aen.* 1.144 *sed et multa alia usus contra antiquitatem uindicauit* ('claim').

putredinem: The word is not attested before Apuleius. It is more commonly used of rotting flesh, but for plant matter, cf. Pallad. 2.15.17 *sic* (sc. *iuglandes*) *beneficio solis et uenti durescunt quae in putredinem transiebant*; see further *TLL* x 2.2773.63–7 (Rey).

4.2.4 Sana igitur et matura debet exponi, quorum alterum spondet de incolumitate perseuerandi, alterum prodest de uirtute generandi: The gist of Palladius' advice (12.7.17, quoted at 4.1.9n. *Quintiliis . . . occasus*) is similar, although he espouses a November rather than September planting. For the stylistic qualities of the words *alterum spondet de incolumitate perseuerandi, alterum prodest de uirtute*, see Introduction §v(a).

Sana . . . matura . . . quorum alterum . . . alterum: The construction *quorum alterum . . . alterum* (very common) is used to pick up many kinds of preceding doublets, not limited to nouns. For close parallels to the present usage, following adjectives, cf. Sall. *Iug.* 7.5 *ac sane, quod difficillumum in primis est, et proelio strenuos erat et bonus consilio, quorum alterum ex prouidentia timorem, alterum ex audacia temeritatem afferre plerumque solet*; Liv. 23.12.9 *si reticeam, aut superbus aut obnoxius uidear, quorum alterum est hominis alienae libertatis obliti, alterum suae.*

spondet de incolumitate perseuerandi: For the notion of a plant (or plant part) 'promising' something of its future state, cf. 1.1.7n. *speciosa exire pollicentur. spondeo* does not as a general rule govern *de*, but the construction is influenced by the symmetry of the sentence: the essence of the thought is contained in the balanced noun phrases *de incolumitate perseuerandi* and *de uirtute generandi*, which evince the respects in which the aforementioned qualities are desirable.

incolumitate perseuerandi: *perseuero* typically means to 'continue' or 'persist' in a particular action, state, condition, vel sim. (*OLD* s.v. 1; see 1.1.7n. *melius perseuerant*), and despite its grammatical dependence on *incolumitate*, the idea here is precisely that of continuing in an 'uncorrupted' or 'healthy state' (for *incolumitas* thus, see *TLL* vii 1.983.29–31 (Buchwald)).

uirtute generandi: *genero* refers to the productivity of a plant also at 3.6.1 *Plurimum generant.* For this usage, from the first century CE on, see *TLL* vi 2.1795.47–55 (G. Meyer). For a relatively close parallel to the present *iunctura*, cf. Col. 3.10.8 (sc. *semen*) *nec ullam generandi uim possidet.*

4.3 After stating that most Greeks (4.3.1 *plerique Graecorum*) believe a spring planting preferable, G. relates instructions for preserving the chestnuts until that time and for testing them to determine their soundness. After desiccation (cf. 4.3.1 *expansa*), the nuts are heaped into a pile and blanketed by sand.

This method of preservation, still used in some parts of the world today, is described by Bounous and Marinoni (2005) 329: 'The "ricciaia", or burr pile, is a very old storage method, almost totally abandoned in Europe, but currently used in Turkey. Chestnuts are harvested while still closed in the burrs and raked into piles about 1 m tall, covered with leaves, burs and soil. Under these conditions, the nuts undergo a form of fermentation, which stablilizes them and allows for their conservation for several months.' The sink-or-float test that G. subsequently mentions (4.3.3–4) is a sound one and is still used today to test the viability of chestnuts as well as other tree seeds. It relies on the fact that disease, insect damage or decay produces minute pockets in the seed that trap air and cause it to float.

Palladius adapts the entire section: 12.7.18 *si uero Februario ponemus, ut usque tunc duret, ita faciendum est: in umbra castaneae siccentur expansae. tunc in angustum et siccum locum translatae cumulum faciunt et eas omnes fluuialis harena diligenter operiat. post dies triginta eas remota harena in aquam frigidam mittis. quae sanae sunt merguntur; supernatat quaecumque uexata est. item quas probasti similiter obrues et post triginta dies aeque probas. hoc cum tertio feceris, usque ad ueris initium serere debebis quae manserint illibatae.*

castaneam melius ex uerna satione procedere: The verb *procedo* is regular of the growth of plants: cf. 2.5.5n. *planta procedit.*

si tamen nux anxie atque sollicite condita seruetur expansa: *tamen* is restrictive (as at 4.6.6 *si tamen limus ex eo stare non soleat*): for weakened uses of *tamen* in *Pom.*, see 2.2.1n. *tamen.* The point is that a spring planting is better only if the nuts are properly preserved in the interim.

anxie atque sollicite condita: Cf. 2.2.1 *medio tamen tempore anxie atque sollicite ossa custodiunt* (with n. on *anxie atque sollicite*).

condita: From *condo*, as the context requires, not *condio* as at 2.12.1 *conditum.*

expansa: Agreeing with the collective singular *nux.* The chestnuts are spread out for an initial period of desiccation before (cf. 4.3.2 *Et tunc*) being transferred into the ricciaia (4.3n.). For *expando* of spreading fruit out for desiccation, see further *TLL* v 2.1598.24–31 (Deicke). Cf. also 2.2.2 *post* (sc. *ossa*) *in umbra explicantur et biduo uentilata.*

4.3.2 Et tunc in angustum locum siccumque translatae eriguntur in cumulum: The ricciaia is described. The detail *angustum* is presumably with a view to the ease of forming the *cumulus.* For the plural *translatae eriguntur* after the collective singular *nux*, see the remarks at 2.4.1n. *persicorum arborem . . . laetissimas.*

supra eas ita <ut> uniuersas obtegat funditur harena fluuialis, quae et pinguis et dulcis est: The restoration of *ut* is required to explain *ita* and *obtegat*; the word could easily have fallen out before *uniuersas.* The

blanket of sand produces a stable, anaerobic environment that aids in the preservation and fermentation of the chestnuts: see 1.1n.

obtegat: See 1.1.6n. *obtecta*.

harena fluuialis, quae et pinguis et dulcis est: A number of sands are distinguished by the Romans, not only river sand (*harena fluuialis* or *fluuiatilis*), but also pit (*fossicia*) and sea (*marina*) sand (cf. Vitr. *De arch.* 2.5.1, Plin. *HN* 36.175). The relative clause *quae et pinguis et dulcis est* seems to define *fluuialis*, detailing the relevant characteristics of this kind of *harena*. That is not to say that the information it contains is superfluous, however: river sand may be coarse as well as fine, and the adjectives *pinguis* et *dulcis* clarify what sort G. has in mind (see the following n.), should there be any doubt.

pinguis . . . dulcis: *pinguis* refers to the fatness or richness of soil (the opposite of *macer*, see 2.4.1n. *macro*), which Columella describes as being closely related to but not identical with its sweetness: cf. 2.2.17–18 *itaque considerandum erit ut solum quod excolere destinamus pingue sit. per se tamen id parum est, si dulcedine caret.* Although the properties are distinct, Columella goes on to describe a means of testing the soil at once for both fattiness and sweetness (2.2.18, with White (1970) 93–4); the qualities are said to depend upon the *pinguitudo* and *naturalis sucus* of the soil (for *sucus* see 2.10.3n.). It is not of course the nutritional content that is relevant here, but because descriptive words for soil may refer to physical texture as well as quality (as remarked at 2.4.1n. *macro*), *pinguis* and *dulcis* should mean that the river sand is thick and rich and will therefore serve as a good barrier for the chestnuts.

4.3.3 trigesimum: The more common spelling is *tricesimus* (Leumann §380.B, *OLD* s.v., de Vaan 262).

eadem harena: For *eadem* here, see the remarks on over-specification in the Introduction §v(b).

[in] nucibus in aquam frigidam missis sanas a corruptis sic licet segregare: With the Paduan reader we should delete *in*, whose intrusion can be explained by the following *in*. This change also gives a point to *sic*, which recapitulates the action expressed in the ablative absolute and anticipates the explanation given by *nam*, etc.

aquam frigidam: The water is cold so as to avoid cooking the chestnuts or simply warming them enough to make them susceptible to rot.

nam corruptae supernatant, sanae residunt: The asyndeton between the clauses is adversative: see further 2.4.6n.

4.3.4 quae placere meruerunt: For the usage of *meruerunt* here, see Introduction §v(a).

post alterum mensem: Providing *uariatio* for 4.3.3 *Post trigesimum diem.*

si quae forsan integrae debent credi, aqua iudicante temptatur: Mazzini's addition of *non* after *forsan* is not unreasonable, but the text gives appropriate sense without emendation: G. leaves unstated the fact that the obviously rotten nuts will be discarded without any need for a sink-or-float test; it is only those that seem sound, i.e. those specified by *si quae,* etc., which will require testing in order to make sure of it.

debent credi: For *debeo,* see 3.6.4n. *debeant germinare.*

aqua iudicante temptatur: *temptatur* is impersonal. *aqua iudicante* (the easier of the proposed emendations from *iudicanto* (**N**)), lit. 'water deciding the matter', refers to the sink-or-float test previously mentioned (4.3.3). But consider that Palladius' version of the line replaces the words *aqua iudicante* with *aeque* (12.7.18 *item quas probasti similiter obrues et post triginta dies aeque probas*) – could the corruption therefore extend to G.'s *aqua* as well?

4.3.5 Hoc idem fit tertio mense donec uernum tempus adueniat: The remarks of 4.2.1–2 presuppose that the fruit are ready for harvesting around the third week of September (see 4.2.1n.). Unless the period of desiccating the chestnuts is imagined to be very long, spring would not arrive immediately after the third month of preserving, so there may be some elision of time between *tertio mense* and *uernum tempus adueniat* which *donec* helps to bridge. The instructions to repeat the monthly test (*Hoc idem fit*) should presumably be taken iteratively for this intervening period.

quo satis nucibus nec includi iam necesse sit nec probari diligentioribus: *includi* takes its point specifically from the ricciaia (4.3.2 *in angustum locum*). It is peculiar that G. says that it is not 'necessary' (*necesse*) for the nuts to be laid up and tested after they have been sown (*satis nucibus*), for strictly speaking it is not even possible. But *necesse* can be given a sense if it is interpreted as referring to the situation more generally, in which we have moved from one period (that of preserving the chestnuts until planting) wherein the cultivator must make provision for storage to another period (planting) where all such necessity has passed.

nec probari diligentioribus: If correct, the substantive *diligentioribus* is obliquely directive (as at 1.1.15 *Diligentioribus rectissime uisum est* (n.)), referring (as it often does) to those exemplary farmers who will take extra care in their labours. Hence it is that *nec probari* applies only to this group, for while everyone who plans for a spring planting time must store up (*includi*) the chestnuts until then, only those who are scrupulous will also conduct periodic tests (*probari*) for their soundness. *diligentioribus* can thus be given a point in the context, but if it is felt to be too vague (it is also rhythmically unattractive), then Reeve's suggestion of *curis* after *diligentioribus* (cretic-trochee *diligentiōrĭbūs cūris*) should be preferred to Scotti's *diligentius*.

probari: Providing *uariatio* for *iudicante* and *temptatur*: see Introduction §v(a).

4.4 The first sentence of the section is an afterthought to the discussion of preservation in 4.3. G. subsequently devotes himself to the question of propagating by nuts versus transplanted suckers (*plantae*; see 4.4.2–3n.). He condemns suckers for the delay in growth that this method produces (thus following Columella and Pliny), but duly reports that some believe that the chestnut grown from a *planta* is better suited to producing nuts. This claim is based on the dubious (cf. 4.4.4 *si tamen fas est*) application of a principle that wood grows into fruit (*lignum in poma proficiat*).

4.4.1 Non in pauimento sed uasibus seruanda, harena ut similiter influat: On the basis of Pallad. 12.7.19 *aliqui in uasculis seruant harena pariter missa*, Antonelli argues that some qualifying word such as *alia* has fallen from the text. That is possible, but G. could equally be providing a forcible correction to the preceding advice; he would thus accept the general principle of preserving by ricciaia, but would prefer packing the chestnuts into a jar to piling them on the floor.

pauimento: In a farm setting often made of clay tiles: cf. Cato, *Agr.* 18.7, Vitr. *De arch.* 7.1.7, 7.4.5, Col. 1.6.13, Pallad. 1.18.2.

harena ut similiter influat: Cf. 4.3.2 *supra eas ita <ut> uniuersas obtegat funditur harena fluuialis, quae et pinguis et dulcis est.* The word *similiter* may refer to the addition of sand in general or to the earlier detail that the nuts should be covered in their entirety (*uniuersas*) or both.

influat: Equivalent in meaning to *infundatur* and providing *uariatio* for 4.3.2 *funditur*. It is unnecessary and obscures the relationship with previous advice to give *influo* the unprecedented sense 'influence', as Mazzini does: 'L'arena non va conservata nel pavimento, ma nei vasi in modo che possa avere simili effetti.'

4.4.2–3 Quidam in plantis iuxta arborem sponte nascentibus castaneas conserendas esse dixerunt. Sed eiusmodi planta tardissime prouenit et continuo biennio languida est, uix ut demum uel illam sucus recreet qui consequitur, quem non habuerat, cum <nuce> dilata non esset: It is clear enough that another method of propagation for the chestnut is mooted here, but it is less certain how to construe the particulars of these difficult sentences. We are aided by Columella and Pliny discussing different methods for planting the chestnut: Col. 4.33.3–4 *ubi frequentanda sunt* (sc. *castaneta*), *melius ex uicino, si competit, mergi more pertica declinata propagatur quam exempta reseritur. haec enim uelut immota sua sede uehementer germinat;*

at quae radicitus exempta et deposita est, biennio reformidat; propter quod compertum est, commodius nucibus, quam uiuiradicibus eiusmodi siluas institui; Plin. *HN* 17.149 *sunt et propagines, nulli quidem faciliores. nudata enim radice tota in sulco prosternitur; tum ex cacumine supra terram relicto renascitur et alia ab radice. sed tralata nescit hospitari pauetque nouitatem biennio fere; postea prosilit. ideo nucibus potius quam uiueradicibus plantaria caedua implentur.* (Compare also Palladius adapting G.: 12.7.17 *sed quae plantis seritur ita aegra est, ut biennio de eius uita saepe dubitetur. serenda est ergo ipsis castaneis, hoc est seminibus suis*, etc.) Both Columella and Pliny mention three techniques for propagating the chestnut: (1) layering (Col. *mergi more* / Plin. *propago*), (2) transplanting a quickset or sucker (*uiuiradix*, i.e. a shoot sent up from the plant's roots) and (3) planting from seed (*nux*). They both recommend without reservation (1) layering, but condemn (2) transplanting a *uiuiradix* as producing a tree that remains weak for two years; if layering is forgone, a better choice than (2) would be (3) planting the nuts (*nux*) themselves and leaving them undisturbed. (Note that the agreement between Columella and Pliny suggests a common source, perhaps Mago.)

This background helps to explain G.'s statements. His first sentence (4.4.2) reports that there are some who grow chestnuts using *plantae* from the base of a tree, and the second (4.4.3) criticizes the method, for such a procedure causes the tree to grow 'very slowly' (*tardissime*) and remain 'weak' (*languida*) for 'an entire two-year period' (*continuo biennio*). It is clear that G. is discussing here the method of transplanting using a quickset or sucker similarly rejected by Columella and Pliny. Note thus G.'s *in plantis iuxta arborem sponte nascentibus castaneas conserendas* = Col. *uiuiradicibus eiusmodi siluas institui* / Plin. *uiueradicibus plantaria . . . implentur.* (For the emendation of *conseruandas* (**N**) to *conserendas* and the instrumental use of *in plantis*, see the following n.) The word *planta* refers in a general way to any slip, shoot, cutting, etc. used for propagation (see 2.1.3n. *nouas plantas*) and hence may stand for the more specialized *uiuiradix* of Columella and Pliny; in order to avoid confusion, G. shows that he has a quickset or sucker in mind with the 'gloss' *iuxta arborem sponte nascentibus*. In the second sentence (4.4.3), note that *tardissime prouenit et continuo biennio languida est* = Col. *biennio reformidat* / Plin. *nescit hospitari pauetque nouitatem biennio fere.*

But what of G.'s obscurely expressed statement *uix ut demum uel illam sucus recreet qui consequitur, quem non habuerat, cum dilata non esset*? (I print the paradosis here: the need for a supplement such as *nuce* is discussed below.) The words are most easily understood as an elaboration on Pliny's *postea prosilit*, which describes the plant's resurgent growth after its two-year period of weakness. Antonelli interprets *dilata* to mean 'transplant', and translates 'sì che a mala pena può alla perfin ravvivarla il sopravvegnente umore, di cui, non essendo trasportata [= *cum dilata non esset*], avea patito per lo innanzi difetto'. Mazzini accepts this interpretation, but it cannot be correct for two reasons: (1) as we noted above, 4.4.2 *in plantis . . . castaneas conserendas* already presupposes transplantation of suckers (*plantae/uiuiradices*) from the base of the tree, so G. could hardly be saying now that such plants had not in fact been transplanted (as implied by Antonelli's interpretation of *cum dilata non esset*); (2) the interpretation implies that propagation through transplanting is somehow beneficial to the chestnut, when to the contrary G. should be condemning the method as harmful, to judge from the passages of Columella and Pliny that he is following. It is therefore necessary to understand *dilata* in an entirely different sense, taking it to refer not to transplanting but to planting out by seed. *differo* can indeed support the latter meaning: in its technical sense among the agronomists of 'setting out (sc. at intervals)', the verb is used both of planting by seed and transplanting young plants (cf. *TLL* v 1.1070.28–42 (Ammann)). Yet precisely because of the ambiguity of *differo* in this respect, a sharper contrast is wanted, and we may achieve it by the addition of a word such as *nuce* (at any event before *dilata*, so as not to spoil the rhythm). Such a supplement has the virtue of establishing a logical relationship between sentences 4.4.2–3: G. will be saying that the transplanted sucker (*eiusmodi planta*) struggles because it did not have the same nourishment (*sucus*) as the plant which had been planted out (*dilata*) from seed (<*nuce*>). Once the nutrition is finally attained by the plant, it restores (*recreet*) the strength even of such a weakling (*uel illam*).

4.4.2 in plantis iuxta arborem sponte nascentibus castaneas conserendas: The transmitted verb *conseruo* is difficult to understand. Antonelli (Mazzini follows) translates 'moltiplicare', a very uncertain extension of its meaning. It is best to accept Antonelli's suggestion *conserendas* for *conseruandas* (so Pallad. 12.7.17 *plantis seritur*); the verb is used from Cato onwards of planting trees (*OLD* s.v. 2a) and is found also at 4.6.3 of chestnuts (*conseri posset*). The deletion of *in* is unnecessary, however, since the preposition can be taken instrumentally: see *Syntactica* II 452–6, *TLL* VII 1.792.16–43 (Hofmann), H–Sz §79.1.g, Pinkster 880–1.

4.4.3 recreet: For the emendation to *recreet*, see 2.1.4n.

4.4.4 de nucibus et hoc genere plantarum audeant aliquod facere discrimen: If *ex hoc genere* of the MS is retained, *plantarum* must refer to plants or trees in general (i.e. as a sort of genus) and *hoc genere* to the chestnut as a particular species: lit. 'nuts of this kind of plant'. But this meaning of *planta* is scarely attested, and the sole example given by Ottink in her *TLL* article (x 1.2326.28–31) is a very poor parallel to the present instance: Boeth. *Top. Arist.* 143a27 *qui uero ipsum tantum superius genus dixit, non dicit et inferius genus; nam qui plantam dicit, non dicit arborem* (where *planta* = φυτόν). The difficulty can be alleviated by emending *ex* to *et* with Mazzini 146, which gives *hoc genere* more point in context, since it will refer specifically to the aforementioned suckers growing under the tree (4.4.2 *plantis iuxta arborem sponte nascentibus*).

nuces . . . caeduis arboribus aptiores, plantas uero nucibus procreandis esse utiliores . . . lignum in poma proficiat, fructus in ligna desistat: For stylistic aspects of the antitheses expressed in these lines, see Introduction §v(a).

caeduis arboribus aptiores: See 4.1–7n., 4.1.5n. *sationem caeduam.*

admittere in animum debeamus: For *debeo*, see 3.6.4n. *debeant germinare.* For *admitto* + *in* + accusative (more common are *in* + ablative or *ad* + accusative), cf. Quint. *Inst.* 4.1.5 *per* (sc. *initia orationis*) *in animum iudicis ut procedere possimus admittimur.*

lignum in poma proficiat, fructus in ligna desistat: For *proficio* of (incremental) plant growth, see *TLL* x 2.1698.16–37 (Kruse). For a reasonably close parallel to the present case, cf. Pelag. *In Rom.* 7.4 (sc. *iustus*) *in florem erumpit, deinde in fructum proficit, postremo ad summam peruenit maturitatem.* It is unnecessary to emend *ligna* to *ligno* with Scotti; the unusual construction of *desisto* + *in* + accusative is influenced by a desire for parallelism (in case, number and rhythm) with the preceding clause: thus *īn pōmă* = *īn lĭgnă.*

4.5 The section is devoted to the question of the right soil type for the chestnut. It is a composite in both doctrine and language from the statements of Columella and Pliny: Col. 4.33.1 *ea* (sc. *castanea*) *pullam terram et resolutam desiderat, sabulonem umidum uel refractum tofum non respuit, opaco et septentrionali cliuo laetatur, spissum solum et rubricosum reformidat*; Plin. *HN* 17.147 *quaerit* (sc. *castanea*) *solum facile nec tamen harenosum, maximeque sabulum umidum aut carbunculum uel tofi etiam farinam . . . recusat eadem glaream, rubricam, cretam omnemque terrae fecunditatem.* Palladius is indebted to G.: 12.7.19 *amant solum molle et solutum, non tamen harenosum. in sabulone proueniunt, sed umecto. nigra terra illis apta est et carbunculus et tofus diligenter infractus. in spisso agro et rubrica uix prouenit; in argilla et glarea non potest nasci.* The laconic report of the *Geoponica* shifts the emphasis and contradicts in some respects the earlier authorities: 10.63.1 τὸ κάστανον . . . ψαμμώδει γῇ καὶ ψυχροῖς τόποις χαίρει.

4.5.1–2 Solum castanea desiderat molle et subactum, non tamen harenosum; patitur et sabulum, si umor adfuerit. Pulla terra uel maxime prodest sed et carbunculus itemque tofus optime seruit diligenter infractus: *pulla terra* aside, the principle for G.'s rejection of *harena* and his endorsement of *sabulum, carbunculus* and crushed *tofus* is neatly summarized by André on Plin. *HN* 17.147 (n. 2): 'Le sol qui convient le mieux au châtaignier est un sol graveleux et siliceux comme celui qui résulte de la décomposition du granit.' Hence the exclusion of *harena,* which is a finer sand than *sabulum* (cf. 2.4.6n. *calculosum . . . proximum*), and the inclusion of *carbunculus* (quarried sand) and *tofus* (pulverized tufa).

4.5.2 Pulla terra: The detail is Columella's (not mentioned by Pliny). The outstanding qualities of this 'black soil' (also called *terra nigra*; cf. Col. 1.pr.24, Pallad. 12.7.19 adapting this passage) are its friability (Cato, *Agr.* 151.2 *terra tenerrima*) and its productivity (Col. 2.2.19 *quae melius prouentu frugum approbatur*). It is connected with the soil of Campania in particular (Cato, *Agr.* 135.2, Col. 1.pr.2, 2.10.18 *putre solum, quod Campani pullum uocant*; cf. Adams (2016) 73). Probably foremost in G.'s mind is the friability of the soil, given his concern that it be *molle et subactum*.

carbunculus: Identified as a kind of quarried sand by Vitr. *De arch.* 2.4.1, which is the meaning here: see André on Plin. *HN* 17.29 (n. 1), Callebat et al. on Vitr. *De arch.* 2.4.1 (n. 3). The fuller investigation of Davidovits (1994) concludes that *carbunculus* is in fact a kind of pozzolana.

itemque: *item* appears only here in *Pom.*, but it is common in *Med.* (25x), where in most cases (17x) it is also paired with *-que*. It is probable that for G. *itemque* is a partially fossilized expression, with a meaning much like *item* (cf. H–Sz §254.c.1) or even *et*: see Maire (2002*a*) xl, and cf. *CG* s.v.

tofus . . . diligenter infractus: I.e. pulverized, as shown by Plin. *HN* 17.147 *tofi . . . farinam*. Though Columella uses *refringo* (4.33.1 *tofum refractum*), G.'s *infringo* is no less apt: 'c. notione frangendi in partes i. q. confringere' (*TLL* VII 1.1492.83–4 (Hofmann)). For tufa in agricultural use, see André on Plin. *HN* 17.29 (n. 2).

4.5.3 Spissum rubricosum locum non uult, uix prouenit in argilla, in glarea cretaque nec nascitur: Although the list as a whole most closely follows Pliny (*HN* 17.147, quoted at 4.5n.), the phrase *spissum rubricosum locum* derives from Columella. *rubricosum locum* points to *rubrica*, which, especially with *spissum*, refers to a heavy red clay, probably the modern *terra rossa* of Italy (*OLD* s.v. 4, *LAT* 22). *argilla* is a white clay or chalk (*OLD* s.v., White (1970) 109–10). We may be tempted to understand *glarea cretaque* as hendiadys, 'gravelly potter's clay' (cf. 2.11.2n. *in glarea argilla*), but the separation of the items in Pliny's list suggests not.

Although the punctuation of this sentence varies among the editors (Mai, for example, has colons where I print commas), only Scotti's affects the construal of the clauses. He places a colon after *uult*, a comma after *argilla* and a comma after *glarea*. The colon is misleading because it implies that what follows *Spissum rubricosum locum non uult* is an explanation or result; the second and third clauses are simply enumerative, however. The comma after *glarea* wrongly separates the noun from *cretaque* (cf. Pallad. 12.7.19, quoted at 4.5n.).

Spissum rubricosum locum: The adjectives are not in asyndeton bimembre, because they differ in rank (see 4.6.2n. *calidas . . . credit* and cf. 2.5.2

siccata erecta). That is, the meaning is 'a dense place of red clay' rather than a 'dense, (red-)clayey place'.

in glarea cretaque nec nascitur: The position of *nec* and the asyndetic parataxis of the preceding clauses argue for taking *nec* without coordinating force, equivalent either to *ne . . . quidem* ('not even', *OLD* s.v. 2b) or perhaps simply *non* (for which see 3.1.5n. *nec ullus*). The three-part asyndeton gives the impression of exhaustively reporting the conditions under which the chestnut will not flourish. (The 'and' added in the translation is for the sake of the English.)

4.6 G. reports various opinions on the climate most suited to the chestnut. The outlier Celsus is reported first. He argues for warm and sunny zones (4.6.2 *calidas apricas regiones*), adducing the produce of Naples by way of justification. Celsus acknowledges, however, that chestnuts can be grown in a cold climate, which most authorities prefer, including Iulius Atticus, Columella, Pliny and the Quintilii (4.6.3 *frigidis locis . . . quae fere omnes anteponenda senserunt*), and which G. himself seems to look favourably on (4.6.3n. *prudentissime . . . senserunt*). There is also a marked preference for planting on slopes (4.6.4–7 *cliuis . . . montana . . . decliuibus locis*), whether because mountainous environments are cooler or for another reason (cf. 1.1.2n.) The Quintilii qualify their recommendation by allowing a level planting field, provided that the soil is thin or near a river.

Palladius follows the gist of G.'s account, but adds the novel statement that the chestnut can handle a warm climate if there is water available: 12.7.19 *diligit caeli statum frigidum, sed et tepidum non recusat si umor assenserit. delectatur cliuis et opacis regionibus ac maxime in septentrionem uersis.* It is possible that Palladius has deduced *sed et tepidum non recusat si humor assenserit* from the Quintilii's advice: if the plains are taken to be warm in contrast to the cold (*frigida*) mountains, then the recommendation to site the orchard near a river (4.6.6 *iuxta uiciniam fluminis*) could be understood to mean that the land must be well watered. The *Geoponica* at any rate recommends without reservation cold and watery places: cf. 2.8.4 δρύες δὲ καὶ καστανέαι, αἱ Διὸς βάλανοι καλούμεναι, ἐν τοῖς συνεχῶς κατομβριζομένοις τόποις κατατιθέσθωσαν; 3.3.5 τούτῳ τῷ μηνὶ (sc. τῷ Μαρτίῳ) φυτεύσομεν πᾶν δένδρον καὶ κάστανον ἀπὸ πασσάλου, μάλιστα ἐν τοῖς ψυχροτέροις καὶ ὑγροτέροις τόποις.

4.6.1 De caeli qualitate nonnulla inter auctores uarietas opinionis exoritur: Cf. 2.3.1 *Interest autem cuiusmodi statum caeli condicio regionis, calidum frigidumne, sortita sit*; for G.'s *caeli qualitate*, cf. Col. 3.1.6, quoted at 2.3.1n.

uarietas opinionis exoritur: For *exorior* here, 'de quaestione, disputatione sim., quales in investigando, explicando sim. surgunt', see the examples at *TLL* v 2.1574.74–1575.4 (Meyer). For *uarietas opinionis*, cf. also *Med.* 22.1 *de uirtute ocimi . . . opinio incerta est*; 22.6 *qui in diuersa opinione sunt*, etc.

4.6.2 pomis: *pomum* refers in general to the edible fruit of a tree ('fructus esculentus arboris', x 1.2602.21 (Grossardt)) and thus picks out not only fleshy fruits (pomes), but also nuts: cf. *TLL* x 1.2602.63–6. The flexibility of the word is well illustrated by Macrob. *Sat.* 3.19.1, where *pomum* is treated as a class that includes both *nuces* and *mala*: *sunt de agri cultura scriptores qui nuces et mala sic diuidunt, ut nuces dicant omne pomum quod foris duro tegatur et intus habeat quod esui est, malum uero*, etc.

calidas apricas regiones castaneae aptissimas credit: Scotti defends the emendation of *apricas* (as transmitted by **N**) to *Africas*: 'Modo [sc. 4.1.2–3] Castaneam *arborem peregrinam* nuncupaverat, dum de Magone *Poeno* ageret'. But G. in fact said that the *arbor* was *peregrina* not to the Italians but to the African Mago (4.1.2 *Poenus et cui peregrinae eiusmodi arboris minus nota cultura sit*)[108] and expected that Celsus, otherwise very knowledgeable about Italian practice (4.1.3 *Cornelium . . . Celsum, Italicae disciplinae peritissimum*), would correct him. It is therefore hardly appropriate that G. would, without comment, now have Celsus recommend African regions for the tree.

The transmitted text *calidas apricas* gives good sense if the asyndeton bimembre is allowed. This kind of two-element asyndeton is generally stylized or formulaic: for discussion and examples see K–S §177.2, H–Sz Stilistik §55.1, Winterbottom on [Quint.] *Decl.* 254.15, Menge et al. (2005) 622–3 and especially Adams (2016) index s.v. 'asyndeton bimembre', with summary remarks at 647–9. Among other possibilities, the elements in asyndeton bimembre may be synonyms or antonyms, stand in a progressively strengthening ('accrescitivo') or disjunctive relationship, or work together to convey a specific emotional or literary effect; they are often alliterative or have multi-part sound effects. In the present instance, the adjectives *calidus* and *apricus* are conceptually associated and can be construed in a strengthening relationship. For another case of asyndeton bimembre in *Pom.*, cf. 2.11.2n. *in glarea argilla*. There are some further possible examples of the construction which I prefer to construe differently. At 2.1.3 *poma ipsa matura integra* (n.), the adjective *integra* is predicated. I also take the participles at 2.5.2 *ossa . . . siccata erecta* and the adjectives at 4.5.3 *Spissum rubricosum locum* (nn.) as differing in rank (Adams (2016) 77–8, 647). Although the judgement is subjective (cf. Adams (2016) 78), the sense is better adapted in each case to the context: 'the dried-out stones that have been set upright' (vs. 'the dried-out, upright stones') and 'dense place of red clay (*terra rossa*)' (vs. 'dense, red-clayey place').

4.6.3 Id illo argumento probat, quod Neapolis, Campaniae ciuitas, eo fructu maxime glorietur: For the fame enjoyed by Neapolitan

[108] Nor does the chestnut originate from Africa: see 4.1.2n.

chestnuts, cf. Plin. *HN* 15.94 *patria laudatissimis* (sc. *castaneis*) *Tarentum et in Campania Neapolis*; Mart. 5.78.14 *et quas docta Neapolis creauit, / lento castaneae uapore tostae.* André on Plin. *HN* 15.94 notes that the esteemed 'Corellian' chestnut (*laudatior Corelliana*) mentioned by Pliny is also Neapolitan, since Pliny (*HN* 17.122) mentions that the Corellius for whom they are named was active in Naples. The detail *Campaniae ciuitas* (cf. Plin. *in Campania*) ensures that there is no confusion about which 'Neapolis', 'the name of several Greek cities' (*OLD* s.v.), is meant, although the word *Neapolis* alone refers especially to the place in Campania.

prudentissime tamen postea adiunxit quod et frigidis locis conseri posset, quae fere omnes anteponenda senserunt: The sentence offers a good example of G.'s practice of 'extensive' didactic layering (Introduction §III(b)): G. does not gainsay Celsus' claim that the chestnut will flourish in warm and sunny climes, but indicates his approval of colder regions both by the detail *prudentissime* and by the statement that *fere omnes* prefer a colder environment. The point is further corroborated by the enumeration of the authorities in the following lines ('intensive' didactic layering).

4.6.4 Iulius Atticus opacis et septentrionem spectantibus cliuis laetissimum fieri castaneae pollicetur: The structure of this sentence is comparable in some respects to 3.7.5 *Illico* [*praesto*] *uidetur ex morsu pecorum omnia haec arbori amaritosa contingere* (n.). The phrase *opacis et septentrionem spectantibus cliuis* is a bare ablative of location, for which see 1.1.8n. *patina noua.*

opacis: This carries the implication that the spot is cool, as suggested both by Pliny's detail *praefrigido* (4.6.5n.) and by G.'s inclusion of Iulius Atticus among those who prefer *frigida loca.*

septentrionem: *septentrio*(*nes*) refers to the seven stars of the constellation Ursa Major, which appear in the northern sky; hence the word can mean simply 'north' (*OLD* s.v. 2).

4.6.5 Columella et Plinius in eadem fere uerba consentiunt: By *eadem fere uerba* G. means that Columella and Pliny agree not only in the substance of their recommendations but also in their language: cf. Col. 4.33.1 *opaco et septentrionali cliuo laetatur*; Plin. *HN* 17.147 *quamlibet opaco septentrionalique et praefrigido situ, uel etiam decliui.*

in eadem fere uerba consentiunt: *consentio* + *in* + accusative becomes common in the first century CE; before then it occurs only rarely, in Pompeius Trogus (*prol.* 30.4) and Livy (27.9.6; cf. Flor. *Ep.* 2.6.15, 4.2.93): see *TLL* IV 398.9–37, 399.70–5 (Burger).

4.6.6 Quintilii montana sine dubio et umida et frigida principaliter probauerunt: As at 3.1.4 *sine dubio et ipse unum latus opponit Aquiloni*, the phrase *sine dubio* marks G.'s own attempt to construe an authority's advice in light of the statements of others (3.1.4n. *aciem . . . Aquiloni*).

principaliter: The adverb is attested only from the first century CE. G. uses the word deliberately here to anticipate the alternative ('second') situation that may be encountered, namely, *in campo*: see *TLL* x 2.1297.49–1298.7 (Spoth), *OLD* s.v. 1.

tenuem locum, et hunc harenosum: Although G. is discussing the planting location and not the soil quality, it is difficult to take *tenuis* as meaning anything other than 'thin' or 'lean' in reference to the soil of the spot (thus synonymous with *macer*; cf. 2.4.1n. *macro*); this meaning of *tenuis* is not uncommon among the agronomists (cf. Col. 2.10.15, 3.5.3, Plin. 16.129, 18.123), and for *locus* so used in reference to the soil quality of the spot, cf. 4.5.3 *Spissum rubricosum locum*. On this interpretation, the following *et hunc harenosum* would be explanatory: it is a lean spot because it is one of fine sand (*harenosum*). The recommendation here is somewhat more emphatic in light of the fact that *harena* is rejected as a planting soil in the fuller discussion of 4.5.1 (cf. *non tamen harenosum*). For explanatory *et hic*, see *TLL* vi 3.2725.5–13 (Schmid-Bulhart-Ehlers); cf. also *TLL* v 2.874.22–875.31 (Hofmann).

iuxta uiciniam fluminis: *iuxta uiciniam* looks redundant, and could be rhetorical *abundantia* (cf. Maire (1999) 435–6), but there is nothing to prevent us from construing it literally: they should be planted not adjacent to the river, but in the land extending around the river.

si tamen limus ex eo stare non soleat: *tamen* is restrictive (2.2.1n. *tamen*). For the meaning of *stare*, see *OLD* s.v. 7c, 9a.

4.6.7 Sunt qui putent et in decliuibus locis castaneas optume prouenire: The sentence seems redundant: why tell us now that 'there are those who' (*sunt qui*) believe that chestnuts grow best on slopes, when we were only just told the names of several authors who agree in this preference (cf. 4.6.4 *cliuis*; 4.6.6 *montana*)? It is at first tempting to delete the statement as a gloss; there is nothing in it uncharacteristic of G.'s language, however, and it may be that whereas the preceding statements were meant to emphasize the consensus regarding a cold environment (cf. 4.6.3n. *prudentissime . . . senserunt*), G. now highlights the fact that the authorities similarly agree on sloping terrain. Note thus that Pliny (quoted at 4.6.5n.) presents a cold environment and sloping terrain as alternatives. (One may in fact wonder whether G.'s whole statement is not an attempt to reproduce the disjunctive *uel etiam decliui* from a Plinian source.)

optume: Elsewhere (2.7.4, 4.5.2) the spelling *optim-* is found. Mazzini 59 singles out this occurrence of the older spelling (*TLL* ii 2079.51–7 (Sinko)) as revealing of G.'s style. I would prefer not to place so much weight on it: cf. 3.7.4n. *quum*.

4.7 This truncated section treats preparation of the soil for the chestnut. G. reports that most prefer trenched soil (4.7.1n. *pastinatos*); the authorities differ, however, on the depth to which the soil should be turned over, with recommendations ranging from 2 to 3 feet. (These depths for trenching agree with the standard range recommended for *pastinatio*: see 3.3.1n. *omnem . . . fodiunt.*) G. intercedes to explain the variation in opinions: probably all of the authorities have a piece of the truth (4.7.4 *Et fortasse singulis satio consistat*), since those who are satisfied with shallower soil are thinking of a nursery, where it is only a matter of coaxing up the plant for two years before it is moved, while those who plant the chestnut to stay (*semina mansura*) rightly employ a deeper bed of soil that will give the roots room to spread.

G.'s mention of two parallel sets of trenching instructions, those for planting with a nursery and those for planting without, may reflect the controversy about the suitability of the chestnut for transplanting that cropped up earlier in *cast.*: see 4.4.2–3n. In the extant portion of 4.7, G. sets out the range of instructions in an ostensibly agnostic way, and it is not certain whether he would have recommended a nursery or not. Even if it is probable that he shared the concerns about transplanting the chestnut (4.4.3 seems to be presented in his own voice), we cannot necessarily infer from this that he would unequivocally recommend against it. Thus Pliny, despite noting the difficulties of transplantation (*HN* 17.149), nevertheless elsewhere states without qualification that the chestnuts are 'transplanted from a nursery after two years' (*HN* 17.148 *ex hoc seminario transferuntur . . . post biennium*).

4.7.1 Maiori parti placet in pastinatos omnino ponere: Cf. Col. 4.33.1 *tum in repastinato nux posita celeriter emicat*; 4.33.2 *seritur . . . sicca terra et repastinata*; Plin. *HN* 17.148 *refringi solum debet sub ea* (sc. *nuce*); Pallad. 12.7.20 *pastinari ergo locus debebit, qui huic destinatur arbusto*. Pliny and Palladius have some further details on the planting site: Plin. *HN* 17.148 *interualla sint pedalia, undique sulco dodrantali. ex hoc seminario*, etc.; Pallad. 12.7.20 *uel totus* (sc. *locus*) *aut sulcis in ordinem destinatis aut certe aratris resolui hinc inde findentibus.*

in pastinatos: For the accusative *pastinatos* where we might expect the ablative *pastinatis* (suggested by Mai), see 3.6.2n. *in locum taedae*.

pastinatos: For remarks on the method of intensive trenching known as (*re*) *pastinatio*, see 2.7.1n. *Scrobes*, 3.3.1n. *omnem . . . fodiunt*. With *pastinatos* here must be supplied *locos*, vel sim.; the more typical way of referring to such trenched ground is with *pastinum* or *pastinatum* used as a neuter substantive (cf. *TLL* x 1.636.14–26, 635.62–75 (Kuhlmann)). The simplex *pastino* and related words *pastinum* and *pastinatio* are not found before Columella, but *repastino* appears already in Cato (but at *or. frg.* 11.1 = 128 Malcovati, not in his agricultural writings) and Varro (*Rust.* 1.18.8, 1.37.4). The family of words derives from *pastinum* (the dibble), a pronged tool for turning over the soil in preparation for planting: see White (1967) 109–10, *TLL* x 1.635.85–636.8 (Kuhlmann).

ita ut terra confracta in duorum pedum altitudinem subleuetur: It was precisely in order to break up the earth (cf. *confracta*) that *pastinatio* was performed. The broken earth is said to be 'raised under the nuts' (*subleuetur*) because it forms a bed on which they are planted. Two feet is not the most common depth for trenching (see 3.3.1n. *omnem . . . fodiunt*), but it is still within the standard range. Besides G., only Palladius mentions this depth specifically in connection with the chestnut, next to the figure of 1½ feet: 12.7.20 *pastinari ergo locus debebit qui huic destinatur arbusto altitudine pedis unius semis uel duorum.*

4.7.2 Columella et Atticus addendum adhuc altitudini semissem putauerunt: Cf. Col. 4.33.2 *terra . . . repastinata in altitudinem dupundii et semissis.*

4.7.3 Sunt aliquot ex Graecis: The substantive use of *aliquot* is rare, but found in all periods of Latin: see *TLL* I 1616.70–9 (Vollmer).

mensura non satis iusta sit nisi quae tribus pedibus excreuerit: *excreuerit* is from *excresco* (not *excerno*). The measure to which something 'grows' or 'comes up' with *excresco* is usually expressed with a preposition (e.g. *in* or *ad*) + accusative: see *TLL* v 2.1284.33–8 (Rehm). *tribus pedibus* is an ablative of extent (Pinkster 827, with comparanda). For the amplification of the phrase, which helps to sustain *uariatio* in the list of trenching depths, cf. 2.5.2 *non altius debent quam quod amplectitur trium mensura palmorum*. For the meaning of *iusta*, see 2.11.2n. *mole iusta.*

4.7.4 singulis satio consistat: *singulis* refers to the three aforementioned *altitudines*: 2 feet (recommended by the *pars maior*), 2½ (Columella and Atticus) and 3 (*aliquot ex Graecis*). The precise meaning of *consisto* will depend on whether *satio* is taken as the act of planting (as translated), even the method, or as what is planted itself (see 2.1.3n. *satio*); with the latter, *consisto* could mean rather 'endure' or 'subsist' (*OLD* s.v. 14c, *TLL* IV 468.27–70 (Remme)).

dipondi: The word *dupondius* (*-um*) shows a bewildering variety of spellings (*TLL* v 1.2285.9–32 (Bannier)). Both *dip-* and the genitive in the single *-i* are attested for the word. In its original meaning *dupondius* refers to the sum of two *asses*, but Columella uses it many times to designate a length of 2 feet; the present instance is the only example outside of Columella's writings with that significance (*TLL* v 1.2285.84–2286.4), and G. is possibly thinking of Col. 4.33.2 (quoted at 4.7.2n.).

nucibus exceptis: The nuts are said to be 'taken up' or 'received' (*exceptis*) by the soil: so cf. Cic. *Sen.* 51 *quae* (sc. *terra*) *cum gremio mollito ac subacto sparsum semen excepit* (with Powell).

alere: Used by the agronomists (and others) of the earth nourishing plants: see *TLL* I 1709.45–52 (v. Mess).

post biennium recessuras: Cf. Plin. *HN* 17.148 *ex hoc seminario transferuntur... post biennium*; Pallad. 12.7.21 *quibus transferre placuerit, bimas plantas transferre debebunt.* G.'s verb *recedo*, however, is not elsewhere so used of transplantation. At Col. 4.15.3, in reference to a slip taken from a plant (*id quod a matre abscisum recessit*), the notion that predominates is that of separation or detachment (*OLD* s.v. 5a).

semina mansura condebant: *condo* is more often used by the agronomists of preserving (noted at 4.2.2n. *condat in terra*); it may here mean nothing more than to bury or conceal the seeds, but in connection with *mansura* it could also appeal to the idea of storing up, in the sense that one is entrusting the seeds to the earth with an eye to the future.

non immerito ampliorem radicibus: Since the shorter depth is identified as 2 feet (cf. *dipondi breuitate*), *ampliorem* (sc. *mensuram*) indicates either 2½ or 3 feet. With *radicibus* the adjective also contains the idea that this depth is 'ample' or 'sufficient' for the roots (*OLD* s.v. *amplus* 1c). For *non immerito*, cf. 4.2.2 *Nec immerito*; for the directive use of litotes, see 1.1.7n. *non leui pondere.*

uti: As for other ostensibly archaizing spellings in *Pom.* (3.7.4 *quum*; 4.6.7 *optume*), one should be cautious in interpreting the significance of *uti* here: see 3.7.4n. *quum.*

quantum*:** The sentence, the last transmitted by **N**, will in all probability have concluded by emphasizing the unfettered growth permitted by the greater depth of trenched soil.

APPENDIX I

THE LATIN TEXT SHARED BY **N** (*DE ARBORIBVS POMIFERIS*) AND THE MANUSCRIPTS OF *MEDICINAE EX HOLERIBVS ET POMIS*

I print below the portion of the Latin text that is shared between **N** and the manuscripts of *Med.* The Latin is given in the edition of Maire (2002*a*) with a selective apparatus. This text represents the first seven sentences and the beginning of the eighth sentence of the chapter on quince medicine in *Med.*, that is, *Med.* 43.1–8 in Maire's edition. I restore the title *Medicina ex cydoneis* from **N** where Maire prints *De cydonio* following the MSS of *Med.* Where **N** breaks off after *madescunt*, the manuscripts of *Med.* finish the sentence and contain another 12 (43.9–20 in Maire's edition). According to the hypothesis offered in the Introduction §i(b), the portion of text shared between **N** and the manuscripts of *Med.* is an important trace of the original form of G.'s work. The purpose of printing this extract is therefore to illustrate the remarks on the relationship between *Pom.* and *Med.* offered in the Introduction §i(b), §vii.

Besides **N** the following nine manuscripts are relevant to the shared text, i.e. to *Med.* 43.1–8. I omit **S** (Florentinus Strozz. lat. LXX), probably a copy of **F** (cf. Reeve (2007) 144) and at all events not departing in this text from the readings of **P** or **F**.

G	Sangallensis 752, s. ix	
V	Leidensis Voss. lat. o. 92, s. x$^{in.}$	
L	Londiniensis Reg. 12.E.XX, s. xii$^{1/2}$	
R	Vaticanus Reg. lat. 1004, s. xii	*consensus GVLRBCW per γ indicatur*
B	Bruxellensis lat. 2419–31, s. xii$^{2/2}$	
C	Cantabrigiensis Sanct. Petr. 222: III, s. xiii–xiv	
W	Vindobonensis Ser. n. 2623, s. xvi	
P	Pragensis lat. XIV.A.12, s. xiv–xv	*consensus PF per δ indicatur*
F	Florentinus Aedil. 165, anno 1433	

Medicina ex cydoneis

1,2 Cydonea plurimi medici eustomacha crediderunt. Eadem
Dioscorides et diuretica existimat austeritatis fortasse beneficio.
3 Ceterum styptica conprimere urinam magis quam ciere consu-
4,5 erunt. Cruda austeriora sunt. Cruda itaque et matura coeliacis,
dysentericis, sanguinem reiectantibus prosunt in cibo sumpta.
6 Coquuntur et in cinere calidissimo farina inuoluta et aegris quo-
7 rum diuturna iam febris est sine periculo dantur. Thoracis et
8 stomachi tormenta mitigant. Aqua etiam caelesti madescunt***

Medicina ex cydoneis *N*: De cydonio *uel sim.* γδ   1 Cydonea *N*: Medicina (-am *R*δ) ex cydoneis γδ   crediderunt *om.* γδ   2 fortasse *om.* *NB*   3 ciere *VLBCW*δ: eicere *GR*: acere *N*   4 Cruda austeriora sunt *om.* γδ   5 itaque et γ: itaque *N*: *om.* δ   6 in cinere] cinere *N*δ   quorum] quarum *N*   7 Thoracis et stomachi tormenta mitigant *om.* γδ

APPENDIX II

MANURE IN THE AGRICULTURAL AUTHORS

Manure (*stercus*) was a subject of the highest importance in Roman agriculture, and we find in the agronomists extensive discussion of its different sorts and the uses to which they could be put. Columella actually waxes poetic in its treatment, calling *stercus* the 'food' whereby a farmer may 'restore the flagging strength' of his fields (2.13.3 *ut stercore adiuues et assumptas uires hoc uelut pabulo refoueas*). His claim that manure remained too lightly examined to his day is only partially true, however (2.13.4 *cum priscis auctoribus quamuis non omissa res, leui tamen admodum cura sit prodita*): Cicero had not put praise of manuring in Cato's mouth for nothing – *Sen.* 54 *quid de utilitate loquar stercorandi?* – as consultation of *De agri cultura* shows (cf. Brehaut (1933) 60 n. 1), and Varro had shown the way that Columella would take (see below). Indeed, all the evidence points to the Romans' appreciation of the advantages of manure at an early time.[1] A word on the agronomists' treatment of manure will be useful, since the information is absent from G.; though it is not certain whether he ever discussed manure, he would at all events have expected a general understanding of the topic from his readers (cf. Introduction §III(b), §IV).

The various treatments of manure in Roman agricultural writing fall into two broad groups: (1) those showing a clear mark of Mago's influence and (2) all others. Varro, Columella, Pliny and the *Geoponica* belong to the Magonian tradition, which is first visible in Varro. Varro (*Rust.* 1.38) reports upon the authority of Cassius Dionysius, the Greek translator of Mago (cf. *Rust.* 1.1.10 with Flach, Domínguez Petit (2004) 186–7), that bird dung is best, especially that of pigeons, followed by human faeces, and

[1] The Greeks had of course already expressed similar opinions: cf. Xen. *Oec.* 20.10 ἀλλὰ καὶ κόπρον λέγουσι μὲν πάντες ὅτι ἄριστον εἰς γεωργίαν ἐστί, κτλ.

thirdly the dung of goats, sheep and donkeys. Columella's treatment (2.14.1–9) is wider ranging than Varro's and incorporates much from his personal experience, but is based on a tripartite division and ranking of manures that fundamentally agrees with Cassius Dionysius' (cf. 2.14.1 *tria genera stercoris, ex auibus, ex hominibus, ex pecudibus*). Pliny (*HN* 17.50–7) also has a fairly extensive discussion of manure explicitly drawing on Cato, Varro and Columella, but at least in the initial summary discussion (50–2) his treatment is oriented towards the same Magonian precedent that informs the accounts of Varro and Columella. Finally, the *Geoponica* (2.21) hews so closely to the tripartite division of Varro and Columella as to reproduce even such details as their prohibition against using the dung of waterfowl. In short, the persistent topical ranking of manures (birds, humans, cattle/ovicaprids etc.) indicates a common thread leading back to Mago. There are of course many apparently un-Magonian elements in the authors' opinions on manure, drawn either from other agricultural authorities or from personal judgement and experience, so it would be unwise to place excessive weight on the practical importance of Mago's rankings; nevertheless, the rankings remain an enduring framework for structuring and communicating knowledge and indicate a shared intellectual heritage underlying the various authors' writings.

Other accounts of manure in antiquity differ from Mago's either in structure or content. Different in content is the advice of Theophrastus (following an authority named Chartodras), who packages his recommendations in a list as Mago did, but with a different assessment of quality: *Hist. pl.* 2.7.4 ἡ δὲ κόπρος οὔτε πᾶσιν ὁμοίως οὔθ' ἡ αὐτὴ πᾶσιν ἁρμόττει· τὰ μὲν γὰρ δριμείας δεῖται τὰ δ' ἧττον τὰ δὲ παντελῶς κούφης. δριμυτάτη δὲ ἡ τοῦ ἀνθρώπου . . . δευτεράν δὲ τὴν ὑείαν, τρίτην δὲ αἰγός, τετάρτην δὲ προβάτου, πέμπτην δὲ βοός, ἕκτην δὲ τὴν λοφούρων. Different in structure is Cato, who endorses bird-, goat-, sheep- and ox-dung (*Agr.* 1.30, 1.36, Harrison (1913) 41 n. 1, White (1970) 134–5) but makes no attempt to organize the different sorts according to the topical ranking of manures common in the other agronomists.[2] Finally, Palladius (1.33) also offers a list, but one that differs in important respects from those found in other agronomists (Martin *ad loc.* (n. 3)): he alone ranks ass's dung best of all; he is not as enthusiastic about the bird dung so highly praised by others (it receives merely the passing remark *auium satis utile*);[3] and he omits human excrement altogether.[4]

Outside of these systematic treatments, the agronomists also recommended other manures for specific purposes. Urine and tanner's refuse, for example, were used as fertilizers for unproductive trees (see further 3.7.3.n.). Additionally, the value of composted vegetable matter and refuse, which were also encompassed by the term *stercus*,[5] was appreciated by the Greek and Roman agronomists, though sometimes presented in ambivalent terms. Theophrastus (*Hist. pl.* 2.7.4) says ἡ δὲ συρματῖτις (sc. κόπρος, i.e. manure mixed with litter, σύρμα) ἄλλη καὶ ἄλλως· ἡ μὲν γὰρ ἀσθενεστέρα ταύτης, ἡ δὲ κρείττων, but declines to go

[2] Two explanations come to mind for Cato's indifference: (1) a chronological impediment – Cato was dead by the time Carthage had fallen and a Latin translation of Mago was commissioned – and (2) his well-known xenophobic sentiments, which might have led him to reject foreign precedents. But on the other hand, scholars (e.g. Boscherini (1970), von Staden (1996) 385–94, Diederich (2007) 15–20) have explored the significant Greek influence on Cato's treatise that is apparent in spite of whatever reservations he expressed publicly; and some have suggested, if on an uncertain basis, that Cato's treatise actually shows some familiarity with Mago and Punic agronomy (Cataudella (2002) 43–4, Krings (2008)). Interpreters interested in Cato's connection with Greek and Punic sources may wish to draw attention to the similarities, rather than differences, between Cato's opinions and those of the Magonian camp.

[3] Note however that some Magonian influence may show in his detail *excepto palustrium*.

[4] It is interesting to compare with the various rankings provided by the ancients the value assigned to different manures in the late nineteenth century in America (Beal (1894) 7): in chemical content, the dung of fowl is ranked as the most valuable ($7.07 per ton), followed by sheep and pigs ($3.30, $3.29), next horses and calves ($2.21, $2.18), last cows ($2.02), with human dung omitted for obvious reasons.

[5] The term *fimum*, also used for manure, has a narrower semantic range: see 3.1.1n. *fimum bubulum*.

into the qualities that would affect the strength of the manure. Cato (*Agr.* 37.2) with his usual pragmatism simply offers a list of greens that could be added to a compost heap. Columella (2.14.6–7) recognizes the benefits of composting, but frames it chiefly as an alternative for those who cannot keep birds or beasts on their farms for droppings. The *Geoponica* (2.22) has fairly elaborate instructions for creating an all-purpose dung and refuse pit. At all events scattered remarks of the agronomists show that the use of fresh and composted vegetable matter was in fact the norm: cf. Spurr (1986*b*) 69–70, 126, Alcock et al. (1994) 150, Jashemski (2018*b*) 433. As implied above, manure of all kinds was stored in pits or heaps (*sterculinum*), from which it could be transported to wherever it was required on the farm (Plate 16).[6] For the agronomists' instructions for storing the manure in heaps, see Cato, *Agr.* 5.7–8, Varro, *Rust.* 1.13.4, Col. 1.6.21–2, 2.14, Plin. *HN* 17.57, Pallad. 1.33.

The bibliography on the usage and value of manure in Roman antiquity is extensive: see e.g. André on Plin. *HN* 17.50–7, White (1970) 125–45, Spurr (1986*b*) 126–32, Flach (1990) 251–2, Alcock et al. (1994), Marcone (1997) 64–7, Meana et al. on *Geop.* 2.21–2, Lelli on *Geop.* 2.21–2, Wilson and Flohr (2011), Kron (2012) 6–7, Jones (2012*b*) 5–9, Halstead (2014) 212–30. For manure in Californian orchards, see Wickson (1891) 192–206. For the importance of manure in the sweep of agricultural history, cf. Jones (2012*a*), (2012*b*). Manure played such a large role in Roman agriculture that it affected the economy of the entire farm: maintaining a supply of it was directly linked with developments in animal husbandry on the steading: cf. Halstead (1987); see also White (1970) 113, 125–6, 273–327, Kron (2005) 293–6, Margaritis and Jones (2008) 160–1, Halstead (2014) 289–94.

[6] For the identification of the *cratis stercoraria* in Plate 16, see the remarks of White (1970) on his plate 30, (1975) 77–8. The men are depicted wearing hoods (*cuculli*) as protection against the cold and wet of winter (Col. 1.8.9; see D–S I 2.1578–9); for the importance of manuring in such conditions, see 2.4.2n. *sic . . . germina*.

APPENDIX III

SPELLING ERRORS IN THE MANUSCRIPT

The following table presents a full register of places where **N** errs, or where I have adopted an orthography different from that of **N**. The register here includes both those errors reported in the apparatus and those not reported there (see Introduction §IX).

Location	N	Text
1.1.5	adfirmant	affirmant
1.1.5	adfixi	affixi
1.1.6	eobtecta	eo obtecta
1.1.7	adfirmamus	affirmamus
1.1.7	policentur	pollicentur
1.1.8	separatasobruunt	separata obruunt
1.1.9	millio	milio
1.1.10	inliniri	illiniri
1.1.11	sicentur	siccentur
1.1.12	inlita	illita
1.1.12	spera	aspera
1.1.13	scrouibus	scrobibus
1.1.13	qui	quo
1.1.13	punicaca	punica
1.1.13	uei (uel **N**[c])	uel
1.1.13	aliam	alia <mala>
1.1.14	Quintilli	Quintilii
1.1.14	iscoue	in scobe
1.1.15	cidoneū	cydoneum
1.1.15	censet	censent
1.1.15	que	quae
1.1.16	conpertum	compertum
2.1.2	Quintilli	Quintilii
2.1.3	sepe	saepe
2.1.3	inlibatum	illibatum
2.1.3	adquirant	acquirat
2.1.4	odiequae	hodieque
2.1.4	putant	putent
2.1.4	que	quae
2.1.4	contecta	contacta

Location	N	Text
2.2.1	persiccum	persicum
2.2.1	anxius	anxie
2.2.1	adque	atque
2.2.2	quidam	quidem
2.2.2	mersi	mergi
2.2.2	uiduo	biduo
2.2.2	conponuntur	componuntur
2.2.2	adque tectum adque oblitum	idque tectum atque oblitum
2.3.1	stantum	statum
2.3.2	mergantur (-untur **N**[c])	merguntur
2.3.2	esterius	exterius
2.3.3	mali	mala
2.3.3	extriare	extra hiare
2.3.4	terre	terrae
2.3.5	aseruentur	asseruentur
2.3.5	temporis	teporis
2.3.5	humoris	umoris
2.3.5	iam iam	iam
2.3.5	gelide	gelidae
2.4.1	letissimas	laetissimas
2.4.1	caeperunt	ceperunt
2.4.2	inriguis	irriguis
2.4.2	aquam (*uix intellegitur*)	?
2.4.3	Quintilli	Quintilii
2.4.3	humidum	umidum
2.4.3	addit	addunt
2.4.4	distentat	distendat
2.4.5	inrigua	irrigua
2.4.5	hortensiad	hortensia
2.4.5	humor	umor
2.4.5	uerti	auerti
2.4.5	et ut	ut et
2.4.6	contrarium tum	tum *del.*
2.4.6	harena	harenae
2.4.6	proximata	proximum
2.4.6	que	quae
2.4.6	gummi	gummis
2.5.2	crebior	crebrior
2.5.2	triumensura	trium mensura
2.5.2	scandente	scandent
2.5.5	praecepit	procedit
2.5.6	dispont	disponant

Location	N	Text
2.5.6	adfirment	affirment
2.5.7	que	quae
2.5.7	frenquens	frequens
2.5.9	inertiam	inertia
2.6.1	frenquens	frequens
2.6.1	adibenda	adhibenda
2.6.1	heruarum	herbarum
2.6.1	condegentium	contegentium
2.6.1	adque	atque
2.6.1	aura	aurae
2.6.2	proprius	propius
2.6.2	adfixo	affixo
2.6.2	uiolentia	uiolenta
2.6.3	conplectitur	complectitur
2.6.3	uima	bima
2.6.3	translatam	translata
2.7.1	essetamnecito uiuit	esset; <et> tamen cito bibit (*dub.*)
2.7.1	contraliarum	contra aliarum
2.7.2	consenescant	consenescat
2.7.3	desidera	desiderat
2.7.3	quarum	quorum
2.7.3	quam scilicet binos pedes	*del.*
2.7.4	si	sic
2.7.4	obtime	optime
2.7.5	inuanti	inuenti
2.7.5	consitas tutas	tutas *del.*
2.7.5	pretegunt	protegunt
2.9.1	adseuerant	asseuerant
2.9.1	satariis (sext- $\mathbf{N}^{m}$)	sextariis
2.9.1	inrigetur	irrigetur
2.9.2	honerata	onerata
2.9.2	sit	sunt
2.9.2	relinquis	reliquis
2.9.2	inperatur	imperatur
2.9.3	sipario (-rlo $\mathbf{N}^{c}$ *dub.*)	sparto
2.9.3	ligata	ligandam
2.9.3	amfora	amphora
2.9.3	extimant	existimant
2.9.3	odieque	hodieque
2.9.3	praemedio (pro remedio $\mathbf{N}^{m}$)	pro remedio
2.9.3	suspendant	suspendunt

Location	N	Text
2.10.1	corticem	cortice
2.10.1	precedente	praecedente
2.10.4	abtissimum	aptissimum
2.10.4	praequoquo	praecoquo
2.10.4	sepe	saepe
2.10.5	abentur	habentur
2.10.5	que	quae
2.10.5	probeniet	proueniet
2.10.6	amigdalum	amygdalum
2.10.7	floriora	floridiora
2.10.7	adsuescunt	assuescunt
2.10.8	ad	at
2.10.8	adque	atque
2.10.8	difficilius	difficilius <recipiunt>
2.10.8	ed	et
2.10.8	sinfirmitatem	sint infirmitatem
2.11.1	oleum	olea
2.11.2	plerunquesse	plerumque esse
2.11.2	arenoso	harenoso
2.12.1	oximelle	oxymelle
2.12.1	squallor	squalor
2.12.2	plerunque	plerumque
2.12.2	temtarunt	temptarunt
2.12.2	adque	atque
2.12.2	rigido	frigido
2.12.2	serbanda	seruanda
2.12.4	Suesatibus	Suessatibus
2.12.4	inpicato	impicato
2.12.4	umpilico	umbilico
2.12.4	adque	atque
2.12.4	sepe	saepe
2.12.4	flacescunt	flaccescunt
3.1.1	fimum	<in> fimum
3.1.1	hiacere	hiascere
3.1.2	tercorum (st- **N**[c] *dub.*) . . . minutissima	terram . . . minutissimam
3.1.2	stercori	stercoris
3.1.2	eateus	eatenus
3.1.3	trina angulari	triangulari
3.1.4	molliuntur	moliuntur
3.1.4	faciem	aciem
3.1.4	id	ad
3.1.4	Fabonium	Fauonium

Location	N	Text
3.1.4	spectet et	et *del.*
3.1.5	nucullus	nec ullus
3.1.5	subministrentur	sumministrentur
3.2.1	adfigitur	affigitur
3.2.1	admiculo	adminiculo
3.2.1	conpletum (*uix intellegitur*)	?
3.2.2	exaceto	exacto
3.3.1	Quintilli	Quintilii
3.3.2	carnalibus	canalibus
3.3.2	inmoretur	immoretur
3.3.3	culturae	cultura
3.3.3	scrobe	scrobes
3.3.4	nouam	nouam <normam>
3.3.4	diffusam	diffissam
3.3.4	nuci	nuce
3.3.4	amigdali	amygdali
3.3.4	adseuerare	asseuerare
3.3.5	sicut	sic et
3.3.5	prestare	praestare
3.3.5	roncinias (Terentinas $\mathbf{N}^{m}$)	Tarentinas
3.4.3	adcommodant	accommodant
3.5.1	sumtuosa	sumptuosa
3.5.1	annui	annuae
3.5.1	prefractos	praefractos
3.5.1	falcae	falce
3.6.1	generent	generant
3.6.2	teda	taeda
3.6.2	illi (ali $\mathbf{N}^{m}$, uli $\mathbf{N}^{c}$)	alii
3.6.2	taede	taedae
3.6.3	perforatisilicem	perforatis silicem
3.6.4	aut pruina	aut *del.*
3.6.4	arenae	harenae
3.6.4	umulo	cumulo
3.6.4	adgregant	aggregant
3.6.4	adque	atque
3.6.4	que	quae
3.6.4	precipiunt	praecipiunt
3.7.1	inrigata	irrigata
3.7.2	stipide	stipite
3.7.2	pruna	pituita
3.7.2	fixserunt	fixerunt
3.7.4	solis	soli
3.7.4	adsignat	assignat

Location	N	Text
3.7.4	crebius	crebrius
3.7.4	dulgior	dulcior
3.7.5	illico praesto (*ex* presto **N**)	praesto *del.*
3.7.5	salilla	saliua
3.7.6	teniora	tenuiora
3.7.6	queque	quaeque
3.8.1	adseruanda	asseruanda
3.8.2	cortes	cortex
3.8.3	liueratae	liberatae
4.1.1	conripiat	corripiat
4.1.1	ille	illae
4.1.1	precinit	praecinit
4.1.1	Amarillas	Amaryllis
4.1.2	brebiter	breuiter
4.1.3	amygdalo eorum	amygdalorum
4.1.4	fit	fuit
4.1.4	uariae	uarie
4.1.4	adfirmat	affirmat
4.1.4	Nobembri	Nouembri
4.1.4	nonciat	nunciat
4.1.4	ut	ut <uix>
4.1.4	quandoque quae	quando quaeque
4.1.4	culpam	culpa
4.1.5	ceduam	caeduam
4.1.5	adsentiatur	assentiatur
4.1.6	Nobembri	Nouembri
4.1.6	indecreuit	decreuit
4.1.7	adsignat	assignat
4.1.8	culturae (*ex* -re **N**) alia	*del.*
4.1.8	rectissimae	rectissime
4.1.8	castaneas serere rentur	castaneae sererentur
4.1.9	Quintilliis	Quintiliis
4.2.1	Grecis	Graecis
4.2.1	actuni	Arcturi
4.2.1	et exortu	et *del.*
4.2.2	inmerito	immerito
4.2.2	tegrumento	tegumento
4.2.2	ut	cum
4.2.2	labantur	labuntur
4.2.3	ad oc	ad hoc
4.2.3	reserbata	reseruata
4.2.4	incolumitatem	incolumitate
4.3.1	sationem	satione

Location	N	Text
4.3.1	anxiae	anxie
4.3.1	adque	atque
4.3.1	sollicitae	sollicite
4.3.2	siccum quae	siccumque
4.3.2	ita uniuersas	ita <ut> uniuersas
4.3.2	arena	harena
4.3.2	flubialis	fluuialis
4.3.2	que	quae
4.3.3	harenam	harena
4.3.3	remobetur	remouetur
4.3.3	in nucibus	in *del.*
4.3.3	missisanas	missis sanas
4.3.3	corruptisic	corruptis sic
4.3.3	segregarere	segregare
4.3.4	qua	quae
4.3.4	iudicanto	iudicante
4.3.5	inclusi	includi
4.4.1	lauimento	pauimento
4.4.2	conserbandas	conserendas
4.4.3	succus	sucus
4.4.3	recreat	recreet
4.4.3	cum dilata	cum <nuce> dilata
4.4.4	inmenti	inuenti
4.4.4	ex hoc	et hoc
4.4.4	uliquod	aliquod
4.4.4	annum	animum
4.5.1	castaneam	castanea
4.5.1	arenosum	harenosum
4.5.3	ult	uult
4.5.3	in argillam glarea	in argilla in glarea
4.6.1	autores	auctores
4.6.3	Neapoli	Neapolis
4.6.3	maximae	maxime
4.6.3	gloriaetur	glorietur
4.6.3	adiuncxit	adiunxit
4.6.3	possae	posset
4.6.3	mones	omnes
4.6.4	letissimum	laetissimum
4.6.6	Quintillis	Quintilii
4.6.6	necessest	necesse est
4.6.6	tunc	hunc
4.7.2	Atticos	Atticus
4.7.3	aliquad (-od **N**^{m})	aliquot

Location	N	Text
4.7.3	Grecis	Graecis
4.7.3	salutis	satis
4.7.3	que	quae
4.7.4	dipundi	dipondi
4.7.4	braebitate	breuitate
4.7.4	mensula	mensura
4.7.4	halerem	alere
4.7.4	mensura	mansura
4.7.4	inmerito	immerito
4.7.4	haberit	haberet

ABBREVIATIONS AND BIBLIOGRAPHY

I ANCIENT WORKS

Editions of Greek texts follow the *Thesaurus Linguae Graecae*. Editions of Latin texts follow those in the corpus provided by the Packard Humanities Institute (http://latin.packhum.org/), or, failing that, those in the Library of Latin Texts provided by BREPOLiS. As important exceptions to these rules, I use Amigues for the texts of Theophrastus, *De causis plantarum* (2012–17) and *Historia plantarum* (1988–2006); André for Pliny the Elder, *Historia naturalis* 15 (1960), 17 (1964), and 18 (1972); Maire for Gargilius Martialis, *Medicinae ex holeribus et pomis* (2002); and Rodgers for Columella (2010) and Palladius (1975): see below, section IV.

Ancient works are abbreviated according to the conventions of the *Oxford Classical Dictionary* (4th edn). For the abbreviation of Latin authors and texts not given in the *Oxford Classical Dictionary*, I follow the conventions of the *Thesaurus Linguae Latinae*. The following list presents a small number of abbreviations that are either an exception to or are not accommodated by these rules:

Arb.	[Columella], *Incerti auctoris liber de arboribus*
Col.	Columella, *Res rustica*
Med.	Gargilius Martialis, *Medicinae ex holeribus et pomis*
Pom.	Gargilius Martialis, *De arboribus pomiferis* (cited by chapter, section and sentence number as given in this edition)
cydon.	<I *De cydoneis*> = *Pom.* 1
persic.	<II *De persicis*> = *Pom.* 2
amygd.	<III *De amygdalis*> = *Pom.* 3
cast.	<IV *De castaneis*> = *Pom.* 4

II ABBREVIATIONS: SELECTED WORKS PERTAINING TO *DE ARBORIBVS POMIFERIS*

In the commentary, references to the following works are to the author's or editor's surname plus page or column number as necessary, e.g. Antonelli, col. 1682 = Antonelli (1848) column 1682; Condorelli xix = Condorelli (1978) xix. Works cited also by author and date, in the Introduction, are included additionally in the respective list (below, section v) for ease of reference.

J. Antonelli, *Gargilii Martialis quae supersunt: frammenti de' libri perduti di Gargilio Marziale volgarizzati per la prima volta con note*. Venice, 1848

S. Condorelli, *Gargilii Martialis quae exstant*. Biblioteca di Helikon Testi e Studi 11. Rome, 1978

A. Mai, *Classicorum auctorum e Vaticanis codicibus editorum tomus 1, complectens Ciceronis De rep. quae supersunt, Gargilii Martialis De arboribus pomiferis, Sallustii Historiarum et Archimedis fragmenta, cum quinque tabulis aeneis*. Rome, 1828 [On the date of publication, see Introduction §vii]

[Mai[2] =] A. Mai, *Gargilii Martialis quae supersunt e codicibus Neapolitano et Vaticanis*. Lüneburg, 1832

[Mai[3] =] A. Mai, *Gargilii Martialis operis deperditi de hortis fragmenta ex codicibus Neapolitano et Romano cum scholiis palaeographiciis et adnotationibus*. Rome, 1846

F. Marx, *A. Cornelii Celsi quae supersunt*. Leipzig and Berlin, 1915

I. Mazzini, *Q. Gargilii Martialis De hortis*. Opuscula philologa 1. Bologna, 1988[2] [1st edn 1978]

A. A. Scotti, 'Memoria del socio A. A. Angelo sopra un codice palimpsesto della Real Biblioteca Borbonica', *Memorie della regale accademia ercolanese di archeologia* 2 (1833) 119–85 [On the date of publication, see Introduction §vii]

M. Winterbottom, review of I. Mazzini, *Q. Gargilii Martialis De hortis* (Bologna 1978), *CR* 30 (1980) 143–4

III ABBREVIATIONS: REFERENCE WORKS

AOM	*Almond orchard management*. Berkeley, CA, 1978
Beekes	R. S. P. Beekes, *Etymological dictionary of Greek*, 2 vols. Leiden Indo-European Etymological Dictionary Series 10. Leiden and Boston, 2010
Chantraine	P. Chantraine, *Dictionnaire étymologique de la langue grecque: histoires des mots*, revised edn with suppl. by A. Blanc, C. de Lamberterie, and J.-L. Perpillou. Paris, 1999
CG	B. Maire, *Concordantiae Gargilianae*. Hildesheim, 2002
CIL	*Corpus inscriptionum Latinarum*. Berlin, 1863–
D–S	C. Daremberg and E. Saglio (eds.), *Dictionnaire des antiquités grecques et romaines d'après les textes et les monuments*. Paris, 1877–1919
de Vaan	M. de Vaan, *Etymological dictionary of Latin and the other Italic languages*. Leiden Indo-European Etymological Dictionary Series 7. Leiden and Boston, 2008
DGRA	W. Smith, W. Wayte, and G. E. Marindin (eds.), *A dictionary of Greek and Roman antiquities*, 2 vols. London, 1890[3]

EFN J. Janick and R. E. Paull (eds.), *The encyclopedia of fruit and nuts*. Cambridge, MA, 2008

E–M A. Ernout and A. Meillet, *Dictionnaire étymologique de la langue latine*. Paris, 1951^{3}

FRH T. J. Cornell (ed.), *The fragments of the Roman historians*, 3 vols. Oxford, 2013

H–Sz J. B. Hofmann, *Lateinische Syntax und Stilistik*, revised by A. Szantyr. Munich, 1965

ILS H. Dessau (ed.), *Inscriptiones Latinae selectae*. Berlin, 1892–1916

K–S R. Kühner and C. Stegmann, *Ausführliche Grammatik der lateinischen Sprache*, 2 vols. Hanover, 1912–14^{2}

LAT M. G. T. Bruno, *Il lessico agricolo latino*. Amsterdam, 1969^{2}

L&S C. T. Lewis and C. Short, *A Latin dictionary founded on Andrews' edition of Freund's Latin dictionary*. Oxford, 1879

Leumann M. Leumann, *Lateinische Laut- und Formenlehre*, revised edn. Munich, 1977

LTBL J. André, *Lexique des termes de botanique en latin*. Études et Commentaires 23. Paris, 1956

Maltby R. Maltby, *A lexicon of ancient Latin etymologies*. ARCA Classical and Medieval Texts, Papers and Monographs 25. Leeds, 1991

NLS E. C. Woodcock, *A new Latin syntax*. Cambridge, 1959

NP H. Cancik and H. Schneider (eds.), *Brill's New Pauly: encyclopaedia of the ancient world. Antiquity*, English edn edited by C. F. Salazar. Boston and Leiden, 2002–11

NPRA J. André, *Les noms de plantes dans la Rome antique*. Paris, 1985

OCD *The Oxford classical dictionary*, online edn. Oxford, 2015–^{5}

OED *The Oxford English dictionary*, online edn. Oxford, 2000–^{2}

OLD *The Oxford Latin dictionary*. Oxford, 2012^{2}

Pinkster H. Pinkster, *The Oxford Latin syntax. Volume 1: The simple clause*. Oxford, 2015

PIR^{2} *Prosopographia imperii Romani saec. I. II. III.* Berlin and Leipzig, 1933–2015^{2}

RE A. Pauly and G. Wissowa (eds.), *Real-Encyclopädie der classischen Altertumswissenschaft*. Stuttgart, 1893–1980

REW W. Meyer-Lübke, *Romanisches etymologisches Wörterbuch*. Heidelberg, 1935^{3}

S–H M. Schanz, *Geschichte der römischen Litteratur bis zum Gesetzgebungswerk des Kaisers Justinian*, revised by C. Hosius, 4 vols. Munich, 1927–59^{4}

Syntactica E. Löfstedt, *Syntactica: Studien und Beiträge zur historischen Syntax des Lateins*, 2 vols. Lund, 1933–56^{2}

TLL *Thesaurus linguae Latinae*. Leipzig/Berlin, 1900–

Wackernagel J. Wackernagel, *Lectures on syntax: with special reference to Greek, Latin, and Germanic*, edited and translated by D. Langslow. Oxford, 2009

IV WORKS CITED BY AUTHOR NAME

References to the following editions, commentaries and translations are to the editor's or translator's surname only. Works cited also by author and date are included additionally in the respective list (below, section v) for ease of reference.

Amigues, S. *Théophraste: Recherches sur les plantes*, 5 vols. Paris, 1988–2006

Théophraste: Les causes des phénomènes végétaux, 3 vols. Paris, 2012–17

André, J. *Pline l'Ancien: Histoire naturelle. Livre XIV*. Paris, 1958

Pline l'Ancien: Histoire naturelle. Livre XV. Paris, 1960

Pline l'Ancien: Histoire naturelle. Livre XVII. Paris, 1964

Pline l'Ancien: Histoire naturelle. Livre XVIII. Paris, 1972

Columelle: De l'agriculture. Livre XII. Paris, 1988

Boldrer, F. *L. Iuni Moderati Columellae rei rusticae liber decimus (carmen de cultu hortorum)*. Testi e studi di cultura classica 17. Pisa, 1996

Callebat, L., P. Gros, and C. Jacquemard. *Vitruve: De l'architecture. Livre II*. Paris, 2003[2]

Clausen, W. *A commentary on Virgil, Eclogues*. Oxford and New York, 1994

Coleman, K. M. *Statius: Silvae IV*. Oxford and New York, 1988

Martial: Liber spectaculorum. Oxford and New York, 2006

Dalby, A. *Cato: On farming (De agricultura). A modern translation with commentary*. Blackawton, 1998

Drossaart Lulofs, H. J., and E. L. J. Poortman. *Nicolaus Damascenus De plantis: five translations*. Verhandelingen der Koninklijke Nederlandse Akademie van Wetenschappen, Afd. Letterkunde, NS 139. Amsterdam and New York, 1989

Dyck, A. R. *Cicero: Catilinarians*. Cambridge, 2008

Fedeli, P. *Properzio: Elegie libro II. Introduzione, testo e commento*. ARCA Classical and Medieval Texts, Papers and Monographs 45. Cambridge, 2005

Fitch, J. G. *Palladius: The Work of farming (Opus agriculturae) and Poem on grafting*. Totnes, 2013

Flach, D. *M. T. Varro: Gespräche über die Landwirtschaft*, 3 vols. Texte zur Forschung 65–7. Darmstadt, 1996–2002

Forster, E. S., and E. H. Heffner. *Columella: On agriculture. Volume 3: Books 10–12, On trees*. Loeb Classical Library 408. Cambridge, MA, 1968

Goujard, R. *Columelle: <Les arbres>*. Paris, 1986

Heraeus, W. *Die Appendix Probi*. Leipzig, 1899

Lachmann, K. *Gromatici veteres: die Schriften der römischen Feldmesser. Erster Band*. Berlin, 1848

Lelli, E. *L'agricoltura antica: i Geoponica di Cassiano Basso*, 2 vols. Altri classici 3. Soveria Mannelli, 2010

Lommatzsch, E. *P. Vegeti Renati digestorum artis mulomedicinae libri*. Leipzig, 1903

Maire, B. *Gargilius Martialis: Les remèdes tirés des légumes et des fruits*. Paris, 2002

Mariotti, I. *Marii Victorini Ars grammatica: introduzione, testo critico e commento*. Biblioteca Nazionale, Serie dei classici greci e latini 6. Florence, 1967

Martin, R. *Palladius: Traité d'agriculture. Tome I: Livres I et II*. Paris, 2003

Meana, M. J., J. I. Cubero and P. Sáez. *Geopónica o extractos de agricultura de Casiano Baso*. Madrid, 1988

Mynors, R. A. B. *Virgil: Georgics*, with a preface by R. G. M. Nisbet. Oxford, 1990

Oakley, S. P. *A commentary on Livy: books VI–X*, 4 vols. Oxford, 1997–2005

Powell, J. G. F. *Cicero: Cato Maior de senectute*. Cambridge Classical Texts and Commentaries 28. Cambridge, 1988

Rodgers, R. H. *Palladii Rutilii Tauri Aemiliani uiri inlustris opus agriculturae. De ueterinaria medicina. De insitione*. Leipzig, 1975

Frontinus de aquaeductu urbis Romae. Cambridge Classical Texts and Commentaries 42. Cambridge, 2004

Columellae res rustica. Incerti auctoris liber de arboribus. Oxford, 2010

Summers, W. C. *Select letters of Seneca edited with introduction and explanatory notes*. London, 1910

Wilkins, A. S. *Cicero: De oratore I–III*. Oxford, 1892

Winterbottom, M. *The minor declamations ascribed to Quintilian*. Texte und Kommentare 13. Berlin and New York, 1984

V WORKS CITED BY AUTHOR NAME AND DATE

References to the following works are to the author's surname and date, e.g. Adams (1976), Zohary et al. (2012).

Abellán, C. A. (2006) (ed.) *Latin vulgaire – latin tardif VII: Actes du VIIème colloque international sur le latin vulgaire et tardif (Séville, 2–6 septembre 2003)*. Colección Actas 54. Seville

Adams, J. N. (1976) *The text and language of a Vulgar Latin chronicle (Anonymus Valesianus II)*. Bulletin of the Institute of Classical Studies Supplement 36. London

(1977) *The Vulgar Latin of the letters of Claudius Terentianus (P. Mich.* VIII, *467–72).* Publications of the Faculty of Arts of the University of Manchester 23. Manchester

(1982) *The Latin sexual vocabulary*. London

(1992*a*) 'Iteration of compound verb with simplex in Latin prose', *Eikasmos* 3: 295–8

(1992*b*) 'Notes on the text, language and content of some new fragments of Pelagonius', *CQ* 42: 489–509

(1994) *Wackernagel's Law and the placement of the copula* esse *in classical Latin*. Cambridge Philological Society, Supplementary volume 18. Cambridge

(1995*a*) *Pelagonius and Latin veterinary terminology in the Roman Empire*. Studies in Ancient Medicine 11. Leiden and New York

(1995*b*) 'The language of the Vindolanda writing tablets: an interim report', *JRS* 85: 86–134

(2003) 'The new Vindolanda writing-tablets', *CQ* 53: 530–75

(2005) 'The accusative + infinitive and dependent *quod-/quia*-clauses: the evidence of non-literary Latin and Petronius', in Kiss et al. (2005) 195–206

(2007) *The regional diversification of Latin: 200 BC–AD 600*. Cambridge

(2011) 'Late Latin', in Clackson (2011) 257–83

(2013) *Social variation and the Latin language*. Cambridge

(2016) *An anthology of informal Latin, 200 BC–AD 900: fifty texts with translations and linguistic commentary*. Cambridge

Adams, J. N., and W. de Melo (2016) '*Ad* versus the dative: from early to late Latin', in Adams and Vincent (2016) 87–131

Adams, J. N., and N. Vincent (2016) (eds.) *Early and late Latin: continuity or change?* Cambridge

Aertsen, H., M. Hannay and R. Lyall (2004) (eds.) *Words in their places: a festschrift for J. Lachlan Mackenzie*. Amsterdam

Ager, B. K. (2010) *Roman agricultural magic*. Diss., University of Michigan

Albrecht, M. von (2003) *Cicero's style: a synopsis followed by selected analytic studies*. Mnemosyne Supplements 245. Leiden and Boston

Alcock, S. E., J. F. Cherry and J. L. Davis (1994) 'Intensive survey, agricultural practice and the classical landscape of Greece', in Morris (1994) 137–69

Ameling, W. (1993) *Karthago: Studien zu Militär, Staat und Gesellschaft*. Vestigia 45. Munich

Andersson, A. (1938) *Studia Vegetiana: commentatio academica*. Uppsala

André, J. (1986) 'Sur la constitution des langues techniques en Latin', *EL* 1: 3–18

Antonelli, J. (1848) *Gargilii Martialis quae supersunt: frammenti de' libri perduti di Gargilio Marziale volgarizzati per la prima volta con note*. Venice

Argentea aetas = *Argentea aetas in memoriam Entii V. Marmorale*. Pubblicazioni dell'Istituto di filologia classica e medievale 37. Genoa, 1973

Astin, A. E. (1978) *Cato the Censor*. Oxford

Atherton, C. (1997) (ed.) *Form and content in didactic poetry*. Nottingham Classical Literature Studies 5. Bari

Axelson, B. (1945) *Unpoetische Wörter: ein Beitrag zur Kenntnis der lateinischen Dichtersprache*. Skrifter utgivna av Vetenskaps-societeten i Lund 29. Lund

Baehrens, W. A. (1912) *Beiträge zur lateinischen Syntax*. Philologus Supplementband 12.2. Leipzig

(1922) *Sprachlicher Kommentar zur vulgärlateinischen Appendix Probi*. Halle an der Saale

Bain, D. (2007) 'Low words in high places: sex, bodily functions, and body parts in Homeric epic and other higher genres', in Finglass et al. (2007) 40–57

Baldi, P., and P. Cuzzolin (2009) (eds.) *New perspectives on historical Latin syntax. Volume 1: Syntax of the sentence*. Trends in Linguistics Studies and Monographs 180.1. Berlin

Baldi, P., and P. Cuzzolin (2011) (eds.) *New perspectives on historical Latin syntax. Volume 4: Complex sentences, grammaticalization, typology*. Trends in Linguistics Studies and Monographs 180.4. Berlin

Baños Baños, J. M. (2009*a*) (ed.) *Sintaxis del latín clásico*. Madrid

(2009*b*) 'Subordinación completiva', in Baños Baños (2009*a*), 523–62

Bartoldus, M. J. (2012) *Palladius Rutilius Taurus Aemilianus: Welt und Wert spätrömischer Landwirtschaft*. Augsburg

Bauer, B. L. M. (2016) 'The development of the comparative in Latin texts', in Adams and Vincent (2016) 313–39

Beagon, M. (1992) *Roman nature: the thought of Pliny the Elder*. Oxford

Beal, W. H. (1894) *Barnyard manure*. Washington, DC

Becker, J. A., and N. Terrenato (2012) (eds.) *Roman republican villas: architecture, context, and ideology*. Papers and monographs of the American Academy in Rome 32. Ann Arbor, MI

Bedon, R., and E. Hermon (2005) (eds.) *Concepts, pratiques, et enjeux environnementaux dans l'empire romain*. Caesarodunum 39. Limoges

Bennett, C. E. (1910–14) *Syntax of early Latin*, 2 vols. Boston

Bergmann, B. (2014) 'The concept of boundary in the Roman garden', with discussion, in Coleman (2014*a*) 245–99

(2018) 'Frescoes in Roman gardens', in Jashemski et al. (2018) 278–316

Berry, D. H. (1996*a*) *Cicero: Pro P. Sulla oratio*. Cambridge Classical Texts and Commentaries 30. Cambridge

(1996*b*) 'The value of prose rhythm in questions of authenticity: the case of *De optimo genere oratorum* attributed to Cicero', in Cairns and Heath (1996) 47–74

Bertholet, F., and K. Reber (2010) (eds.) *Jardins antiques: Grèce – Gaule – Rome*. Regards sur l'Antiquité 2. Gollion

Bertocchi, A., and M. Maraldi (2003) 'Some concessive expressions in the passage from classical to late Latin', in Solin et al. (2003) 459–77

Bertocchi, A., and A. Orlandini (1996) 'Quelques aspects de la comparaison en latin', *IF* 101: 195–232

Bilderback, T., R. E. Bir and T. G. Ranney (2014) 'Grafting and budding nursery crop plants', *NC Cooperative Extension Resources* [Available at http://content.ces.ncsu.edu/grafting-and-budding-nursery-crop-plants/]

Birley, A. R. (2005) *The Roman government of Britain*. Oxford

Biville, F., M.-K. Lhommé and D. Vallat (2012) (eds.) *Latin vulgaire – latin tardif IX: Actes du IXe colloque international sur le latin vulgaire et tardif (Lyon, 2–6 septembre 2009)*. Collection de la Maison de l'Orient et de la Méditerranée 49. Lyons

Blanc, N. (2015) 'Paradis et *hortus conclusus*: formes et sens de la clôture', in Morvillez (2015) 105–30

Blomgren, S. A. (1937) *De sermone Ammiani Marcellini quaestiones variae*. Uppsala Universitets Årsskrift 1937, 6. Uppsala

(1944) 'In Venantii Fortunati carmina adnotationes', *Eranos* 42: 100–34

Boll, F. (1911) *Griechische Kalender II: der Kalender der Quintilier und die Überlieferung der* Geoponica. Sitzungsberichte der Heidelberger Akademie der Wissenschaften, Philosophisch-historische Klasse 1911, 1. Heidelberg

Boscherini, S. (1970) *Lingua e scienza greca nel* De agri cultura *di Catone*. Ricerche di Storia della Lingua Latina 8. Rome

Bounous, G., and D. T. Marinoni (2005) 'Chestnut: botany, horticulture, and utilization', *Horticultural Reviews* 31: 291–347

Bovey, M. (1999) 'La greffe de l'olivier sur le figuier: Columelle, Virgile et la greffe des arbres', *REL* 77: 184–204

Bowman, A. K., and A. I. Wilson (2013) (eds.) *The Roman agricultural economy: organization, investment, and production*. Oxford and New York

Brehaut, E. (1933) *Cato the Censor on farming*. Records of Civilization, Sources and Studies 17. New York

Brenni, G. M. R. (1985) *The dolia and sea-borne commerce of Imperial Rome*. M.A. thesis, Texas A&M University

Brouquier-Reddé, V. (1991) 'La sandale en fer du boeuf romain ou bousandale', *SJ* 46: 41–56

Bruère, R. T. (1956) 'Pliny the Elder and Virgil', *CPh* 51: 228–46

Bruun, C. (2015) 'Water use and productivity in Roman agriculture', in Erdkamp et al. (2015) 132–49

Buckley, E., and M. Dinter (2013) (eds.) *A companion to the Neronian age*. Malden, MA

Burton, P. (2000) *The Old Latin Gospels: a study of their texts and language*. New York and Oxford

Cabrillana, C. (2011) 'Purpose and result clauses', in Baldi and Cuzzolin (2011) 19–92

Cairns, F., and M. Heath (1996) (eds.) *Papers of the Leeds international Latin seminar 9: Roman poetry and prose, Greek poetry, etymology, and historiography*. ARCA Classical and Medieval Texts, Papers and Monographs 34. Leeds

Calboli, G. (1989) (ed.) *Subordination and other topics in Latin: Proceedings of the third colloquium on Latin linguistics (Bologna, 1–5 April 1985)*. Studies in language companion series 17. Amsterdam and Philadelphia

(1990) (ed.). *Latin vulgaire – latin tardif II: Actes du IIème colloque international sur le latin vulgaire et tardif (Bologne, 29 août – 2 septembre 1988)*. Tübingen

(1995) 'La frase consecutiva graeca: rapporto con la sua sovraordinata e uso dei modi', in Cuzzolin (1995) 57–73

(2003) *Marci Porci Catonis oratio pro Rhodiensibus: Catone, l'Oriente greco e gli imprenditori romani*, 2nd edn. Bologna

(2005) (ed.) *Papers on grammar IX. Latina lingua! Nemo te lacrimis decoret neque funera fletu faxit. Cur? Volitas viva per ora virum: Proceedings of the twelfth international colloquium on Latin linguistics (Bologna, 9–14 June 2003)*. Rome

(2012) 'Syntaxe nominale et subordination en latin tardif', in Biville et al. (2012) 439–51

Callebat, L. (1982) 'La prose du "De architectura" de Vitruve', *ANRW* II 30.1: 696–722

(1990) 'Langages techniques et langue commune', in Calboli (1990) 45–56

(1992) 'Problèmes formels de la vulgarisation scientifique et technique', in Iliescu and Marxgut (1992) 63–73

Callender, M. H. (1965) *Roman amphorae with index of stamps*. London and New York

Campbell, B. (2000) *The writings of the Roman land surveyors: introduction, text, translation and commentary*. Journal of Roman Studies monographs 9. London

Cataudella, M. (2002) 'Agronomia', in Santini et al. (2002) 31–62

Cennamo, M. (1998) 'The loss of the voice dimension between late Latin and early Romance', in Schmid et al. (1998) 77–100

Chahoud, A. (2011) 'The language of Latin verse satire', in Clackson (2011) 367–83

Champlin, E. (1981) 'Serenus Sammonicus', *HSCP* 85: 189–212

Chantraine, P. (1963) *Grammaire homérique II: syntaxe*. Paris

Chastagnol, A. (1994) *Histoire Auguste: les empereurs romains des IIe et IIIe siècles. Édition bilingue latin-français*. Paris

Chiesa, P., A. M. Fagnoni and R. E. Guglielmetti (2017) (eds.) *Ingenio facilis: per Giovanni Orlandi (1938–2007)*. Millennio Medievale 111 / Strumenti e Studi 43. Florence

Chouquer, G., and F. Favory (2001) *L'arpentage romain: histoire des textes, droit, techniques*. Paris

Christmann, E. (1982) 'Zur antiken Georgica-Rezeption', *WJA* 8: 57–67

Cichorius, C. (1887) 'Gargilius Martialis und die Maurenkriege unter Gallienus', *Leipziger Studien zur classischen Philologie* 10: 319–27

(1922) *Römische Studien: Historisches, Epigraphisches, Literargeschichtliches aus vier Jahrhunderten Roms*. Leipzig

Cipolla, C. (1907) *Codici bobbiesi della Biblioteca nazionale universitaria di Torino*. Collezione paleografica bobbiese 1. Milan

Clackson, J. (2011) (ed.) *A companion to the Latin language*. Malden, MA

Clackson, J., and G. Horrocks (2007) *The Blackwell history of the Latin language*. Malden, MA and Oxford

Clausen, W. (1955) '*Silva coniecturarum*', *AJPh* 76: 47–62

(1965) review of P. J. Enk, *Sex. Propertii Elegiarum liber secundus* (Leiden 1962), *AJPh* 86: 95–101

Clément-Tarantino, S. (2006) 'La poétique romaine comme hybridation féconde: les leçons de la greffe (Virgile, *Géorgiques*, 2, 9–82)', *Interférences, Ars scribendi* 4: 1–26

Coleman, K. M. (2012) 'Bureaucratic language in the correspondence between Pliny and Trajan', *TAPhA* 142: 189–239

(2014*a*) (ed.) *Le jardin dans l'antiquité*. Entretiens sur l'Antiquité Classique 60. Vandoeuvres

(2014*b*) 'Melior's plane tree: an introduction to the ancient garden', in Coleman (2014*a*) 1–26

Coleman, R. G. G. (1975) 'Greek influence on Latin syntax', *TPhS* 74: 101–56

(1993) 'Vulgar Latin and proto-Romance: minding the gap', *Prudentia* 25: 1–14

(1999) 'Vulgarism and normalization in the text of *Regula Sancti Benedicti*', in Petersmann and Kettemann (1999) 345–56

Collins, D. (2008) *Magic in the ancient Greek world*. Malden, MA

Collura, P. (1943) *Studi paleografici: la precarolina e la carolina a Bobbio*. Fontes Ambrosiani 22. Milan

Condorelli, S. (1978) *Gargilii Martialis quae exstant*. Biblioteca di Helikon Testi e Studi 11. Rome

(1985) 'Gargiliana', *NAFM* 3: 1025–33

(1995–8) 'Su alcuni spunti Pliniani nella c.d. *Medicina* di Gargilio Marziale', *Helikon* 35–38: 241–82

Cool, H. E. M. (2006) *Eating and drinking in Roman Britain*. Cambridge and New York

Cooper, F. T. (1895) *Word formation in the Roman sermo plebeius: an historical study of the development of vocabulary in vulgar and late Latin, with special reference to the Romance languages*. Boston and London

Courtney, E. (1988) review of M. Pulbrook, *Ovid: Nux* (Maynooth, Co. Kildare 1985), *AJPh* 109: 274–8

Craik, E. M. (1990) (ed.) *Owls to Athens: essays on classical subjects presented to Sir Kenneth Dover*. Oxford

Cubberly, A. (1995) 'Bread-baking in Ancient Italy. *Clibanus* and *sub testu* in the Roman world: further thoughts', in Wilkins et al. (1995) 55–68

Cubberly, A., J. A. Lloyd and P. C. Roberts (1988) '*Testa* and *clibani*: the baking covers of classical Italy', *PBSR* 56: 98–119

Cuomo, S. (2011) 'A Roman engineer's tales', *JRS* 101: 143–65

Curtis, R. I. (2001) *Ancient food technology*. Technology and change in history 5. Leiden and Boston

(2016) 'Food storage technology', in Irby (2016) II 587–604

Cuzzolin, P. (1994) *Sull'origine della costruzione* dicere quod*: aspetti sintattici e semantici*. Pubblicazioni della Facoltà di Lettere e Filosofia dell'Università di Pavia 72. Florence

(1995) (ed.) *Studi di linguistica greca*. Materiali linguistici 16. Milan

(2011) 'Comparative and superlative', in Baldi and Cuzzolin (2011) 549–659

Dalby, A. (2003) *Food in the ancient world from A to Z*. London and New York

(2011) *Geoponika: Farm work. A modern translation of the Roman and Byzantine farming handbook*. Totnes

Danckaert, L. (2014) '*Quidem* as a marker of emphatic polarity', *TPhS* 112: 97–138

(2015) 'Spelling out the obvious: Latin *quidem* and the expression of presuppositional polarity', *Journal of Historical Pragmatics* 16: 109–41

Davidovits, F. (1994) 'A la découverte du *carbunculus*', *Voces* 5: 33–46

De Angelis, A. (1995) *La coltivazione delle piante da frutto nella letteratura agronomica latina: dalla produzione alla conservazione dei frutti*. Academia Latinitati fovendae, Varia 6. Rome

(1997*a*) 'La coltivazione delle piante da frutto: un'occasione di rilettura dei testi agronomici latini', *QUCC* 55: 71–88

(1997*b*) 'Il termine tecnico *bipalium* e la sua interpretazione nei trattati rustici', *RCCM* 39: 205–30

Decker, M. (2007) 'The authorship and context of early Byzantine farming manuals', *Byzantion* 77: 106–15

De Meo, C. (1986) *Lingue tecniche del latino*, 2nd edn. Testi e Manuali per l'Insegnamento Universitario del Latino 16. Bologna

Deroux, C. (2002) (ed.) *Hommages à Carl Deroux II: prose et linguistique, médecine.* Collection Latomus 267. Brussels

Deschamps, L. (2000) 'Le vin dans les *Satires Ménippées* de Varron', *Pallas* 53: 29–37

Devine, A. M., and L. D. Stephens (2006) *Latin word order: structured meaning and information.* Oxford and New York

Dickey, E. (2002) *Latin forms of address: from Plautus to Apuleius.* Oxford and New York

Dickey, E., and A. Chahoud (2010) (eds.) *Colloquial and literary Latin.* Cambridge and New York

Dickie, M. W. (2001) *Magic and magicians in the Greco-Roman world.* London and New York

Diederich, S. (2007) *Römische Agrarhandbücher zwischen Fachwissenschaft, Literatur und Ideologie.* Untersuchungen zur antiken Literatur und Geschichte 88. Berlin and New York

Diggle, J. (2004) *Theophrastus: Characters.* Cambridge Classical Texts and Commentaries 43. Cambridge

Domínguez Petit, R. (2004) 'Fuentes literarias para la agricultura cartaginesa: el tratado de Magon', *Habis* 35: 179–92

van Dommelen, P., and C. Gómez Bellard (2008) (eds.) *Rural landscapes of the Punic world.* Monographs in Mediterranean Archaeology 11. London and Oakville

Doody, A. (2007) 'Virgil the farmer? Critiques of the *Georgics* in Columella and Pliny', *CPh* 102: 180–97

(2009) 'Pliny's *Natural History: enkuklios paideia* and the ancient encyclopedia', *JHI* 70: 1–21

(2010) *Pliny's encyclopedia: the reception of the* Natural History. Cambridge

Drossaart Lulofs, H. J., and E. L. J. Poortman (1989) *Nicolaus Damascenus De plantis: five translations.* Verhandelingen der Koninklijke Nederlandse Akademie van Wetenschappen, Afd. Letterkunde, NS 139. Amsterdam and New York

Duke, J. A. (1989) *CRC handbook of nuts.* Boca Raton, FL

Dumont, J. C. (2008) 'Columella and Vergil', *Vergilius* 54: 49–58

Eichholz, D. E. (1965) *Theophrastus: De lapidibus.* Oxford

(1967) 'Some mineralogical problems in Theophrastus' *De lapidibus*', *CQ* 17: 103–9

Erdkamp, P., K. Verboven and A. Zuiderhoek (2015) (eds.) *Ownership and exploitation of land and natural resources in the Roman world.* Oxford and New York

Eriksson, N. (1934) *Studien zu den Annalen des Tacitus.* Lund

Ernout, A., and F. Thomas (1953) *Syntaxe latine,* 2nd edn. Nouvelle Collection à l'usage des classes 38. Paris

Espinilla, E., P. J. Quetglas and E. Torrego (2002) (eds.) *La comparación en latín*. Colección textos i estudis del GLLUB 4. Barcelona and Madrid

Fabretti, R. (1702) *Inscriptionum antiquarum quae in aedibus paternis asseruantur explicatio et additamentum, una cum aliquot emendationibus Gruterianis & indice rerum & uerborum memorabilium*. Rome

Fantar, M. H., and M. Ghaki (1995) (eds.) *Actes du IIIe congrès international des études phéniciennes et puniques: Tunis, 11–16 novembre 1991*. Tunis

Faraone, C. A. (1991) 'The agonistic context of early Greek binding spells', in Faraone and Obbink (1991) 3–32

Faraone, C. A., and D. Obbink (1991) (eds.) *Magika hiera: ancient Greek magic and religion*. Oxford

Feltenius, L. (1977) *Intransitivizations in Latin*. Acta Universitatis Upsaliensis, Studia Latina Upsaliensia 9. Stockholm

Ferraces Rodríguez, A. (1994) 'El Pseudo-Dioscórides *De herbis femininis*, los *Dynamidia* e Isidoro de Sevilla, *Etym.* XVII, 7–11', in Vázquez Buján (1994) 183–203

(1999) *Estudios sobre textos latinos de fitoterapia entre la antigüedad tardía y la alta edad media*. Monografías (Universidade de Coruña) 73. A Coruña

(2000) 'Le *Ex herbis femininis*: traduction, réélaboration, problèmes stylistiques', in Pigeaud and Pigeaud (2000) 77–89

(2006) 'Texto y tradición indirecta: reflexiones sobre una edición reciente de Gargilio Marcial', *Athenaeum* 94: 707–15

Finglass, P. J., C. Collard and N. J. Richardson (2007) (eds.) *Hesperos: studies in Ancient Greek poetry presented to M. L. West on his seventieth birthday*. Oxford and New York

Fischer, K.-D. (1997*a*) 'Q. Gargilius Martialis', in Sallmann (1997) 269–73

(1997*b*) 'Pseudo-Gargilius Martialis, *Curae boum*', in Sallmann (1997) 273

Fitch, J. G. (2013) *Palladius: The Work of farming (Opus agriculturae) and Poem on grafting*. Totnes

Flach, D. (1990) *Römische Agrargeschichte*. Handbuch der Altertumswissenschaften 3.9. Munich

Flobert, P. (1975) *Les verbes déponents latins: des origines à Charlemagne*. Publications de la Sorbonne, Série N.S. Recherches 17. Paris

Fögen, T. (2003) 'Metasprachliche Reflexionen antiker Autoren zu den Charakteristika von Fachtexten und Fachsprachen', in Horster and Reitz (2003) 31–60

(2005) (ed.) *Antike Fachtexte / Ancient technical texts*. Berlin and New York

(2009) *Wissen, Kommunikation und Selbstdarstellung: zur Struktur und Charakteristik römischer Fachtexte der frühen Kaiserzeit*. Zetemata 134. Munich

(2011) 'Latin as a technical and scientific language', in Clackson (2011) 445–63

Formisano, M. (2005) 'Veredelte Bäume und kultivierte Texte: Lehrgedichte in technischen Prosawerken der Spätantike', in Horster and Reitz (2005) 295–312

(2006) 'Frontinus on water supply', review of R. H. Rodgers, *Frontinus: De aquaeductu urbis Romae* (Cambridge Classical Texts and Commentaries 42; Cambridge 2004), *CR* 56: 132–5

Formisano, M., and P. van der Eijk (2017) (eds.) *Knowledge, text and practice in ancient writing*. Cambridge and New York

Foxhall, L. (1996) 'Feeling the earth move: cultivation techniques on steep slopes in classical antiquity', in Shipley and Salmon (1996) 44–67

Frayn, J. M. (1974) 'Subsistence farming in Italy during the Roman period: a preliminary discussion of the evidence', *G&R* 21: 11–18

Frederiksen, M. (1980) 'Plinio il vecchio e l'agricoltura in età imperiale romana: gli aspetti tecnici ed economici', in *Tecnologia* (q.v.) 81–97

Gaertner, J. F. (2007) '*tum* und *tunc* in der augusteischen Dichtersprache', *RhM* 150: 211–24

Gaide, F., and F. Biville (2003) (eds.) *Manus medica: actions et gestes de l'officiant dans les textes médicaux latins. Questions de thérapeutique et de lexique: Actes du colloque tenu à l'Université Lumière-Lyon II, les 18 et 19 septembre 2001*. Aix-en-Provence

Galdi, G. (2014) 'Some considerations on the apodotic uses of *atque* and *et* (2nd c. BC–2nd c. AD)', *Journal of Latin Linguistics* 13: 63–91

(2016) 'On *coepi/incipio* + infinitive: some new remarks', in Adams and Vincent (2016) 246–64

Gale, M. R. (2000) *Virgil on the nature of things: the* Georgics, *Lucretius and the didactic tradition*. Cambridge and New York

Garcia, D., and D. Meeks (1997) (eds.) *Techniques et économie antiques et médiévales: le temps de l'innovation. Colloque international (C.N.R.S) (Aix-en-Provence 21–23 mai 1996)*. Travaux du Centre Camille Jullian 21. Paris

Gautherie, A. (2017) *Rhétorique et thérapeutique dans le* De medicina *de Celse*. Recherches sur les Rhétoriques Religieuses 25. Turnhout

Gemoll, W. (1883) *Untersuchungen über die Quellen, den Verfasser und die Abfassungszeit der* Geoponica. Berlin

Georgoudi, S. (1990) *Des chevaux et des boeufs dans le monde grec: réalités et representations animalières à partir des livres XVI et XVII des* Géoponiques. Paris

Gibson, R. K. (1997) 'Didactic poetry as "popular" form: a study of imperatival expressions in Latin verse and prose', in Atherton (1997) 67–98

Gleason, K. L., and M. A. Palmer (2018) 'Constructing the ancient Roman garden', in Jashemski et al. (2018) 369–401

Goodyear, F. D. R. (1983) 'The *Oxford Latin Dictionary*', *Proceedings of the African Classical Associations* 17: 124–36

Gordon, R. L. and F. Marco Simón (2010) (eds.) *Magical practice in the Latin West: papers from the international conference held at the University of Zaragoza 30 Sept.–1 Oct. 2005*. Religions in the Graeco-Roman World 168. Leiden and Boston

Goujard, R. (1975) *Caton: De l'agriculture*. Paris

(1979) 'Le *De arboribus* de Columelle: problèmes de l'authenticité', *RPh* 53: 7–28

(1984) 'Le *De arboribus* de Columelle: difficultés et solutions proposées', *RPh* 58: 217–22

(1986*a*) *Columelle: <Les arbres>*. Paris

(1986*b*) 'Encore apropos de l'authenticité du *De arboribus*', *Latomus* 45: 612–18

Goulet, R. (ed.) (2003) *Dictionnaire des philosophes antiques: supplément*. Paris

Gowers, E. (2000) 'Vegetable love: Virgil, Columella, and garden poetry', *Ramus* 29: 127–48

(2011) 'Trees and family trees in the *Aeneid*', *ClAnt* 30: 87–118

Grandgent, C. H. (1907) *An introduction to Vulgar Latin*. Boston

Greene, J. A., and D. P. Kehoe (1995) 'Mago the Carthaginian', in Fantar and Ghaki (1995) 110–17

Grocock, C., and S. Grainger (2006) *Apicius: a critical edition with an introduction and an English translation of the Latin recipe text* Apicius. Totnes

Halla-aho, H. (2009) *The non-literary Latin letters: a study of their syntax and pragmatics*. Commentationes Humanarum Litterarum 124. Helsinki

(2012) 'What does "Latin" mean? A terminological pamphlet', in Leiwo et al. (2012) 13–23

Hallén, M. (1941) *In scriptores Historiae Augustae studia: commentatio academica*. Uppsala

Halstead, P. (1987) 'Traditional and ancient rural economy in Mediterranean Europe: *plus ça change*?', *JHS* 107: 77–87

(2014) *Two oxen ahead: pre-mechanized farming in the Mediterranean*. Malden, MA

Handford, S. A. (1947) *The Latin subjunctive: its usage and development from Plautus to Tacitus*. London

Hardy, G., and L. Totelin (2016) *Ancient botany*. Abingdon

Harrison, F. [A Virginia Farmer] (1913) *Roman farm management: the treatises of Cato and Varro, done into English, with notes of modern instances*. New York

Hartmann, H. T., D. E. Kester, F. T. Davies, Jr. and R. L. Geneve (2013) *Hartmann & Kester's plant propagation: principles and practices*, 8th edn. Upper Saddle River, NJ

Haverling, G. (2000) *On* sco-*verbs, prefixes, and semantic functions: a study in the development of prefixed and unprefixed verbs from early to late Latin*. Studia Graeca et Latina Gothoburgensia 64. Gothenburg

(2006) 'On cases and prepositions in Vulgar and literary Late Latin', in Abellán (2006) 345–60

(2012) 'Literary Late Latin and the development of the spoken language', in Leiwo et al. (2012) 153–76

(2014) 'Il latino letterario della tarda antichità', in Molinelli et al. (2014) III 845–72

(2015) (ed.) *Latin linguistics in the early 21st century: Acts of the 16th international colloquium on Latin linguistics (Uppsala, June 6th–11th, 2011)*. Acta Universitatis Upsaliensis, Studia Latina Upsaliensia 35. Uppsala

Havers, W. (1911) *Untersuchungen zur Kasussyntax der indogermanischen Sprachen*. Untersuchungen zur indogermanischen Sprach- und Kulturwissenschaft 3. Strasbourg

Heberlein, F. (2011) 'Temporal clauses', in Baldi and Cuzzolin (2011) 235–371

Henderson, J. (2002) 'Columella's living hedge: the Roman gardening book', *JRS* 92: 110–33

(2004) *The Roman book of gardening*. London and New York

Henkel, J. (2014) 'Vergil talks technique: metapoetic arboriculture in *Georgics* 2', *Vergilius* 60: 33–66

Hentz, G. (1983) 'Le *De arboribus*: œuvre de jeunesse de Columelle ou compilation d'un anonyme?', in Hentz and Zehnacker (1983) 303–13

Hentz, G., and H. Zehnacker (1983) (eds.) *Hommages à Robert Schilling*. Collection d'Études latines 37. Paris

Herman, J. (1963) *La formation du système roman des conjonctions de subordination*. Deutsche Akademie der Wissenschaften zu Berlin, Veröffentlichungen des Instituts für romanische Sprachwissenschaft 18. Berlin

(1989) 'Accusativus cum infinitivo et subordonnée à *quod*, *quia* en latin tardif – nouvelles remarques sur un vieux problème', in Calboli (1989) 133–52

(1994) (ed.) *Linguistic studies on Latin: selected papers from the 6th international colloquium on Latin linguistics (Budapest, 23–27 March 1991)*. Amsterdam and Philadelphia

(2000) *Vulgar Latin*, trans. R. Wright. University Park, PA

Hesnard, A. (1997) 'Entrepôts et navires à dolia: l'invention du transport de vin en vrac', in Garcia and Meeks (1997) 130–1

Heurgon, J. (1976) 'L'agronome carthaginois Magon et ses traducteurs en latin et en grec', *CRAI* 120: 441–56

Hilgers, W. (1969) *Lateinische Gefässnamen: Bezeichnungen, Funktion und Form römischer Gefäße nach den antiken Schriftquellen*. Düsseldorf

Hine, H. M. (2011) '*Discite . . . agricolae*: modes of instruction in Latin prose agricultural writing from Cato to Pliny the Elder', *CQ* 61: 624–54

Hoffmann, M. E. (1987) *Negatio contrarii: a study of Latin litotes*. Studies in Greek and Latin Linguistics 4. Assen and Wolfeboro, NH

(1989) 'Litotes expressions in Latin', in Calboli (1989) 589–610

Hofmann, J. B. (1926) 'Beiträge zur Kenntnis des Vulgärlateins', *IF* 43: 80–122

(1978) *Lateinische Umgangssprache*, 4th edn. Heidelberg

Horster, M., and C. Reitz (2003) (eds.) *Antike Fachschriftsteller: literarischer Diskurs und sozialer Kontext*. Palingenesia 80. Wiesbaden

(2005) (eds.) *Wissensvermittlung in dichterischer Gestalt*. Palingenesia 85. Stuttgart

Houston, G. W. (1989) 'Review. The state of the art: current work in the technology of ancient Rome', *CJ* 85: 63–80

Howe, N. P. (1985) 'In defense of the encyclopedic mode: on Pliny's preface to the *Natural History*', *Latomus* 44: 561–76

Hugonnard-Roche, H. (2003) 'Pseudo-Aristote, *De plantis*', in Goulet (2003) 499–505

Hunt, A. (2010) 'Elegiac grafting in Pomona's orchard: Ovid, *Metamorphoses* 14.623–771', *MD* 65: 43–58

Hus, A. (1965) Docere *et les mots de la famille de* docere*: étude de sémantique latine*. Paris

Huxley, A., and M. Griffiths (1992) (eds.) *The new Royal Horticultural Society dictionary of gardening*, 4 vols. New York

Iliescu, M., and W. Marxgut (1992) (eds.) *Latin vulgaire, latin tardif III: Actes du IIIème colloque international sur le latin vulgaire et tardif (Innsbruck, 2–5 septembre 1991)*. Tübingen

Irby, G. L. (2016) (ed.) *A companion to science, technology, and medicine in ancient Greece and Rome*, 2 vols. Malden, MA

Janakat, S., A. A. R. Al-Nabulsi, S. Allehdan, A. N. Olaimat and R. A. Holley (2015) 'Antimicrobial activity of *amurca* (olive oil lees) extract against selected foodborne pathogens', *Food Science and Technology* 35: 259–65

Jansen, G. C. M. (2018) 'Water and water technology in Roman gardens', in Jashemski et al. (2018) 402–31

Jansen, G. C. M., O. Koloski-Ostrow and E. M. Moorman (2011) (eds.) *Roman toilets: their archaeology and cultural history*. BABESCH, Supplement 19. Leuven

Janson, T. (1979) *Mechanisms of language change in Latin*. Acta Universitatis Stockholmiensis, Studia Latina Stockholmiensia 23. Stockholm

Jashemski, W. F. (1979–93) *The gardens of Pompeii, Herculaneum, and the villas destroyed by Vesuvius*, 2 vols. New Rochelle, NY

(2018*a*) 'Produce gardens', in Jashemski et al. (2018) 121–51

(2018*b*) 'Gardening practices and techniques', in Jashemski et al. (2018) 432–54

Jashemski, W. F., K. L. Gleason, K. J. Hartswick and A.-A. Malek (2018) (eds.) *Gardens of the Roman Empire*. Cambridge and New York

Jeanneret, M. (1918) *La langue des tablettes d'exécration latines*. Diss., Neuchâtel

(1922) 'La langue des tablettes d'exécration latines (Notes complémentaires)', *RPh* 46: 14–25

Johnston, S. I. (2007*a*) (ed.) *Ancient religions*. Cambridge, MA

(2007*b*) 'Magic', in Johnston (2007*a*) 139–52

Jones, R. (2012*a*) (ed.) *Manure matters: historical, archaeological and ethnographic perspectives*. Burlington, VT

(2012*b*) 'Why manure matters', in Jones (2012*a*) 1–11

Jordan, D. R. (1985) 'A survey of Greek *defixiones* not included in the special corpora', *GRBS* 26: 151–97

Josephson, A. (1940) *Casae litterarum: Studien zum corpus agrimensorum Romanorum*. Uppsala

Juniper, B. E., and D. J. Mabberley (2006) *The story of the apple*. Portland, OR

Kajanto, I. (1965) *The Latin cognomina*. Rome

Kaufman, D. B. (1931) 'Horseshoes in antiquity', *CJ* 26: 619–20

Keeline, T. (2013) 'The literary and stylistic qualities of a Plinian letter: a commentary on Plin. *Ep*. 7.9', *HSCP* 107: 229–64

Kiss, S. (1982) *Tendances évolutives de la syntaxe verbale en latin tardif*. Studia Romanica, Series linguistica 4. Debrecen

Kiss, S., L. Mondin and G. Salvi (2005) (eds.) *Latin et langues romanes: études de linguistique offertes à József Herman à l'occasion de son 80ème anniversaire*. Tübingen

Knox, P. E. (1986) 'Adjectives in *-osus* and Latin poetic diction', *Glotta* 64: 90–101

Kolendo, J. (1980) *L'agricoltura nell'Italia romana: tecniche agrarie e progresso economico dalla tarda repubblica al principato*. Biblioteca di storia antica 10. Rome

Kottmann, M. (1903) *De elocutione L. Junii Moderati Columellae*. Rottweil

König, J., and G. Woolf (2013*a*) (eds.) *Encyclopaedism from antiquity to the Renaissance*. Cambridge and New York

(2013*b*) 'Introduction', in König and Woolf (2013*a*) 1–20

(2013*c*) 'Encyclopaedism in the Roman Empire', in König and Woolf (2013*a*) 23–63

Krenkel, W. A. (1973). 'A. Cornelius Celsus', in *Argentea aetas* (q.v.) 17–28

Krings, V. (2008) 'Rereading Punic agriculture: representation, analogy and ideology in the classical sources', in van Dommelen and Gómez Bellard (2008) 22–43

Kron, G. (2005) 'Sustainable Roman intensive mixed farming methods: water conservation and erosion control', in Bedon and Hermon (2005) 285–308

(2012) 'Food production', in Scheidel (2012) 156–174 [Expanded version at http://web.stanford.edu/~scheidel/CCRE, where it is paginated 1–47; I adopt the pagination of the expanded version]

Kronenberg, L. (2009) *Allegories of farming from Greece and Rome: philosophical satire in Xenophon, Varro, and Virgil*. Cambridge and New York

Kroon, C. (2004) 'Latin *quidem* and the structure of the move', in Aertsen et al. (2004) 199–209

(2005) 'The relationship between grammar and discourse: evidence from the Latin particle *quidem*', in Calboli (2005) 577–90

(2009) 'Latin linguistics between grammar and discourse: units of analysis, levels of analysis', in Rieken and Widmer (2009) 143–58

(2011) 'Latin particles and the grammar of discourse', in Clackson (2011) 176–95

Langslow, D. R. (2000) *Medical Latin in the Roman Empire*. Oxford and New York

(2005) '"Langues réduites au lexique"? The languages of Latin technical prose', in Reinhardt et al. (2005) 287–302

Lausberg, H. (1998) *Handbook of literary rhetoric: a foundation for literary study*, trans. M. T. Bliss et al. Leiden and Boston

Lawrence, R. M. (1899) *The magic of the horse-shoe: with other folk tales*. Boston

Layne, D. R., and D. Bassi (2008) (eds.) *The peach: botany, production and uses*. Cambridge, MA

Le Gall, J. (1959) 'Les falces et la faux', *Études d'archéologie classique* 2: 55–72

Lehoux, D. (2007) *Astronomy, weather, and calendars in the ancient world: parapegmata and related texts in classical and near eastern societies*. Cambridge and New York

(2012) *What did the Romans know? An inquiry into science and worldmaking*. Chicago

Leiwo, M. (2010) 'Petronius' linguistic resources', in Dickey and Chahoud (2010) 281–91

Leiwo, M, H. Halla-aho and M. Vierros (2012) (eds.) *Variation and change in Greek and Latin*. Papers and Monographs of the Finnish Institute at Athens 17. Helsinki

Lelli, E. (2010) *L'agricoltura antica: i Geoponica di Cassiano Basso*, 2 vols. Altri classici 3. Soveria Mannelli

Leumann, M. (1940) 'Das lat. Suffix *-āneŭs*', *IF* 40: 116–23

Lieberg, G. (1982) *Poeta creator: Studien zu einer Figur der antiken Dichtung*. Amsterdam.

Linde, H. (1936) *Studier till Columellas nionde bok*. Doktorsavhandlingar i Latinsk Filologi vid Göteborgs Högskola 12. Diss., Gothenburg

Linderbauer, B. (1922) *S. Benedicti Regula monachorum*. Metten

Lloyd, G. E. R. (1979) *Magic, reason and experience: studies in the origin and development of Greek science*. Cambridge and New York

Löfstedt, B. (1963) 'Zum lateinischen possessiven Dativ', *Zeitschrift für vergleichende Sprachforschung auf dem Gebiete der indogermanischen Sprachen* 78: 64–83

Löfstedt, E. (1907) *Beiträge zur Kenntnis der späteren Latinität*. Uppsala Universitets Årsskrift 1907. Diss., Uppsala

(1911) *Philologischer Kommentar zur* Peregrinatio Aetheriae*: Untersuchungen zur Geschichte der lateinischen Sprache*. Uppsala

(1920) *Zur Sprache Tertullians*. Lunds Universitets Årsskrift NS 1.16.2. Lund

(1936) *Vermischte Studien zur lateinischen Sprachkunde und Syntax*. Lund

(1948) 'On the style of Tacitus', *JRS* 38: 1–8

(1959) *Late Latin*. Instituttet for Sammenlignende Kulturforskning, Serie A, Forelesninger 25. Oslo

Loreti, F., and S. Morini (2008) 'Propagation techniques', in Layne and Bassi 2008, 221–43

Lowe, D. (2010) 'The symbolic value of grafting in ancient Rome', *TAPhA* 140: 461–88

Lowe, E. A. (1934–71) *Codices latini antiquiores: a palaeographical guide to Latin manuscripts prior to the ninth century*, 12 vols. Oxford

Luck, G. (2006) *Arcana mundi: magic and the occult in the Greek and Roman worlds. A collection of ancient texts*, 2nd edn. Baltimore, MD

Lundström, S. (1948) *Neue Studien zur lateinischen Irenäusübersetzung*. Acta Universitatis Lundensis NS 1.44.8. Lund

Mahaffy, J. P. (1889) 'The work of Mago on agriculture', *Hermathena* 7: 29–35

Mai, A. (1828) *Classicorum auctorum e Vaticanis codicibus editorum tomus I, complectens Ciceronis De rep. quae supersunt, Gargilii Martialis De arboribus pomiferis, Sallustii Historiarum et Archimedis fragmenta, cum quinque tabulis aeneis*. Rome

(1831) *Classicorum auctorum e Vaticanis codicibus editorum tomus III, complectens mythographos tres, fabulas Phaedri ut aiunt nouas, Boethii opuscula duo, Cassiodori supplementum, epigrammata vetera, geographum veterem, Gargilii Martialis fragmentum de pomis, Placidi glossas, et alia quaedam*. Rome

Maiden, M., J. C. Smith and A. Ledgeway (2011) (eds.) *The Cambridge history of Romance languages. Volume I: Structures*. Cambridge

Maire, B. (1997) 'La *variatio* dans le lexique des *Medicinae ex oleribus et pomis* de Gargilius Martialis', *RFIC* 125: 306–18

(1999) 'Éléments vulgaires dans les *Medicinae ex holeribus et pomis* de Gargilius Martialis', in Petersmann and Kettemann, 429–39

(2000) 'Les *Medicinae* de Gargilius: un manuel pratique aux ambitions littéraires?', in Pigeaud and Pigeaud (2000) 153–64

(2002*a*) *Gargilius Martialis: Les remèdes tirés des légumes et des fruits*. Paris

(2002*b*) 'L'esprit ou la lettre: les sources des *Medicinae* de Gargilius Martialis', in Deroux (2002) 514–35

(2003*a*) 'Actions thérapeutiques ou gestes littéraires: le lexique des *Medicinae* de Gargilius Martialis', in Gaide and Biville (2003) 147–60

(2003*b*) 'Quand la littérature vulgaire enfante herbes, légumes et fruits', in Solin et al. (2003) 361–8

(2010) 'Le jardin qui soigne', in Bertholet and Reber (2010) 87–113

Malouta, M., and A. Wilson (2013) 'Mechanical irrigation: water-lifting devices in the archaeological evidence and in the Egyptian papyri', in Bowman and Wilson (2013) 273–305

Maltby, R. (2008) 'The language, style and origins of Pseudo-Dioscorides *De herbis femininis*', in Wright (2008) 392–9

(2016) 'Analytic and synthetic forms of the comparative and superlative from early to late Latin', in Adams and Vincent (2016) 340–69

Marcone, A. (1997) *Storia dell'agricoltura romana: dal mondo arcaico all'età imperiale*. Studi Superiori NIS 338. Rome

Margaritis, E., and M. K. Jones (2008) 'Greek and Roman agriculture', in Oleson (2008) 158–74

Marini, R. P., and G. L. Reighard (2008) 'Crop load management', in Layne and Bassi (2008) 289–302

Martin, R. (1971) *Recherches sur les agronomes latins et leurs conceptions économiques et sociales*. Paris

(1985) 'État présent des études sur Columelle', *ANRW* II 32.3: 1959–79

Marx, F. (1915) *A. Cornelii Celsi quae supersunt*. Leipzig and Berlin

Mazzini, I. (1977) 'Dati biografici ed opera di un minore: Quinto Gargilio Marziale', *AMArc* 7: 99–121

(1988) *Q. Gargilii Martialis De hortis*, 2nd edn. Opuscula philologa 1. Bologna [1st edn 1978]

Meader, C. L., and E. Wölfflin (1900) 'Zur Geschichte der Pronomina demonstrativa I', *Archiv für lateinische Lexikographie und Grammatik* 11: 369–93

Meech, W. W. (1888) *Quince culture: an illustrated hand-book for the propagation and cultivation of the quince, with descriptions of its varieties, insect enemies, diseases and their remedies*. New York

Meiggs, R. (1982) *Trees and timber in the ancient Mediterranean world*. Oxford

Menge, H., T. Burkard and M. Schauer (2005) *Lehrbuch der lateinischen Syntax und Semantik*, 4th edn. Darmstadt

Michelini, A. (1978) '"Ὕβρις and plants', *HSCP* 82: 35–44

Molinelli, P., P. Cuzzolin and C. Fedriani (2014) (eds.) *Latin vulgaire – latin tardif X: Actes du Xe colloque international sur le latin vulgaire et tardif (Bergamo, 5–9 septembre 2012)*, 3 vols. Biblioteca di Linguistica e Filologia 1. Bergamo

Moraux, P. (1951) *Les listes anciennes des ouvrages d'Aristote*. Louvain

(1973) *Der Aristotelismus bei den Griechen: von Andronikos bis Alexander von Aphrodisias. Erster Band: Die Renaissance des Aristotelismus im I. Jh. v. Chr.* Philologisch-historische Studien zum Aristotelismus 5.1. Berlin

von Morawski, C. (1881) 'Bemerkungen zu den sogenannten quintilianeischen Declamationen', *Zeitschrift für die österreichischen Gymnasien* 32: 1–12

Morel, J.-P., and A. M. Mercuri (2009) (eds.) *Plants and culture: seeds of the cultural heritage of Europe*. Studio, tutela e fruizione dei beni culturali 3. Bari

Morris, I. (1994) (ed.) *Classical Greece: ancient histories and modern archaeologies*. Cambridge and New York

Morvillez, E. (2015) (ed.) *Paradeisos, genèse et métamorphose de la notion de paradis dans l'Antiquité: Actes du colloque international d'Avignon (Palais des Papes – Vaison-la-Romaine, 22 mars 2009)*. Orient & Méditerranée 15. Paris

Mudge, K., J. Janick, S. Scofield and E. E. Goldschmidt (2009) 'A history of grafting', *Horticultural Reviews* 35: 437–93

Mugione, E., C. Giordano and A. Ciarallo (2012) 'Gli elementi architettonici nella pittura di giardino a Pompei', in Senatore and Perrotta (2012) 213–16

Nehring, A. (1923) 'Zum Namen der Quitte', *Glotta* 13: 11–16

Nelsestuen, G. A. (2015) *Varro the agronomist: political philosophy, satire, and agriculture in the late Republic*. Columbus, OH

(2016) 'Storing produce and staging dinner parties: fruit-galleries and genre in Varro's *De re rustica* 1', *ICS* 41: 21–40

Nelson, H. L. W. (1981) *Überlieferung, Aufbau und Stil von Gai* Institutiones. Studia Gaiana 6. Leiden

Niaounakis, M., and C. P. Halvadakis (2006) *Olive processing waste management: literature review and patent survey*, 2nd edn. Waste Management 5. Amsterdam

Nisbet, R. G. M. (1961) *Cicero: In L. Calpurnium Pisonem oratio*. Oxford

(1990) 'Cola and clausulae in Cicero's speeches', in Craik (1990) 349–59

Nutting, H. C. (1922) 'The *si*-clause in substantive use', *University of California Publications in Classical Philology* 7: 129–42

(1923) 'Subjunctive conditions in Tacitus', *University of California Publications in Classical Philology* 7: 143–95

(1925) 'The Latin conditional sentence', *University of California Publications in Classical Philology* 8: 1–185

Obbink, D., and R. Rutherford (2011) (eds.) *Culture in pieces: essays on ancient texts in honor of Peter Parsons*. Oxford and New York

Oberhelman, S. M. (2003) *Prose rhythm in Latin literature of the Roman Empire: first century B.C. to fourth century A.D.* Studies in Classics 27. Lewiston, NY

Olcott, G. N. (1898) *Studies in the word formation of the Latin inscriptions, substantives and adjectives, with special reference to the Latin sermo vulgaris*. Rome

Oleson, J. P. (2000*a*) 'Irrigation', in Wikander (2000) 183–215

(2000*b*) 'Water lifting', in Wikander (2000) 217–302

(2008) (ed.) *The Oxford handbook of engineering and technology in the classical world.* Oxford and New York

Önnerfors, A. (1956) *Pliniana: in Plinii Maioris Naturalem historiam studia grammatica semantica critica.* Uppsala Universitets Årsskrift 1956, 6. Uppsala

(1963) *In Medicinam Plinii studia philologica: de memoria et verborum contextu opusculi, de elocutione et aetate deque iis operibus, quibus medio aevo conceptum est.* Lunds Universitets Årsskrift NS 2.55.5. Lund

(1993) 'Das medizinische Latein von Celsus bis Cassius Felix', *ANRW* II 34.1: 227–341

Orlandini, A., and P. Poccetti (2008) 'Three types of negation: different values of *nec* in Latin and in the languages of ancient Italy', in Wright (2008) 1–12

Panagl, O. (1975) 'Präsuppositionen und die Syntax der lateinischen Komparation', *Salzburger Beiträge zur Linguistik* 1: 361–75

Paratore, E. (1948) *Apulei Metamorphoseon libri IV–VI (La favola di Amore e Psiche): introduzione e testo critico.* Biblioteca di Studi Superiori, Filologia Latina 2. Florence

Parker, A. J. (1992) 'Cargoes, containers and stowage: the ancient Mediterranean', *IJNA* 21: 89–100

Pascucci, G. (1968) 'Aspetti del latino giuridico', *SIFC* 40: 3–43

Pease, A. S. (1933) 'Notes on ancient grafting', *TAPhA* 64: 66–76

Penney, J. W. H. (2005) 'Connections in archaic Latin prose', in Reinhardt et al. (2005) 37–51

Petersmann, H. (1977) *Petrons urbane Prosa: Untersuchungen zu Sprache und Text (Syntax).* Österreichische Akademie der Wissenschaften, Philosophisch-historische Klasse, Sitzungsberichte 323. Vienna

Petersmann, H., and R. Kettemann (1999) (eds.) *Latin vulgaire – latin tardif V: Actes du Ve colloque international sur le latin vulgaire et tardif (Heidelberg, 5–8 septembre 1997).* Heidelberg

Pigeaud, A., and J. Pigeaud (2000) (eds.) *Les textes médicaux latins comme littérature: Actes du VIe colloque international sur les textes médicaux latins du 1er au 3 septembre 1998 à Nantes.* Nantes

Pinkster, H. (2005) 'The language of Pliny the Elder', in Reinhardt et al. (2005) 239–56

Pomoell, R. (1931) *Textkritiska Studier till Columellas femte Bok.* Gothenburg

Praun, J. (1885) *Bemerkungen zur Syntax des Vitruv mit eingehender Darstellung der Substantivsätze.* Bamberg

Pruvost-Versteeg, G. (2015) '*Dativus sympatheticus* et tour possessif chez Plaute: concurrence ou complémentarité?', in Haverling (2015) 264–75

Ragep, F. J., and S. P. Ragep (1996) (eds.) *Tradition, transmission, transformation: Proceedings of two conferences on pre-modern science held at the University of Oklahoma*. Collection des travaux de l'Académie Internationale d'Histoire des Sciences 37. Leiden and Boston

Ramina, A., P. Tonutti and W. McGlasson (2008) 'Ripening, nutrition and postharvest physiology', in Layne and Bassi (2008) 550–74

Reay, B. (2005) 'Agriculture, writing, and Cato's aristocratic self-fashioning', *ClAnt* 24: 331–61

Reeve, M. D. (2007) 'The editing of Pliny's *Natural History*', *RHT* NS 2: 107–79

Regenbogen, O. (1937) 'Eine Polemik Theophrasts gegen Aristoteles', *Hermes* 72: 469–75

Reighard, G. L., and F. Loreti (2008) 'Rootstock development', in Layne and Bassi (2008) 193–220

Reinhardt, T. (2010) 'Syntactic colloquialism in Lucretius', in Dickey and Chahoud (2010) 203–28

Reinhardt, T., M. Lapidge and J. N. Adams (2005) (eds.) *Aspects of the language of Latin prose*. Proceedings of the British Academy 93. Oxford

Reitz, C. (2013) 'Columella, *De re rustica*', in Buckley and Dinter (2013) 275–87

(2017) '*Auctoritas* in the garden: Columella's poetic strategy in *De re rustica* 10', in Formisano and van der Eijk (2017) 217–30

Reitzenstein, R. (1884) *De scriptorum rei rusticae qui intercedunt inter Catonem et Columellam libris deperditis*. Berlin

Renehan, R. (1977) 'Compound–simplex verbal iteration in Plautus', *CPh* 72: 243–8

Richardson, W. F. (2004) *Numbering and measuring in the classical world*, 2nd edn. Bristol

Richter, W. (1972) *Der* liber de arboribus *und Columella*. Sitzungsberichte der Bayerischen Akademie der Wissenschaften 1972, 1. Munich

(1978) *Gegenständliches Denken, archaisches Ordnen: Untersuchungen zur Anlage von Cato* De agri cultura. Bibliothek der klassischen Altertumswissenschaften NS 2.62. Heidelberg

(1981–3) *Columella: Zwölf Bücher über Landwirtschaft*. Munich

Riddle, J. M. (1984) 'Gargilius Martialis as a medical writer', *JHM* 39: 408–29

Rieken, E., and P. Widmer (2009) (eds.) *Pragmatische Kategorien: Form, Funktion, und Diachronie. Akten der Arbeitstagung der indogermanischen Gesellschaft vom 24. bis 26. September 2007 in Marburg*. Wiesbaden

Risselada, R. (1993) *Imperatives and other directive expressions in Latin: a study in the pragmatics of a dead language*. Amsterdam studies in classical philology 2. Amsterdam

(1994) '*modo* and *sane*, or what to do with particles in Latin directives', in Herman (1994) 319–43

(1998) 'The discourse functions of *sane*: Latin marker of agreement in description, interaction, and concession', *Journal of Pragmatics* 30: 225–44

Roby, C. (2016) *Technical ekphrasis in Greek and Roman science and literature: the written machine between Alexandria and Rome.* Cambridge

Rodgers, R. H. (1975) *An introduction to Palladius.* Bulletin of the Institute of Classical Studies Supplement 35. London

(2010) *Columellae Res rustica. Incerti auctoris liber de arboribus.* Oxford

Roelcke, T. (2005) *Fachsprachen*, 2nd edn. Berlin

Rose, H. J. (1933) 'The folklore of the *Geoponica*', *Folklore* 44: 57–90

Rose, V. (1863) *Aristoteles pseudepigraphicus: pars prima fragmenta Aristotelis continens.* Leipzig

(1864–70) *Anecdota graeca et graecolatina: Mitteilungen aus Handschriften der Geschichte der griechischen Wissenschaft*, 2 vols. Berlin

(1875) *Plinii Secundi quae fertur una cum Gargilii Martialis medicina.* Leipzig

(1886) *Aristotelis qui ferebantur librorum fragmenta.* Leipzig

Rosén, H. (1983) 'The mechanisms of Latin nominalization and conceptualization in historical view', *ANRW* II 29.1: 178–211

(2009) 'Coherence, sentence modification, and sentence-part modification: the contribution of particles', in Baldi and Cuzzolin (2009) 317–441

(2012*a*) '*Coepi* + infinitif dans une sélection de traductions en latin tardif', in Biville et al. 2012, 365–75

(2012*b*) 'Two phasal verbs: Lat. *coepi* and Gk. ἔβαλον/ἔβαλα, βάλλω/βάζω/βάνω', *IF* 117: 119–72

Rosengarten, F., Jr. (1984) *The book of edible nuts.* New York

Ross, D. O., Jr. (1969) *Style and tradition in Catullus.* Cambridge, MA

(1980) '*Non sua poma*: Varro, Virgil, and Grafting', *ICS* 5: 63–71

(1987) *Virgil's elements: physics and poetry in the* Georgics. Princeton, NJ

Sadori, L. et al. (2009) 'The introduction and diffusion of peach in ancient Italy', in Morel and Mercuri (2009) 45–61

Sallmann, K. (1997) (ed.) *Die Literatur des Umbruchs: von der römischen zur christlichen Literatur 117 bis 284 n. Chr.* Handbuch der lateinischen Literatur 4 = Handbuch der Altertumswissenschaft 8.4. Munich

Salvi, G. (2011) 'Morphosyntactic persistence', in Maiden et al. (2011) 318–81

Santini, C. (2000) 'Le traduzioni dei libri di Magone', in Sconocchia and Toneatto (2000) 367–80

Santini, C., I. Mastrorosa and A. Zumbo (2002) (eds.) *Letteratura scientifica e tecnica di Grecia e Roma.* Studi Superiori (Carocci Editore) 401. Rome

Scheidel, W. (2012) (ed.) *The Cambridge companion to the Roman economy.* Cambridge and New York

Schiefsky, M. J. (2005) 'Technical terminology in Greco-Roman treatises on artillery construction', in Fögen (2005) 253–70

Schmalz, J. H. (1909) '*Si tamen*', *Glotta* 1: 333–9

Schmid, M. S., J. R. Austin and D. Stein (1998) (eds.) *Historical linguistics 1997: selected papers from the 13th international conference on historical linguistics (Düsseldorf, 10–17 August 1997)*. Amsterdam studies in the theory and history of linguistic science 4.164. Amsterdam and Philadelphia

Schneider, J. G. (1794) *Scriptorum rei rusticae ueterum Latinorum*, 4 vols. Leipzig

Schrickx, J. (2011) *Lateinische Modalpartikeln:* nempe, quippe, scilicet, *und* nimirum. Amsterdam studies in classical philology 19. Leiden and Boston

(2015) 'Reflexionen über lateinische *re*-Komposita', *Glotta* 91: 264–80

Schulze, C. (2001) *Celsus*. Studienbücher Antike 6. Hildesheim

(2002) *Die pharmazeutische Fachliteratur in der Antike: eine Einführung*. Göttinger Forum für Altertumswissenschaften 10. Göttingen

Sconocchia, S., and L. Toneatto (1993) (eds.) *Lingue tecniche del greco e del latino: Atti del 1. seminario internazionale sulla letteratura scientifica e tecnica greca e latina*. Trieste

(2000) (eds.) *Lingue tecniche del greco e del latino III: Atti del 3. seminario internazionale sulla letteratura scientifica e tecnica greca e latina*. Edizioni e Saggi Universitari di Filologia Classica 6. Bologna

Scotti, A. A. (1833) 'Memoria del socio A. A. Angelo sopra un codice palimpsesto della Real Biblioteca Borbonica', *Memorie della regale accademia ercolanese di archeologia* 2: 119–85

Senatore, M. R., and A. Perrotta (2012) *Scienze naturali e archeologia: il paesaggio antico. Interazione uomo/ambiente ed eventi catastrofici: Museo Archeologico Nazionale (Napoli 14–16 ottobre 2010)*. Scienze della terra 35. Rome

Senn, G. (1929) 'Hat Aristoteles eine selbständige Schrift über Pflanzen verfaßt?', *Philologus* 85: 113–40

Sharrock, A. (1997) '*Haud mollia iussa*', response to Gibson (1997), in Atherton (1997), 99–115

Sherwin-White, A. N. (1966) *The letters of Pliny: a historical and social commentary*. Oxford

Shipley, G., and J. Salmon (1996) (eds.) *Human landscapes in classical antiquity: environment and culture*. Leicester–Nottingham studies in ancient society 6. London and New York

Shipp, G. P. (1944) 'A Latin idiom and Pliny *Ep.* iv.2.2', *CPh* 39: 117–18

Skydsgaard, J. E. (1968) *Varro the scholar: studies in the first book of Varro's* De re rustica. Analecta Romana Instituti Danici 4. Copenhagen

Smith, A. E., and D. M. Secoy (1975) 'Forerunners of pesticides in Classical Greece and Rome', *Journal of Agricultural and Food Chemistry* 23: 1050–5

Solin, H., M. Leiwo and H. Halla-aho (2003) (eds.) *Latin vulgaire – latin tardif VI: Actes du VIe colloque international sur le latin vulgaire et tardif (Helsinki, 29 août – 2 septembre)*. Hildesheim and New York

Solmsen, F. (1911) 'Zur Geschichte des Namens der Quitte', *Glotta* 3: 241–5

Solodow, J. (1978) *The Latin particle* quidem. American Classical Studies 4. University Park, PA

Sorrento, L. (1949) *Sintassi romanza: ricerche e prospettive.* Varese

Spanier, E. (2010) *The good farmer in ancient Rome: war, agriculture and the elite from the republic to early empire.* Diss., University of Washington (Seattle)

Speranza, F. (1974) *Scriptorum Romanorum de re rustica reliquiae: volumen prius.* Biblioteca di Helikon Testi e Studi 8. Messina

Spevak, O. (2010) *Constituent order in classical Latin prose.* Studies in language companion series 117. Amsterdam and Philadelphia

Spurr, M. S. (1986*a*) 'Agriculture and the *Georgics*', *G&R* 33: 164–87

(1986*b*) *Arable cultivation in Roman Italy c.200 B.C. – c. A.D. 100.* Journal of Roman studies monographs 3. London

von Stackelberg, K. T. (2014) 'Garden hybrids: hermaphrodite images in the Roman house', *ClAnt* 33: 395–426

von Staden, H. (1990) 'Women and dirt', *Helios* 19: 7–30

(1996) 'Liminal perils: early Roman receptions of Greek medicine', in Ragep and Ragep (1996) 369–418

Stefenelli, A. (1962) *Die Volkssprache im Werk des Petron: im Hinblick auf die romanischen Sprachen.* Wiener romanistische Arbeiten 1. Stuttgart

Stok, F. (1993) 'Gargilio Marziale: un epigono dell'enciclopedismo', in Sconocchia and Toneatto (1993) 220–34

Stover, J. (forthcoming) (ed.) *The Oxford guide to the transmission of the Latin classics.* Oxford

Svennung, J. (1927) 'De auctoribus Palladii 1', *Eranos* 25: 123–78

(1935) *Untersuchungen zu Palladius und zur lateinischen Fach- und Volkssprache.* Uppsala

Syme, R. (1971) *Emperors and biography: studies in the* Historia Augusta. Oxford

(1983) *Historia Augusta papers.* Oxford

Tapper, R. M. (1980) *The Materia medica of Gargilius Martialis.* Diss., University of Wisconsin (Madison)

Tarriño, E. (2011) 'Comparative clauses', in Baldi and Cuzzolin (2011) 373–427

Taub, L. (2003) *Ancient meteorology.* New York

Tavenner, E. (1916) *Studies in magic from Latin literature.* New York

Teall, J. L. (1971) 'The Byzantine agricultural tradition', *DOP* 25: 33–59

Tecnologia = *Tecnologia, economia e società nel mondo romano: Atti del convegno di Como 27/28/29 settembre 1979.* Como, 1980

Terrenato, N. (2012) 'The enigma of "Catonian" villas: the *De agri cultura* in the context of second-century BC Italian architecture', in Becker and Terrenato (2012) 69–93.

Thielscher, P. (1963) *Des Marcus Cato Belehrung über die Landwirtschaft*. Berlin

Thomas, R. F. (1987) 'Prose into poetry: tradition and meaning in Virgil's *Georgics*', *HSCP* 91: 229–60

Thomas, R. G., and A. T. Hodge (2000) 'Water-supply', in Wikander (2000) 3–99

Thurmond, D. L. (2006) *A handbook of food processing in classical Rome: for her bounty no winter*. Technology and change in history 9. Leiden and Boston

Tidner, E. (1922) *De particulis copulativis apud scriptores historiae Augustae quaestiones selectae*. Uppsala Universitets Årsskrift 1922, Filosofi, Språkvetenskap och Historiska Vetenskaper 3. Uppsala.

(1938) *Sprachlicher Kommentar zur lateinischen Didascalia Apostolorum*. Kungl. Vitterhets Historie och Antikvitets Akademiens Handlingar 45.1. Stockholm

Tölle-Kastenbein, R. (1990) *Antike Wasserkultur*. Munich

Torrego, E. (2002) 'Los SN comparativos: el segundo término de la comparación', in Espinilla et al. (2002) 251–79

(2009) 'Coordination', in Baldi and Cuzzolin (2009) 443–87

Ullmann, M. (1972) *Die Natur- und Geheimwissenschaften im Islam*. Handbuch der Orientalistik, Abt. 1: Der Nahe und der Mittlere Osten 6.2. Leiden

Väänänen, V. (1981) *Introduction au latin vulgaire*, 3rd edn. Paris

(1987) *Le journal-épître d'Égérie (Itinerarium Egeriae): étude linguistique*. Annales Academiae Scientiarum Fennicae, Series B, 230. Helsinki

(1990) '*Plicare/applicare* "se diriger vers": simplex pro composito?', in Calboli (1990) 239–47

Vázquez Buján, M. E. (1994) (ed.) *Tradición e innovación de la medicina Latina de la antigüedad y de la Alta Edad Media: Actas del IV coloquio internacional sobre los textos médicos latinos antiguos*. Cursos y congresos de la Universidade de Santiago de Compostela 83. A Coruña

Versnel, H. S. (1991) 'Beyond cursing: the appeal to justice in judicial prayers', in Faraone and Obbink (1991) 60–106

Victorius, P. [Piero Vettori] (1542) *Explicationes suarum in Catonem, Varronem, Columellam castigationum*. Lyons

Vollmer, F. (1918) 'Lesungen und Deutungen II', *Sitzungsberichte der Königlich Bayerischen Akademie der Wissenschaften* 4: 1–28

Wahlén, S. (1930) *Studia critica in declamationes minores quae sub nomine Quintiliani feruntur: commentatio academica*. Uppsala

Watkins, C. (1966) 'An Indo-European construction in Greek and Latin', *HSCP* 71: 115–19

Waugh, F. A. (1913) *The American peach orchard: a sketch of the practice of peach growing in North America at the beginning of the twentieth century*. New York

Wehr, B. (1984) *Diskurs-Strategien im Romanischen: ein Beitrag zur romanischen Syntax*. Romanica Monacensia 22. Tübingen

(2008) 'Spätlatein aus der Sicht der Romanistik: zu apodosis-einleitendem *et*', in Wright (2008) 179–90

Wellmann, M. (1908) 'Palladius und Gargilius Martialis', *Hermes* 43: 1–31

Westerath, H. (1938) *Die Fachausdrücke des Ackerbaues bei den römischen Agrarschriftstellern*. Osnabrück

White, K. D. (1963) 'Wheat-farming in Roman times', *Antiquity* 37: 207–12

(1967) *Agricultural implements of the Roman world*. Cambridge

(1970) *Roman farming*. Ithaca, NY

(1973) 'Roman agricultural writers I: Varro and his predecessors', *ANRW* I 4: 439–97

(1975) *Farm equipment of the Roman world*. Cambridge

(1984) *Greek and Roman technology*. London

Whitton, C. (2013) *Pliny the Younger: Epistles book II*. Cambridge

Wickson, E. J. (1891) *The California fruits and how to grow them: a manual of methods which have yielded greatest success; with lists of varieties best adapted to the different districts of the state*, 2nd edn. San Francisco, CA

Wikander, O. (2000) (ed.) *Handbook of ancient water technology*. Technology and change in history 2. Leiden and Boston

Wilkins, J., D. Harvey and M. Dobson (1995) (eds.) *Food in antiquity*. Exeter

Wilkinson, L. P. (1963) *Golden Latin artistry*. Cambridge

Wilson, A. I. (2000) 'Land drainage', in Wikander (2000) 303–17

(2008) 'Hydraulic engineering and water supply', in Oleson (2008) 285–318

(2012) 'Raw materials and energy', in Scheidel (2012) 133–55

Wilson, A. I., and M. Flohr (2011) 'The economy of ordure', in Jansen et al. (2011) 147–56

Winterbottom, M. (1980) review of I. Mazzini, *Q. Gargilii Martialis De hortis* (Bologna 1978), *CR* 30: 143–4

(2011) 'On ancient prose rhythm: the story of the dichoreus', in Obbink and Rutherford (2011) 262–76

(2017) 'The *Tribunus Marianus* and the development of the *cursus*', in Chiesa et al. (2017) 231–46

Wistrand, E. K. (1933) *Vitruvius-studier*. Doktorsavhandlingar i Latinsk Filologi vid Göteborgs Högskola 1926, 8. Diss., Gothenburg

(1941) *Über das Passivum*. Göteborgs Kungl. Vetenskaps- och Vitterhets-Samhälles Handlingar 6.A.1.1. Gothenburg

Wöhrle, G. (1985) *Theophrasts Methode in seinen botanischen Schriften*. Studien zur antiken Philosophie 13. Amsterdam

Wright, R. (2008) (ed.) *Latin vulgaire – latin tardif* VIII: *Actes du* VIIIe *colloque international sur le latin vulgaire et tardif (Oxford, 6–9 septembre 2006)*. Hildesheim and New York

Zainaldin, J. L. (forthcoming 1) 'Gargilius Martialis: *Medicinae ex holeribus et pomis*', in Stover (forthcoming)

(forthcoming 2) 'Ps.-Gargilius Martialis: *Curae boum*', in Stover (forthcoming)

(forthcoming 3) 'Gargilius Martialis: *De hortis / De arboribus pomiferis*', in Stover (forthcoming)

Zimmerman, M. (2012) *Apulei Metamorphoseon libri* XI. Oxford

Zohary, D., M. Hopf, and E. Weiss (2012) *Domestication of plants in the Old World: the origin and spread of domesticated plants in south-west Asia, Europe, and the Mediterranean basin*, 4th edn. Oxford

INDEX LOCORVM

For abbreviations and editions used, see Abbreviations and Bibliography §I.

INDEX NOMINVM ET RERVM

INDEX VERBORVM